W0042222

The fungal nucleus

The fungal nucleus

SYMPOSIUM OF
THE BRITISH MYCOLOGICAL SOCIETY
HELD AT QUEEN ELIZABETH COLLEGE
LONDON, APRIL 1980

EDITED BY
K.GULL & S.G.OLIVER

CAMBRIDGE UNIVERSITY PRESS
CAMBRIDGE
LONDON   NEW YORK   NEW ROCHELLE
MELBOURNE   SYDNEY

CAMBRIDGE UNIVERSITY PRESS
Cambridge, New York, Melbourne, Madrid, Cape Town,
Singapore, São Paulo, Delhi, Tokyo, Mexico City

Cambridge University Press
The Edinburgh Building, Cambridge CB2 8RU, UK

Published in the United States of America by Cambridge University Press, New York

www.cambridge.org
Information on this title: www.cambridge.org/9780521279215

© British Mycological Society 1981

This publication is in copyright. Subject to statutory exception
and to the provisions of relevant collective licensing agreements,
no reproduction of any part may take place without the written
permission of Cambridge University Press.

First published 1981
First paperback edition 2011

A catalogue record for this publication is available from the British Library

Library of Congress Catalogue Card Number: 81-6079

ISBN 978-0-521-23492-4 Hardback
ISBN 978-0-521-27921-5 Paperback

Cambridge University Press has no responsibility for the persistence or
accuracy of URLs for external or third-party internet websites referred to in
this publication, and does not guarantee that any content on such websites is,
or will remain, accurate or appropriate.

Contents

Contributors

B. W. Bainbridge, *Microbiology Department, Queen Elizabeth College, Campden Hill Road, London W8 7AH, UK.*

G. R. Banks, *National Institute for Medical Research, The Ridgeway, Mill Hill, London NW7 1AA, UK.*

G. Beakes, *Department of Plant Biology, University of Newcastle upon Tyne, Newcastle upon Tyne NE1 7RU, UK.*

A. Beckett, *Department of Botany, Bristol University, Bristol BS8 1UG, UK.*

E. M. Bradbury, *Department of Biological Chemistry, School of Medicine, University of California, Davis, California 95616, USA.*

C. E. Caten, *Department of Genetics, University of Birmingham, Birmingham B15 2TT, UK.*

O. H. M. de Vries, *Department of Developmental Plant Biology, Biological Centre, University of Groningen, Haren, The Netherlands.*

J. J. M. Dons, *Department of Developmental Plant Biology, Biological Centre, University of Groningen, Haren, The Netherlands.*

K. Gull, *Biological Laboratory, University of Kent, Canterbury, Kent CT2 7NJ, UK.*

I. B. Heath, *Biology Department, York University, 4700 Keele Street, Downsview, Toronto, Ontario M3J 1P3, Canada.*

J. H. C. Hoge, *Department of Developmental Plant Biology, Biological Centre, University of Groningen, Haren, The Netherlands.*

L. H. Johnston, *National Institute for Medical Research, The Ridgeway, Mill Hill, London NW7 1AA, UK.*

B. C. Lamb, *Department of Botany, Imperial College, London SW7 2BB, UK.*

E. B. Lane, *Department of Zoology, University College, Gower Street, London WC1E 6BT, UK.*

H. R. Matthews, *Department of Biological Chemistry, School of Medicine, University of California, Davis, California 95616, USA.*

P. Nurse, *School of Biological Sciences, University of Sussex, Falmer, Brighton, Sussex BN1 9QC, UK.*

S. G. Oliver, *Applied Molecular Biology Group, Department of Biochemistry, University of Manchester Institute of Science and Technology, P.O. Box 88, Manchester M60 1QD, UK.*

J. Springer, *Department of Developmental Plant Biology, Biological Centre, University of Groningen, The Netherlands.*

J. G. H. Wessels, *Department of Developmental Plant Biology, Biological Centre, University of Groningen, Haren, The Netherlands.*

D. H. Williamson, *National Institute for Medical Research, The Ridgeway, Mill Hill, London NW7 1AA, UK.*

A. Zantinge, *Department of Developmental Plant Biology, Biological Centre, University of Groningen, Haren, The Netherlands.*

D. Zickler, *Laboratoire de Génétique, Bât 400, Université de Paris-Sud, 91405–Orsay, France.*

Preface

The fungi have become favoured model organisms for the study of eukaryotic genetics and molecular biology. Although some may consider the fungi to be on a branch-line of evolution, it is becoming increasingly evident that they are typically eukaryotic in most essential features. The fungal nucleus, the subject of this volume, has been intensively studied. The small size of the fungal nuclear genome is a great attraction to geneticists, but the correspondingly small size of the nucleus has posed problems for classical cytologists. It is very difficult, and in some species impossible, to see fungal chromosomes with the light microscope. Cytologists have been forced to turn to the electron microscope and have made a great virtue out of this necessity. The small size of the fungal nucleus makes it practicable to take serial sections through the whole nucleus and this has permitted very detailed analyses of nuclear structure during both mitosis and meiosis.

The study of the ultrastructure of fungal nuclei has been carried out with a wide range of species and has established a number of important general features. This broad survey is now, to a large extent, complete, and the first four chapters of this volume summarize its outcome. The diversity of approach is reflected in the range of terms that are used to describe similar structures, e.g. spindle pole bodies, centriolar plaques, centrioles, nucleus-associated organelles, microtubule organizing centres. There is no consensus among the authors as to which terms are the most appropriate and we, as editors, have made no attempt to impose uniformity.

While the ultrastructural studies have ranged over the whole fungal kingdom, those on the genetics, biochemistry and molecular biology of fungal nuclei have concentrated on only a few species. Pre-eminent among these, perhaps, is the yeast *Saccharomyces cerevisiae*. This concentration of effort has fostered its own brand of arrogance and insularity. In this volume, and elsewhere, the word yeast is taken to mean *S. cerevisiae* unless it is otherwise qualified. Such is the effect of this propaganda machine that one author, who usually works on the fission yeast *Schizosaccharomyces pombe*, appears to have been browbeaten into writing his chapter on *S. cerevisiae*!

In the fields of nuclear division and chromosome structure, yeast has a serious rival in the slime mould *Physarum polycephalum* and that organism is well represented in this volume. Finally, the delightful and useful sexual versatility of the filamentous fungi has not been overlooked. For instance, we feel that the review of the parasexual cycle presented here is both valuable and long overdue.

This book is based on the symposium on the Fungal Nucleus, which was organized by the Physiology Group of the British Mycological Society and held at Queen Elizabeth College, London, during April, 1980. Most of the chapters were written after the meeting and we have done our best to update the references at the proof stage. If this volume has appeared later than we would have liked it is because geographical separation is as inhibitory to the exchange of knowledge as it is to the exchange of genetic information.

S. G. Oliver
K. Gull
Canterbury, 1980

Ultrastructure of the Phycomycete nucleus

G. BEAKES

Department of Plant Biology,
University of Newcastle upon Tyne,
Newcastle upon Tyne NE1 7RU, UK

Introduction

The understanding of nuclear behaviour in the lower fungi has been severely hampered by the small size of most of their nuclei (Table 1). Although the use of modern phase optical systems has improved observation of nuclei in living hyphae (McCully & Robinow, 1973; McNitt, 1973; Powell, 1975; Tanaka, 1970), squash preparations of stained chromatin require very critical interpretation since, even when resolved, most Phycomycete chromosomes are very near the limits of optical resolution (Robinow & Bakerspigel, 1965; Sansome, 1980; Win-Tin & Dick, 1975). It is not surprising that the application of transmission electron microscopy (TEM), with its greatly extended resolving capabilities, has vastly increased knowledge of nuclear structure in fungi (compare review of Robinow & Bakerspigel, 1965 with that of Fuller, 1976). However, even TEM is not without its own limitations which should always be kept in mind when evaluating current knowledge of nuclear fine structure. It is often extremely difficult to reconstruct dynamic cellular events from static electron micrographs. The complex changes associated with nuclear division in fungi are usually extremely rapid (times for completion of mitosis: 25–35 min in *Basidiobolus*, Tanaka, 1970; 3–5 h in *Phlyctochytrium*, with events from metaphase onwards taking 4 min or less, McNitt, 1973; and 30 min in *Allomyces*, Olson, 1974a), although, fortunately, divisions that occur are often reasonably synchronous (Howard & Moore, 1970; Powell, 1975; Tanaka, 1970). The very thinness (60 nm) of sections required for TEM means that even with Phycomycete nuclei, extensive serial sectioning is required to obtain a complete three-dimensional picture of the nucleus and its associated organelles. In spite of their small size,

Table 1. *Organelles associated with interphase nuclei*

Species		Type of organelle	No.	Dimensions: diameter × length (nm)	Pocket	Orientation	Associated cytoplasmic microtubules
Hydromyxomycetes	*Labyrinthula* sp.	none	–	–	–	–	–
	Thraustochytrium	centriole	2	180 × 600	+	100°–130°	+
	Plasmodiophora brassicae	centriole	2	200 × 200	+	180°	+
Plasmodiophoromycetes	*Sorophaera veronica*	centriole	2	200 × 200	+	180°	+
Chytridiomycetes	*Blastocladiella emersonii*	centriole	2	180 × 200 180 × 120	+	parallel	–
	Catenaria anguillulae	centriole	2	180 × 250	+	90°	–
	Entophlyctis sp.	centriole	2	180 × 200 180 × 130		parallel	+
	Haprochytrium hedinii	centriole	2	180 × 200	+	90°	+
	Phlyctochytrium sp.	centriole	2	215 × 215	+	90°–parallel	+
Oomycetes	*Albugo* sp.	centriole	2	200 × 200	+	180°	–
	Phytophthora spp.	centriole	2	200 × 200	+	180°	–
	Saprolegnia spp.	centriole	2	200 × 200	+	180°	±
	Thraustotheca clavata	centriole	2	200 × 200	+	180°	±
Mucorales	*Mucor hiemalis*	plaque	1	c. 200	–	–	–
Entomophthorales	*Ancyclistes* sp.	plaque	1–2	c. 250	–	–	–
	Basidiobolus ranarum	electron-dense body	2(?)	20 × 120	+	separate	–

complete reconstruction of Phycomycete nuclei by painstaking serial sectioning has only been undertaken by Heath (1974*a*). Fixation and preparation procedures necessary for TEM may introduce artifacts, such as undulating nuclear membranes in osmium-fixed material (Beakes & Gay, 1977), or fail to preserve structures such as microtubules and centrioles (Heath 1974*b*). Fixation problems are particularly acute when trying to follow nuclear behaviour in thick-walled, dormant, structures like resting sporangia (Olson, 1974*b*; Olson & Reichle, 1978 *a, b*), oospores (Beakes & Gay, 1977), and zygospores. Indeed, because of the problems outlined above, and the overall complexity of TEM techniques, only a minute sample of species and isolates of Phycomycete fungi have been examined at the ultrastructural level (Table 2). This clearly imposes severe limitations as to the generalizations that can be drawn from these studies and makes it virtually impossible to assess the normal range of phenotypic and genotypic variation.

This review, in addition to describing nuclear structure in the traditional Phycomycete subdivisions of the largely aquatic Mastigomycotina and the terrestrial Zygomycotina will also include some Orders (Labyrinthules and Thraustochytriales) and Classes (Plasmodiophormycetes) now usually placed in the Myxomycota (plasmodial thalli). However, the true slime moulds (Acrasiomycetes and Myxomycetes) are outside the scope of the present review, although their nuclear structure is probably better documented and understood than any other group of lower fungi (see review by Mohberg, 1974). This present review will concentrate on those papers specifically dealing with nuclear structure, many of which have been summarized in Table 2.

Interphase nuclei

With the exception of the extraordinarily large nuclei (25 × 10 μm) of the Entomophthoralean Zygomycete *Basidiobolus ranarum* (Tanaka, 1970; Sun & Bowen, 1972; Gull & Trinci, 1974*a*), Phycomycete nuclei are extremely small (2.5–3.5 μm × 1.5–2.5 μm). Analysis of nuclear deoxyribonucleic acid (DNA) shows there is generally good correlation between observed nuclear volume at different stages of the organisms' life cycle and DNA content (Bryant & Howard, 1969; Howard & Moore, 1970; Olson & Reichle, 1978*a, b*). However, this generalization need not always apply. For instance, male and female gametic nuclei although of similar ploidy are often of different size (Pommerville & Fuller, 1976; Beakes & Gay, 1977). Similarly, the

Table 2. A summary of ultrastructural accounts of nuclear division in the lower fungi

Species	Mitosis (MT) or meiosis (ME)	Class of study	Material examined	Authority
MYXOMYCOTINA				
Hydromyxomycetes				
Labyrinthula sp.	ME/MT	**/*	reticulate sori	Perkins & Amon, 1969; Perkins, 1970
Thraustochytrium sp.	MT	***	vegetative cells	Kazama, 1974
Plasmodiophormycetes				
Plasmodiophora brassicae	MT/ME	***	plasmodia	Garber & Aist, 1979a, b
Polymyxa betae	MT	*	plasmodia	Keskin, 1971
Sorosphaera veronicae	ME/MT	*/***	plasmodia	Braselton & Miller, 1973; Braselton et al., 1975; Dylewski et al., 1978
MASTIGOMYCOTINA				
Chytridiomycetes				
Allomyces macrogynus	MT	*	vegetative thalli	Robinow & Bakerspigel, 1965
	ME	**	resting sporangia	Olson, 1974b
Allomyces neo-moniliformis	MT	***	germinating spores	Olson, 1974a
Blastocladiella emersonii	MT	*	zoosporangia	Lessie & Lovett, 1968
	ME	**	resting sporangia	Olson & Reichle, 1978a
Catenaria anguillulae	MT	***	zoosporangia	Ichida & Fuller, 1968
	ME	**	resting sporangia	Olson & Reichle, 1978b
Coelomyces indicus	MT	*	sporangia	Madelin & Beckett, 1972
Entophlyctis sp.	MT	***	sporangia	Powell, 1975
Harpochytrium hedinii	MT	***	vegetative thalli	Whisler & Travland, 1973
Phlyctochytrium irregulare	MT	***	sporangia	McNitt, 1973

Oomycetes

Achlya ambisexualis	ME/MT	**/*	antheridia/hyphae	Ellzey, 1974; Ellzey & Huizar, 1977
Albugo spp.	MT	**	hyphae	Khan, 1976
Aphanomyces astaci	MT	*	hyphae	Heath, 1974b
Aphanomyces euteiches	MT	*	germinating cysts	Hoch & Mitchell, 1972
Lagenisma coscinodisci	MT/ME	*/*	sporangia	Schnepf & Deichgräber, 1978
Phytophthora palmivora	MT	***	secondary sporangia	Hemmes & Hohl, 1973
Pythium aphanidermatum	MT	*	germinating cysts	Grove & Bracker, 1978
Saprolegnia ferax	MT	***	hyphae	Heath & Greenwood, 1968, 1970
	ME	***	oogonia	Beakes, 1980
Saprolegnia furcata	ME	***	oogonia & antheridia	Beakes & Gay, 1977
Thraustotheca clavata	MT/ME	***	hyphae/oogonia	Heath, 1974a

ZYGOMYCOTINA

Zygomycetes

Ancyclistes sp.	MT	**	hyphae	Moorman, 1976
Basidiobolus ranarum	MT	***	hyphae	Gull & Trinci, 1974; Sun & Bowen, 1972; Tanaka, 1970
Mucor hiemalis	MT	***	germinating spores	McCully & Robinow, 1973
Phycomyces blakesleeanus	MT	***	sporangiophores	Franke & Reau, 1972
Pilobus crystallinus	MT	***	sporangia	Bland & Lunney, 1975

* Usually only single stage described;
** detailed account, but with substantial gaps in divisional stages;
*** detailed account covering all, or most of divisional stages.

presumed diploid restitution nuclei in oospheres of an emasculate isolate of *Saprolegnia ferax* are over twice the volume of pre-meiosis nuclei (Beakes, 1980).

Examination of nuclei at different stages of the fungal life cycle usually reveals a much more varied picture of nuclear morphology than that indicated in Table 1. In hyphae and other stages, such as sporangiophores, where there is active cytoplasmic streaming, nuclei have elongate, often spindle-shaped profiles (Ellzey, 1974; Heath, 1974*a*; McCully & Robinow, 1973; Tanaka, 1970), whereas in zoosporangia and zoospores they have pyriform or conical profiles (Thraustochytriales: Kazama, 1974; Chytridiomycetes: Lessie & Lovett, 1968; Olson & Rønne, 1975; Oomycetes: Grove & Bracker, 1978; Heath & Greenwood, 1971; Hoch & Mitchell, 1972). Such variations in shape often appear to be the result of interactions between the nuclear envelope and cytoplasmic microtubules (Grove & Bracker, 1978; Heath, 1974*a*; Heath & Greenwood, 1971; Lessie & Lovett, 1968; Tanaka, 1970).

In addition, nuclear shape may be modified by environmental conditions. Hyphal nuclei in *Saprolegnia ferax* treated with a range of microtubule inhibitors acquire a variety of irregularly looped (enclosing islands of cytoplasm) and horned (nuclear projections) profiles (Heath, 1975*a*). Somewhat similar profiles have also been described in some oospore germlings of this species, which may be the result of genotypic variation (Beakes, 1980).

Nuclear envelope

Nuclei are delimited by a cisternal envelope which is part of the cellular endomembrane system (for comprehensive review of envelope structure and function, see Franke & Scheer, 1974). The nuclear envelope in Phycomycete fungi has a slightly uneven profile and is perforated by characteristic 80–100 nm pores. In *Basidiobolus* the pores appear in an orderly array throughout the envelope (Sun & Bowen, 1972), whereas in *Saprolegnia* they often occur in localized clusters (Beakes, 1976). Tangential (grazing) sections of the envelope show the pore has an electron-dense rim, composed of a number of 20 nm subunits and plugged by an electron-dense granule (Garber & Aist, 1979*a*; Grove & Bracker, 1978; Tanaka, 1970), as reported in other eukaryotes (Franke & Scheer, 1974). Glancing profiles of vegetative nuclei of *Saprolegnia* do not reveal a clear rim structure (Beckett, Heath & McLaughlin, 1974) although comparable sections of meiotic nuclei show a particulate rim and possible spoke-like threads connecting it to the central granule (Howard & Moore, 1970). It is therefore possible that the nuclear pore complex may vary during differentiation.

Ribosomes are usually attached to the outer face of the nuclear envelope, although they have rarely attracted comment. The inner face of the nuclear envelope frequently has small patches of electron-dense material adhering to it, believed to be inactive heterochromatin. These patches have been reported in the Labyrinthulales (Perkins & Amon, 1969) Plasmodiophoromycetes (Braselton, Miller & Pechak, 1975) and widely within the Blastocladiales (Ichida & Fuller, 1968; Olson, 1974a), and Mucorales (Bland & Lunney, 1975; Franke & Reau, 1972; McCully & Robinow, 1973). In many species the nucleolus is also attached to the inner face of the nuclear envelope. As in other organisms (Franke & Scheer, 1974), the nuclear envelope often shows connections with cisternae of rough endoplasmic reticulum (Table 1). More rarely, the envelope may be connected to larger vesicles, which in *Labyrinthula* contain flocculent (slime) material (Perkins & Amon, 1969) and in *Saprolegnia*, bundles of flimmer tubules (Heath, Greenwood & Griffiths, 1970). In the Hydromyxomycetes and Oomycetes the nuclear envelope is also associated with one or more Golgi dictyosomes with the inner face contributing small vesicles to the dictyosome-forming face. Therefore, in aquatic Phycomycetes the nuclear envelope may have important secretory functions.

Nuclear-associated organelles (NAO)

Interphase nuclei of phycomycete fungi are usually associated with some form of nuclear associated body (NAO, see Heath, 1978) which forms a focus for spindle development during nuclear division (Table 1). In their non-motile phases the nuclei of zoosporic species are usually associated with a pair of microtubular centrioles, located some 20–100 nm from the nuclear envelope. The dimensions and orientation of these centrioles with respect to each other have been summarized in Table 1, and these clearly vary with the taxonomic groupings (see Heath, 1974b, 1975c for phylogenetic speculations).

The interphase nuclei of Plasmodiophoromycetes (Braselton & Miller, 1973; Garber & Aist, 1979a) and Oomycetes (Heath, 1974b; Heath & Greenwood, 1968, 1970) are associated with two short (200 nm) centrioles orientated in a characteristic, and relatively unusual, end-to-end fashion whereas those of Chytridiomycetes are often of uneven length (Lessie & Lovett, 1968; Olson, 1974a, b; Whisler & Travland, 1973) and orientated at right angles to each other. However, as Table 1 shows, there is some variability within the Chytridiomycetes with many species having equal centrioles and showing a range of orientations. As pointed out by Heath (1975*b*), variations in centriole length and orientation may sometimes be associated with the transformation of centrioles into the elongate basal bodies of the flagellar apparatus (kinetosomes) prior to the assumption of the motile phase (Heath & Greenwood, 1971; Lessie & Lovett, 1968).

Centrioles may not always be associated with interphase nuclei, however, and may disappear and reform during the fungal life cycle. In *Labyrinthula*, centrioles are not associated with interphase nuclei but reform prior to mitosis and meiosis (Perkins & Amon, 1969; Perkins, 1970). In *Plasmodiophora brassicae*, the single centriole associated with the post-meiotic nucleus has been shown to disintegrate following cytokinesis (Garber & Aist, 1979*b*). In *Saprolegnia* also there is good evidence that the centriole may revert to some pro-centriolar state following karyogamy (Beakes & Gay, 1977; Beakes, 1980). It should be pointed out, however, that the reported absence of centrioles with interphase nuclei must always be viewed with some caution, since centrioles are not always well-preserved and may be overlooked without extensive serial sectioning (see Heath, 1974*b*).

Interphase centrioles often are associated with Golgi dictyosomes and certain amounts of electron-dense amorphous material. In *Thraustochytriales* (Kazama, 1974), *Plasmodiophoromycetes* (Brasleton *et al.*, 1975; Garber & Aist, 1979*a*) and Chytridiales (McNitt, 1973; Powell, 1975) the centrioles (in particular the associated electron-dense material) are the focus for numerous radiating aster microtubules, whereas in the Blastocladiales and Oomycetes such tubules are relatively scarce.

The centrioles are often located in shallow depressions, or pockets in the nuclear envelope (Table 1). However, only in Oomycete fungi is this pocket associated with differentiation of the envelope membranes (Heath, 1974*a, b*; Heath & Greenwood, 1970). In these fungi, the pocket membranes show increased electron density and thickness, and the intervening cisternal space is of uniform diameter. The inner face of the Oomycete pocket envelope is, in addition, associated with an array of short microtubules, terminated by either electron-dense globules (Heath, 1974*a*; Beakes, 1980) or more diffuse material (Hemmes & Hohl, 1973). These are believed to be nascent spindle kinetochores (Heath, 1974*b*, 1978) and have not been observed in other Orders.

The interphase nuclei of the non-motile Zygomycetes are not associated with microtubular centrioles (Table 1). However a localized (200 nm) region of differentiated nuclear envelope has been reported in *Mucor hiemalis* (McCully & Robinow, 1973), and *Ancyclistes* (Moorman, 1976). Since this region is associated with the development of the spindle during mitosis, it has been termed the spindle pole body (see McCully & Robinow, 1973). The giant nuclei of the *Basidiobolus* are associated with two separate NAOs which, although termed centrioles by Sun & Bowen (1972), do not appear to have the conventional centriolar structure (ring of nine triplet microtubules). Fuller (1976) has suggested they resemble certain annular spindle pole bodies reported in some red algae. However in common with centrioles, the NAOs in *Basidiobolus* are located in pockets and are associated with cytoplasmic microtubules (Sun & Bowen, 1972).

Nucleoplasm: chromatin and nucleoli

Detailed understanding of the organization of nucleoplasm requires the application of rigorous cell biological techniques, including the use of inhibitors and high resolution autoradiography (see reviews by Bouteille, Laval & Dupuy-Coin, 1974; Smetana & Busch, 1974). No such studies have been carried out on Phycomycete nuclei and most descriptions of nuclear contents are based on comparison with other organisms. In most Phycomycetes the nucleoplasm is composed of a uniformly fibrillar material (diffuse chromatin?) although Oomycetes often have a conspicuous particulate fraction as well (Beakes, 1980; Heath & Greenwood, 1970; Hemmes & Bartnicki-Garcia, 1975; Hickey & Coffey, 1977). Since these particles are ribosome-sized and are particularly clearly resolved in osmium-fixed material (Beakes & Gay, 1977) they probably represent pro-ribosomes (see Smetana & Busch, 1974). Indeed, in mature oospheres and oospores of *Saprolegnia*, the nuclear matrix is uniformly fibrillar in glutaraldehyde-fixed material (Beakes, 1980; Beakes & Gay, 1977). This is consistent with the likely cessation of RNA and ribosome synthesis at these stages. As already mentioned, many nuclei contain small regions of dense (heterochromatic?) material which, except in the Mucorales, are also distributed throughout the nucleoplasm. Localization of interphase DNA by Feulgen staining in *Allomyces* indicates that it occurs throughout the non-nucleolar nucleoplasm (Olson, 1974a). This is likely to be the general pattern in most aquatic Phycomycetes. In contrast, similar studies in Mucorales show that chromatin is largely associated with the nuclear envelope (McCully & Robinow, 1973) and in *Basidiobolus*, it occurs in patches within the extensive nucleolar region (Tanaka, 1970).

The most conspicuous nuclear inclusion in all Phycomycetes is the relatively electron-dense nucleolus, which may be centrally or eccentrically placed within the nucleoplasm or attached to the nuclear envelope (Table 1). Phycomycete nucleoli are usually small (0.8–1.0 μm), well-defined, spherical or hemispherical bodies although in Peronosporalean oomycetes often occur as a number of separate blocks (*Peronospora pisi*, Hickey & Coffey, 1977; G. Beakes, unpublished observations). As in higher organisms (Smetana & Busch, 1974) the nucleolus is composed of both fibrillar (6–8 nm) and particulate (15–20 nm) components. These are clearly seen in the giant nucleoli of *Basidiobolus*, where the particulate component forms a dense reticulum (Gull & Trinci, 1974*a*; Sun & Bowen, 1972; Tanaka, 1970).

The fibrillar and particulate components of the nucleolus are not always randomly intermixed. In *Saprolegnia*, for instance, the fibrillar part tends to be centrally concentrated, with the particulate fraction around the periphery (Beakes, 1980). In addition, lighter regions frequently occur within the nucleolus which resemble either nucleolar vacuoles (as in *Basidiobolus*, Gull & Trinci, 1974*a*; Jordan, Birkett & Trinci, 1980) or nucleolini (as in zoospore nuclei of *Saprolegnia diclina*, Beakes, unpublished) reported in other eukaryotes (Smetana & Busch, 1974).

In fact the nucleolus in most Phycomycetes shows a good deal of variation in structure, although these nucleoli have not so far been analysed in detail. For instance in zoospores the nucleolus is often more discrete than in vegetative stages, and in the Blastocladiales it is located at the nuclear apex, adjacent to the flagellar base (Lessie & Lovatt, 1969; Olson & Rønne, 1975), whereas in Oomycetes it is located on the side away from the flagella (Grove & Bracker, 1978; Heath & Greenwood, 1971; Hoch & Mitchell, 1972). Recently, Jordan *et al.* (1980) have shown that nucleolar size in *Basidiobolus* decreases with increasing temperature. At low temperatures (<10 °C) the nucleolus is large and diffuse, and contains many nucleolar vacuoles, whereas at high temperatures (>20 °C) it is smaller and more compact. Clearly, nucleolar behaviour in Phycomycete nuclei deserves more careful examination.

In addition to the nucleolus, Phycomycete nuclei also contain a variety of other inclusions, the significance of which is poorly understood. Aggregates of 7–10 nm microfilaments have been reported in both colchicine-treated (Heath *et al.*, 1970) and untreated nuclei

(Ellzey, Huizar & Yanez, 1976; Gleason, 1973) in the *Saprolegniales*. Aggregates of spherical bodies, 40 nm in diameter, have also been seen in the nuclei of an unidentified (asexual) *Saprolegnia* isolate (Beakes, unpublished) and undulating 20 nm rodlets within the oosphere nuclei of *Phytophthora capsici*. Both of these aggregates are not unlike intranuclear virus particles. Recently, arrays of membrane inclusions, containing nuclear pores similar to annulate lamellae (see Franke & Scheer, 1974), have been described in nuclei in hyphae of *Basidiobolus* grown at low temperature (Jordan *et al.*, 1980), and in mature oospore germlings of *Saprolegnia* (Beakes, 1980). Finally, elaborate invaginations of nuclear envelope (often enclosing vesicular arrays) have been reported in nuclei of inhibitor-treated hyphae and gemmae and in untreated oospore germlings of *Saprolegnia*, and probably represent a response to stress conditions (Beakes, 1976, 1980; Heath, 1975*a*).

Nuclear fusion and autolysis

In species undergoing a normal sexual life cycle there will come a stage when male and female nuclei will fuse. This has only been observed at the ultrastructural level in *Allomyces* (Olson & Rønne, 1975; Pommerville & Fuller, 1976) and in *Saprolegnia* (Beakes, 1976). In the former, motile anisomorphic male and female gametes fuse to give rise to the sporophytic (2n) generation (Olson & Rønne, 1975), whereas in the latter, male and female nuclei, derived from gametangial meiosis, unite within differentiated oospheres (Beakes, 1980; Beakes & Gay, 1977). Following gamete fusion, both in normal (Pommerville & Fuller, 1976) and multiple unions of *Allomyces* (Olson & Rønne, 1975), the nuclei become closely aligned, and a localized region of nuclear envelope fuses. As envelope fusion progresses the apical nucleoli also coalesce (Olson & Rønne, 1975). In *Saprolegnia*, comparable lobed fusion profiles have only been observed in atypical tap-water-induced spores (Beakes, 1976).

In addition to changes in nuclear numbers as a result of divisions or fusions, loss of interphase nuclei by autolysis has also been reported in Oomycete fungi (Beakes, 1976, 1980; Heath, 1975*a*; Hemmes & Hohl, 1973). There appear to be two types of nuclear death. The first type is exemplified in *Saprolegnia*, by the nucleolus and parts of the nucleoplasm becoming intensely stained and the nuclear envelope becoming locally inflated, as reported in colcemid- (Heath, 1975*a*) and streptomycin- (Beakes & Gay, 1980) grown hyphae and in post-fertilization antheridia (Beakes & Gay, 1977). In the second type, the nuclei first become surrounded by one or more cisternae of smooth endoplasmic reticulum, and/or vesicles, to form a lysosomal compartment (Beakes, 1980; Hemmes & Hohl, 1973). The nucleoplasm then starts to disintegrate and the nuclear envelope or investing endoplasmic reticulum inflates to form a vesicular compartment. The final stages of this type of autolysis result in the nuclear remnants disintegrating into multivesicular bodies (Beakes, 1976; Hemmes & Hohl, 1973). This type of autolysis appears widespread in *Saprolegnia* and has been reported in low-temperature-grown hyphae (Heath, 1975a), in mature oospore germlings (Beakes, 1980), and in streptomycin-induced gemmae (Beakes, 1976). It is also the mechanism by which supernumerary nuclei are eliminated after mitosis in secondary sporangia in *Phytophthora* (Hemmes & Hohl, 1973) and after meiosis in *Saprolegnia* oogonia (Beakes, 1980) and *Phytophthora* antheridia (Hemmes & Bartnicki-Garcia, 1975). It is perhaps significant that in nearly all these cases exogenous nutrients are limiting and, therefore, this is probably a mechanism by which nuclear components are recycled (Beakes, 1980; Hemmes & Hohl, 1973).

Mitosis

Of all aspects of Phycomycete nuclear behaviour, mitosis has received most scrutiny from electron microscopists (Table 2). In addition to resolving mechanistic aspects of division, these studies were given additional significance by the recognition of the phylogenetic importance of the mitotic apparatus following the pioneering speculations of Pickett-Heaps (1969, 1972). This is of particular relevance to

Explanation of Figures

Figs. 1–8 summarize mitotic divisions and Figs. 9–10 meiotic divisions, based on the micrographs in the following papers. Fig. 1: Kazama (1975). Fig 2: Garber & Aist (1979a); Dylewski *et al.* (1978). Fig. 3: Powell (1975). Fig. 4: McNitt, 1975. Fig. 5: Ichida & Fuller (1968). Fig. 6: Heath (1974a); Heath & Greenwood (1968, 1970); Hemmes & Hohl (1973); Khan (1976). Fig. 7: Bland & Lunney (1975). Fig. 8: Gull & Trinci (1974a); Şun & Bowen (1972); Tanaka (1970). Fig. 9: Garber & Aist, 1979b. Fig. 10: Beakes (1980); Beakes & Gay (1977); Howard & Moore (1970). The stages of division have been abbreviated as follows: I, pre-divisional interphase; P (P1 early, P2 late) prophase; M (M1 early, M2 late) metaphase; A (A1 early, A2 late) anaphase; T (T1 early, T2 late) telophase; I2 post divisional interphase. In meiotic divisions the numerals I and II refer respectively to first and second divisions of meiosis.

such a diverse and complex assemblage of organisms as the Phycomycete fungi and has already been comprehensively reviewed (Fuller, 1976; Heath, 1974*c*, 1975*c*, 1978; Kubai, 1974, 1978).

Nuclear division represents a complex flow of structural events that are difficult to describe concisely. A selection of Phycomycete mitoses have, therefore, been illustrated diagrammatically in Figs. 1–8 and

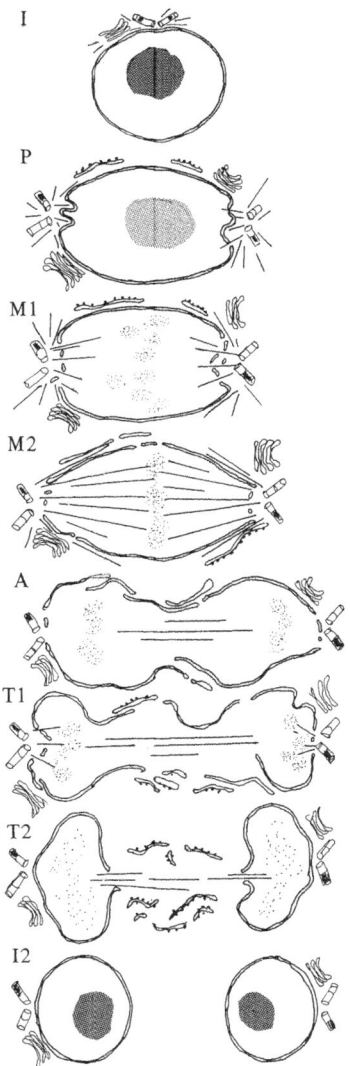

Fig. 1. Mitosis in *Thraustochytrium* sp.

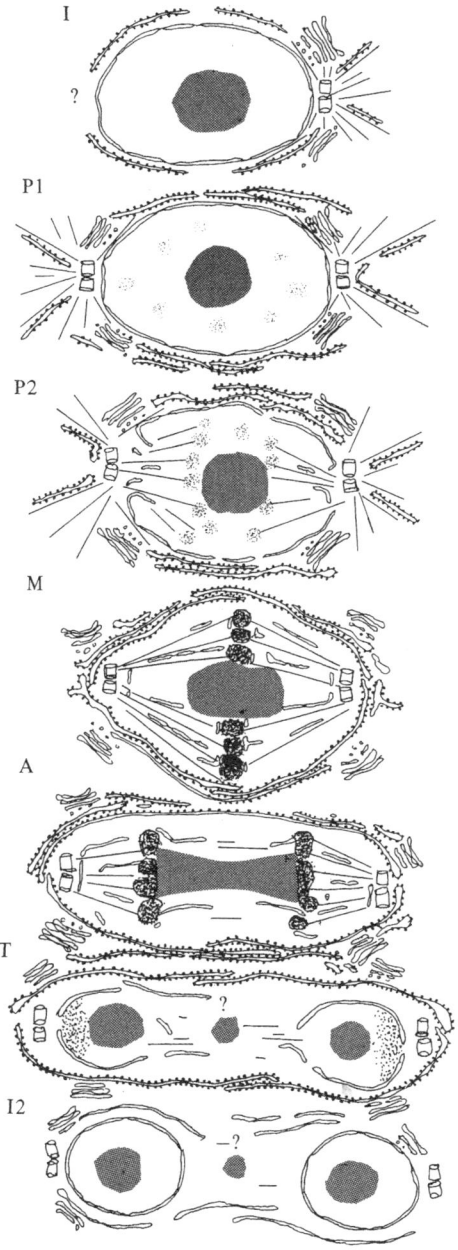

Fig. 2. Mitosis in Plasmodiophoromycetes

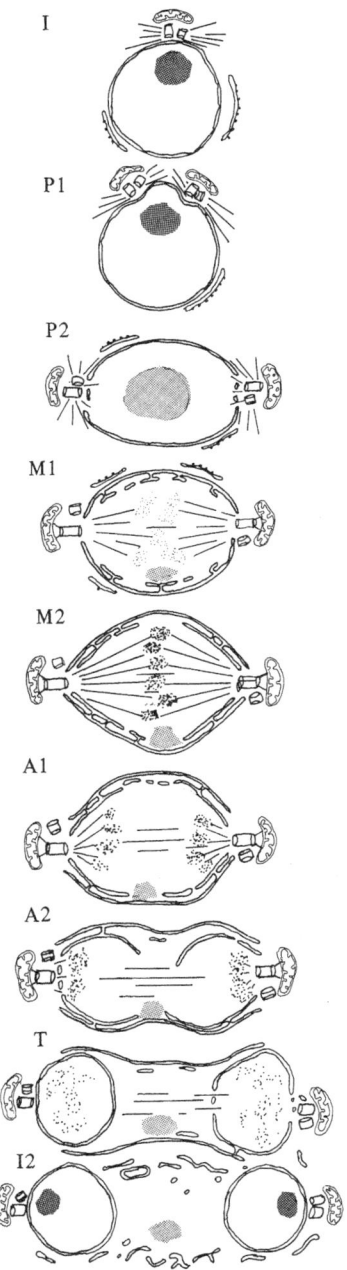

Fig. 3. Mitosis in *Entophlyctis* sp.

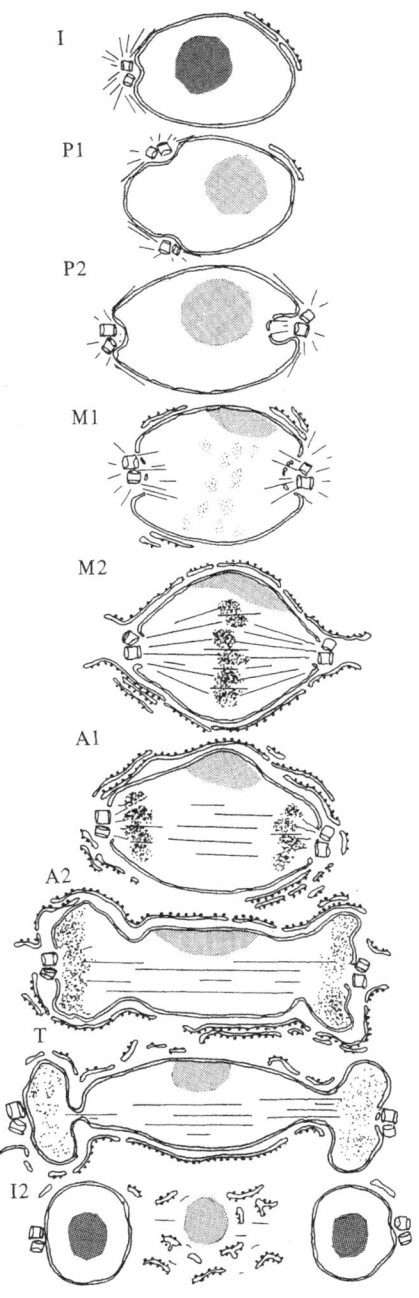

Fig. 4. Mitosis in *Phlyctochytium irregulare*.

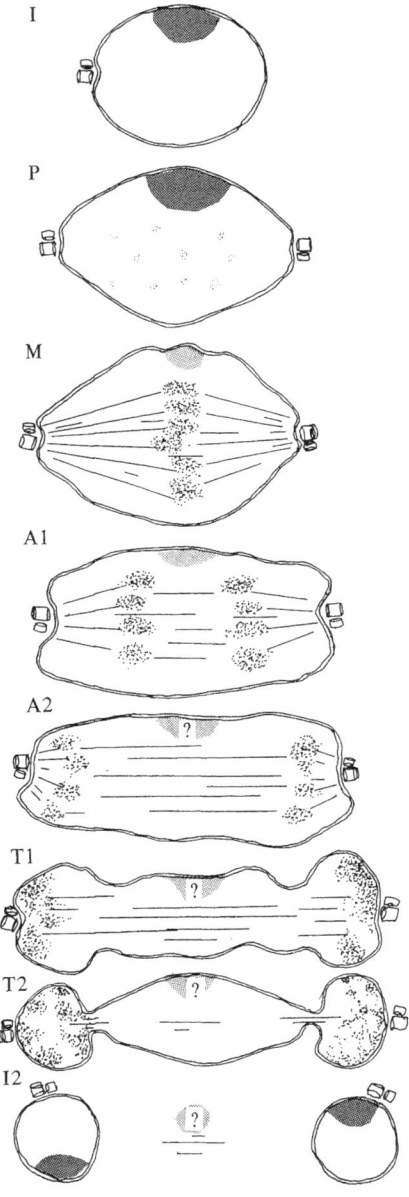

Fig. 5. Mitosis in *Catenaria anguillulae*.

summarized in Table 4. There is greater variation in mitotic behaviour in the Chytridiomycetes (Figs. 3–5) than the Oomycetes (Fig. 6), which suggests they are of more diverse origin. Mitosis in the Zygomycete classes so far examined, the Mucorales (Fig. 7) and Entomophthorales (exemplified by *Basidiobolus*, Fig. 8) is so different that it is difficult to imagine such variation arising in closely related organisms.

Nuclear envelope

The lower fungi can be divided into several groups on the basis of the behaviour of the nuclear envelope during mitosis (Table 4). In the

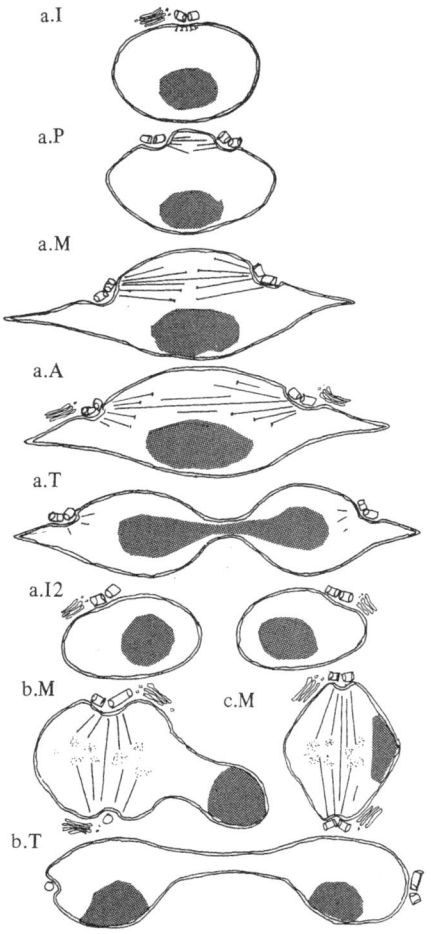

Fig. 6. Mitosis in *Saprolegnia* (*a*), *Phytophthora* (*b*) and *Albugo* (*c*).

Hydromyxomycetes (Fig. 1), Plasmodiophoromycetes (Fig. 2) and Chytridiales (Figs. 3 and 4) the nuclear envelope undergoes a localized breakdown to form polar fenestrae as the spindle develops at metaphase (Figs. 1–4, Table 3). These fenestrae are usually confined to the regions immediately adjacent to the polar centrioles (Figs. 2–4) but are more extensive in the Hydromyxomycetes (Kazama, 1974; Perkins & Amon, 1969; Perkins, 1970). In the Zygomycete *Basidiobolus* there is a much more widespread disruption of the nuclear envelope during metaphase (Fig. 8; Gull & Trinci, 1974*a*; Sun & Bowen, 1972; Tanaka, 1970). The disrupted envelope does not, however, break down, but forms a complex sponge-like reticulum capping late metaphase profiles before reassembling during late anaphase (Tanaka, 1970). In contrast, in the

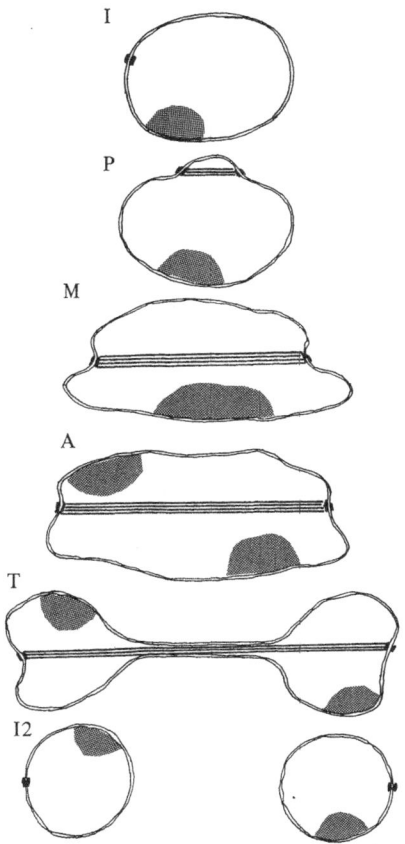

Fig. 7. Mitosis in *Pilobus crystallinus*.

Blastocladiales (Fig. 5), Oomycetes (Fig. 6) and most Zygomycetes (Fig. 7; Moorman, 1976), the nuclear envelope remains intact throughout nuclear division and the spindle apparatus is truly intranuclear (Table 3).

In many of the fenestrate species, the nuclear envelope becomes associated with cisternae of rough endoplasmic reticulum during mitosis (Table 3). These arrays are particularly extensive in the Plasmodiophoromycetes (Fig. 2, Braselton *et al.*, 1975; Dylewski, Braselton &

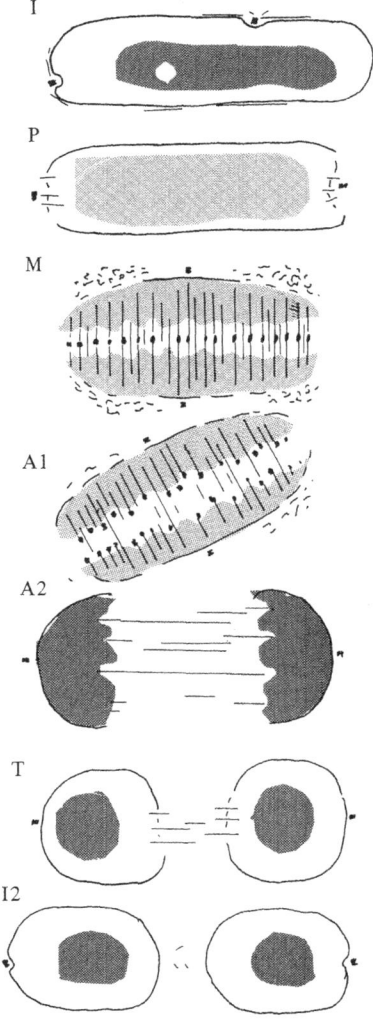

Fig. 8. Mitosis in *Basidiobolus ranarum*.

Miller, 1978; Garber & Aist, 1979*a*) and *Phlyctochytrium* (Fig. 4; McNitt, 1973). More unusually, in *Entophlyctis* (Fig. 3) intranuclear vesicles and cisternae are generated during metaphase from the inner face of the nuclear envelope and form the new envelope around the daughter nuclei (Fig. 3, Powell, 1975). In Plasmodiophoromycetes extensive intranuclear cisternae and vesicles also appear during metaphase (Fig. 2, Braselton *et al.*, 1975; Dylewski *et al.*, 1978; Garber & Aist, 1979*a*). Indeed, Garber & Aist (1979*a*) suggest that most of these cisternae are derived from the original nuclear envelope (they possess nuclear pores) and the nucleoplasm is in fact delimited by cisternae of rough endoplasmic reticulum during metaphase and anaphase. At telophase the nuclear envelope is assembled around the separated chromatin from the residual cisternae of the original nuclear envelope. Therefore, although the nuclear envelope was thought to remain largely intact throughout nuclear division in Plasmodiophoromycetes (Braselton *et al.*, 1975) it seems more probable that it breaks down and reassembles during mitosis.

At telophase in the Hydromyxomycetes (Fig. 1), Plasmidiophoromycetes (Fig. 2), Chytridiomycetes (Figs. 3–5) and *Basidiobolus* (Fig. 8), some portions of the original envelope and residual spindle apparatus are excluded in the intervening mid-piece region (Table 3). These fragments rapidly disintegrate, forming cisternae of endoplasmic reticulum between the separated daughter nuclei (Kazama, 1974; McNitt, 1973; Powell, 1975). In Oomycete fungi (Fig. 6) and in other Zygomycetes (Fig. 7) mitosis is completed by the fusion of the attenuated envelope in the mid-region without loss of nuclear material (Bland & Lunney, 1975; Heath, 1974*a*; Heath & Greenwood, 1970; McCully & Robinow, 1973; Moorman, 1976).

Nuclear associated organelles and spindle apparatus

During prophase polar nuclear associated organelles (NAOs) replicate and move to the poles of the developing spindle apparatus (Figs. 1–7). Mitotic centrioles, although generally similar in size to the interphase ones (Table 1), may show changes in their orientation (*Thraustochytrium*, Kazama, 1974; *Phlyctochytrium*, McNitt, 1973; *Entophlyctis*, Powell, 1975). Centriole duplication and migration is often associated with a proliferation of aster microtubules, many of which radiate over the surface of the nucleus (Heath, 1974*a*; Heath & Greenwood, 1970; Kazama, 1974; McNitt, 1973; Powell, 1975). In fenestrate species aster microtubules are reduced in number as the

Table 3. *Summary of mitosis in the lower fungi*

Species and taxonomic groupings	Nuclear envelope				Nucleolus			Spindle apparatus		Metaphase chromatin		Telophase		Perinuclear endoplasmic reticulum
	Intact	Polar fenestrae	Disperses	Intranuclear cisternae	Absent	Disperses	Persistent	Continuous and interdigitating microtubules	Chromosomal microtubules	Usually distinct well-defined metaphased plate	Usually indistinct no metaphase plate	Nuclear material excluded	Division by construction	
Hydromyxyomycetes														
Labyrinthula sp.		+			+			?	+	+		+		+
Thraustochytrium sp.		+			+			+	+	+		+		+
Plasmodiophormycetes														
Plasmodiophora brassicae			?	+			+	?	+	+		+		+
Sorosphaera veronicae			?	+			+	?	+	+		+		+
Chytridiomycetes														
Entophylctis sp.		+				+		+	+	+		+		+
Harpochytrum hedinii		+		+		+		+	+	+		+		+
Phlyctochytrium sp.		+				+		?	+	+		+		+
Allomyces neo-moniliformis	+						+	?	+	+		+		
Blastocladiella emersonii	+					?		+	+	+		+		
Catenaria anguillulae	+					+		+	+	+		+		

Taxon							
Oomycetes							
Saprolegnia ferax	+	+	+	+	+	+	+
Thraustotheca clavata	+	+	+	+	+	+	+
Albugo sp.	+	+	?	+	+	+	+
Phytophthora palmivora	+	+	?	+	+	+	+
Mucorales							
Mucor hiemalis	+	+	+	+	+	+	+
Phycomyces blakesleeanus	+	+	+	+	+	+	+
Pilobus crystallinus	+	+	+	?	+	+	+
Entomophthorales							
Basidiobolus ranarum	+				+		
Ancylistes sp.	+	+	+	+	+	+	+

? Denotes probable but unconfirmed behaviour.

spindle develops during metaphase (McNitt, 1973; Powell, 1975; Whisler & Travland, 1973), suggesting an interconversion of subunits from aster to spindle microtubules.

In the Hydromyxomycetes (Fig. 1) and Chytridiomycetes (Figs. 3–5) centriole migration precedes spindle development whilst in Oomycetes (Fig. 6, except Leptomitales, Heath, 1975c) the two events are concomitant, as is true of spindle pole body migration in the Mucorales (Fig. 7, Bland & Lunney, 1975; McCully & Robinow, 1972). In *Basidiobolus* (Fig. 8) the two separate NAOs replicate during prophase prior to spindle development (Sun & Bowen, 1972), as seems likely of the plaque-like NAOs in *Ancyclistes* (Moorman, 1976).

During metaphase and anaphase the pairs of centrioles (except in *Coelomyces* where single centrioles were reported, Madelin & Beckett, 1972) are located at either end of the elongating nuclei (Braselton *et al.*, 1975; Hemmes & Hohl, 1973; Ichida & Fuller, 1969; Kazama, 1974; Khan, 1976; McNitt, 1973; Olson, 1974a; Powell, 1975). The main exceptions to this are the Saprolegniales, where the centriolar pockets are sub-apical, and the nuclei have characteristic horns, resulting from interactions between aster microtubules and the nuclear envelope (Fig. 6, Heath, 1974a, b, 1975c, 1978; Heath & Greenwood, 1968, 1970).

The spindle apparatus which forms between the polar centrioles is composed of characteristic 20 nm microtubules which terminate in a 50–100 nm zone of amorphous electron-dense material (microtubule organizing centre) associated with either the centrioles in fenestrate species or the nuclear envelope where the spindle is internal (Table 3). However, in *Basidiobolus*, spindle microtubules terminate in the broad band of polar nucleolar material (Sun & Bowen, 1972) and in *Sapromyces* the intranuclear spindle seems to terminate within the nucleoplasm rather than at the nuclear envelope (Heath, 1978). Although complete reconstructions of mitotic spindles from serial sections have only been reported in *Thraustotheca clavata* (Heath, 1974a) it seems probable that the majority of Phycomycete spindles contain both chromosomal and a variety of non-chromosomal microtubules (Heath, 1978). The former are recognized by their termination in some form of kinetochore, attached to the chromatin (see Heath, 1974a, 1975c, 1978 for full details). The major exceptions are the Mucorales (Fig. 7, Table 3) which produce a spindle composed of a relatively small number (10–20) of apparently continuous microtubules (Bland & Lunney, 1972; Franke & Reau; McCully & Robinow, 1973). Although development of the spindle is reported to be inhibited at metaphase by colchicine in

Allomyces (Olson, 1972, 1974*a*) and griseofulvin in *Basidiobolus* (Gull & Trinci, 1974*b*), the Saprolegnian spindle appears curiously resistant to a wide range of antimicrotubule agents (Heath, 1975*a*, 1978). The mechanisms of genome separation in fungi will be described elsewhere in this volume (Heath, pp. 85–112).

Nucleoplasm: chromatin and nucleoli

Chromatin is often very difficult to see in most Phycomycete nuclei (Heath, 1974*a*; Kazama, 1974) and individual chromosomes can rarely be distinguished directly. Metaphase chromatin is most easily resolved in the Plasmodiophoromycetes (Braselton *et al.*, 1975; Garber & Aist, 1979*a*), many of the Chytridiomycetes (*Catenaria*, Ichida & Fuller, 1973; *Phlyctochytrium*, McNitt, 1973; *Entophyctis*, Powell, 1975; *Harpochytrium*, Whisler & Travland, 1973), and *Basidiobolus* (Gull & Trinci, 1974*a*; Sun & Bowen, 1972; Tanaka, 1970) where it is aligned along equatorial metaphase plates. The metaphase chromatin appears largely uncondensed in Oomycete fungi and the kinetochores are not aligned on well-defined metaphase plates (Beakes, 1980; Heath, 1974*a*, 1978; Heath & Greenwood, 1970; Hemmes & Hohl, 1973; Khan, 1976). Location of chromatin within dividing Mucoralean nuclei is also problematical, and it appears to be associated with the envelope (Franke & Reau, 1972; McCully & Robinow, 1973) or central spindle apparatus (Bland & Lunney, 1975; Franke & Reau, 1972).

Except in the Hydromyxomycetes, the nucleolus is normally retained throughout mitosis (Table 3). It is probably most striking in the Plasmodiophoromycetes where it is aligned along the metaphase spindle axis, surrounded by an annulus of chromatin to form characteristic cruciate profiles (Fig. 2, Braselton *et al.*, 1975; Garber & Aist, 1979*a*; Keskin, 1971). In most Chytridiomycetes the nucleolus is partially dispersed during prophase and is probably excluded in the mid-piece region at telophase (McNitt, 1973; Powell, 1975). In *Allomyces* the nucleolus remains condensed and is excluded at telophase (Olson, 1974*a*).

In the Oomycete genera *Albugo* and *Saprolegnia* and the Zygomycete, *Ancyclistes*, the conspicuous nucleoli are retained in their condensed state during metaphase, located to one side of the spindle apparatus (Heath & Greenwood, 1968; Khan, 1976; Moorman, 1976). In *Phytophthora*, the metaphase nucleolus is located in a curious pocket-like extension of the nucleus (Hemmes & Hohl, 1973). During anaphase the Oomycete nucleolus disperses somewhat as it segregates between the two daughter nuclei (Heath & Greenwood, 1968; Hemmes & Hohl, 1973).

In the Mucorales the nucleolus either segregates along the spindle apparatus (*Mucor hiemalis*, McCully & Robinow, 1973) or attaches to the nuclear envelope (Fig. 7, *Pilobus crystallinus*, Bland & Lunney, 1975). In *Basidiobolus* the large nucleolus disperses throughout the nucleoplasm during prophase and at metaphase segregates to form two broad diffuse bands at each pole of the spindle (Fig. 8, Gull & Trinci, 1974*a*; Sun & Bowen, 1972; Tanaka, 1970).

Meiosis

There have been relatively few ultrastructural accounts of meiosis in Phycomycete fungi and the number of orders covered is very small (Table 2). This is largely due to difficulty in satisfactorily fixing the thick-walled structures in which zygotic meiosis usually occurs (Olson, 1974*b*; Olson & Reichle, 1978*a, b*). However, even with Oomycetes, which have much more amenable gametangial meiosis, there have been no ultrastructural accounts of meiosis in the Peronosporales, in spite of increasing numbers of corresponding light microscopic studies (Sansome, 1980; Win-Tin & Dick, 1975). Meiosis in *Plasmodiophora brassicae* (Garber & Aist, 1979*b*) and *Saprolegnia* (Beakes, 1980; Beakes & Gay, 1977; Howard & Moore, 1970) is summarized in Figs. 9 and 10, respectively. Although prophase I events are usually easy to follow ultrastructurally, due to their relatively slow speed and high degree of synchrony (Beakes, 1980; Garber & Aist, 1979*b*; Howard & Moore, 1970; Olson, 1974*b*; Olson & Reichle, 1978*a, b*), later stages of meiosis occur much more rapidly and have often been incompletely described (Table 2). In *Labyrinthula* (Perkins & Amon, 1969), *Plasmodiophora* (Fig. 9, Garber & Aist, 1979*b*) and the Blastocladiales (Olson, 1974*b*, Olson & Reichle, 1978*a, b*) the first (meiosis I) and second (meiosis II) divisions of meiosis occur independently (Fig. 9), whereas in the Saprolegniales, both take place within the same envelope (Fig. 10). This results in the characteristic cruciate (clover-leaf) profiles at telophase II (Howard & Moore, 1970).

Nuclear envelope

Species that have fenestrate mitotic envelopes (Table 3) also develop polar fenestrae during metaphase I of meiosis (*Labyrinthula*, Perkins & Amon, 1969; Perkins, 1970; *Sorosphaera*, Braselton & Miller, 1973). Similarly, the nuclear envelopes of both *Allomyces*

(Olson, 1974*b*) and *Saprolegnia* (Beakes, 1980; Beakes & Gay, 1977; Ellzey, 1974; Howard & Moore, 1970) remain intact at this stage (Fig. 10). Although not present during mitosis, the meiosis I nuclei of *Achlya* (Ellzey, 1974) and meiosis II nuclei of *Saprolegnia* (Beakes, 1980) become associated with arrays of perinuclear endoplasmic reticulum.

Centrioles and spindle apparatus

In *Labyrinthula* the centrioles undergo replication during both prophase I and prophase II (Perkins & Amon, 1969; Perkins, 1970). However, in *Plasmodiophora* (Garber & Aist, 1979*b*) and *Saprolegnia*

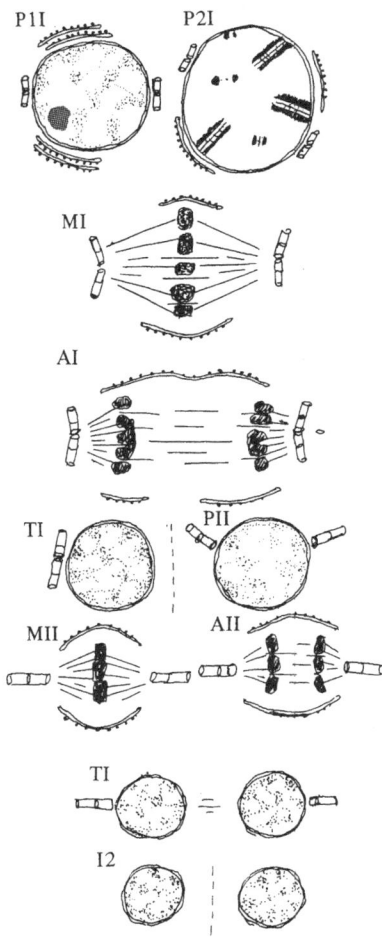

Fig. 9. Meiosis in *Plasmodiophora brassicae.*

(Beakes, 1980; Beakes & Gay, 1977; Howard & Moore, 1970) the centrioles only replicate during prophase I. This results in the poles of the second divisional spindles being associated with single centrioles, which characteristically are orientated at 90° to the nuclear envelope (Figs. 9 and 10).

Another phenomenon shared by *Plasmodiophora* (Garber & Aist, 1979*b*) and some members of the Saprolegniales (*Saprolegnia ferax*, Beakes, 1980; *S. furcata*, Beakes & Gay, 1977, but not *Achlya*

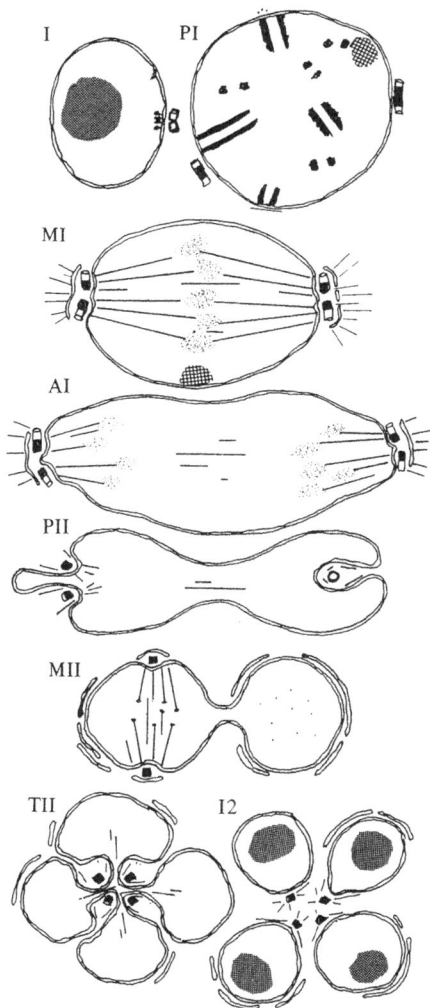

Fig. 10. Meiosis in *Saprolegnia*.

ambisexualis, Ellzey, 1974 and *S. terrestris*, Howard & Moore, 1970) is the elongation of centrioles (from 200 to 400 μm) during meiosis I. In *Saprolegnia* elongate centrioles develop during prophase I and persist only as far as anaphase I (Beakes, 1980; Beakes & Gay, 1977) whereas in *Plasmodiophora*, elongation occurs during metaphase I and persists throughout meiosis II (Garber & Aist, 1979*b*). It has been suggested that this phenomenon may represent the vestiges of a suppressed motile phase (Beakes & Gay, 1977; Garber & Aist, 1979*b*).

The meiotic centrioles in *Saprolegnia* and *Achlya* also differ from their mitotic counterparts by becoming ensheathed in a layer of electron-dense material and a cap of pericentriolar endoplasmic reticulum, both of which appear during prophase I (Beakes, 1980; Beakes & Gay, 1977; Ellzey, 1974; Howard & Moore, 1970). In contrast, centriole behaviour in *Sorosphaera* and *Allomyces* appears identical to that described during mitosis (Braselton & Miller, 1973; Olson, 1974*b*). In all species in which meiosis has been described (Table 2), including members of the Saprolegniaceae, spindle development during metaphase I occurs after centriole duplication and migration (Figs. 9 and 10). In *Saprolegnia* kinetochores were not seen in metaphase I nuclei, where chromatin is quite well differentiated, but were observed in metaphase II spindles (Beakes, 1980).

Nucleoplasm: synaptonemal complexes, chromatin and nucleoli
The nucleoplasm of prophase I nuclei is characterized by the appearance of synaptonemal complexes (SC) between pairs of condensing bivalents during leptotene and zygotene (reviewed by Westegaard & von Wettstein, 1972). The appearance and dimensions of these complexes in Phycomycete fungi are summarized in Table 4, together with those of the well-defined SCs that occur in the cup fungus, *Neotiella rutilans* (Westergaard & von Wettstein, 1972). The best documented Phycomycete SCs are those that occur in members of the Blastocladiales (Table 4). They have amorphous 30 nm lateral elements and an electron-dense 20 nm core (Olson, 1974*b*; Olson & Reichle, 1978*a, b*). Although the 100 nm region of electron-dense chromatin is clearly defined in both *Allomyces* (Olson, 1974*a*) and *Blastocladiella* (Olson & Reichle, 1978*b*) it does not stain well in *Catenaria*, and the lateral elements are more easily resolved (Olson & Reichle, 1978*a*). In *Labyrinthula* (Perkins & Amon, 1969), *Plasmodiophora* (Garber & Aist, 1979*b*) and *Sorosphaera* (Braselton & Miller, 1973) the chromatin obscures the lateral elements, but the electron-dense core can be seen.

Table 4. *Synaptonemal complex dimensions*

Group	Species	Mean dimensions (nm)					Termination at nuclear envelope
		Chromatin	Lateral element	Central zone	Core	Cross fibres	
Hydromyxomycetes	*Labyrinthula* sp.	91		70	20	+	+
Plasmodiophoromycetes	*Plasmodiophora brassicae*	60–70		75–80	15–20	–	+
	Sorosphaera veronicae	80	c. 100	c. 20	+	+	
Blastocladiales	*Allomyces macrogynus*	200	110	20	–	+	
	Blastocladiella emersonii	100	30	110	20	–	+
	Catenaria anguillulae	?	30	110	20	–	+
Saprolegniales	*Achlya ambisexualis*	160		200	20	–	+
	Saprolegnia ferax	75–90		150	–	–	+
	Saprolegnia furcata	80–90		160	–	–	+
Lagenidiales	*Lagenidium callinectes*[a]	–	25	50	10	–	–
Leptomitales	*Sapromyces androgynus*	119		80	30	+	+
Ascomycetes	*Neotiella rutilans*	350	50	100	20	+	+

[a] polycomplexes

The central region of the complex is usually electron-transparent, although ill-defined cross fibres have been reported in *Labyrinthula* (Perkins & Amon, 1969) and *Sorosphaera* (Braselton & Miller, 1973). In *Blastocladiella*, mispairing of the SCs has been described which suggests this organism is probably polyploid (Olson & Reichle, 1978*b*).

Although well-defined parallel regions of condensed bivalent chromatin also appear in prophase I nuclei in the Saprolegniales (Beakes, 1980; Beakes & Gay, 1977; Ellzey, 1974; Howard & Moore, 1970) and Lagenidiales (Schnepf & Deichgräber, 1978), an intervening and rather ill-defined SC has only been found in *Achlya* (Ellzey & Huizar, 1977). The central zone in the Saprolegniales is also wider than for typical SCs (Table 4). However, the bivalent chromatin shows the usual termination at the nuclear envelope (Table 4) and serial reconstructions of prophase I nuclei have been used by Heath (1974*a*) to determine chromosome number in *Thraustotheca*. However, typical SCs with well-defined central cores have been reported in two other Saprolegnian orders (Leptomitales, Gotelli, 1979; Lagenidiales, Amerson & Bland, 1973). The latter was a curious report of polycomplexes in an encysted spore of *Lagenidium callinectes*, which presumably represent residual SCs in a zoomeiospore (Schnepf & Deichgräber, 1978).

In the Saprolegniales, metaphase I chromatin is much more conspicuous than in mitosis, and there is recognizable equatorial alignment (Beakes, 1980; Beakes & Gay, 1977; Ellzey, 1974; Howard & Moore, 1970). However, following meiosis I the chromatin once again becomes virtually impossible to see.

Unlike in mitosis the nucleolus is not usually retained during meiosis. In *Plasmodiophora* (Garber & Aist, 1979*b*), *Sorosphaera* (Braselton & Miller, 1973), *Allomyces* (Olson, 1974*b*) and *Saprolegnia* (Beakes & Gay, 1977; Howard & Moore, 1970) it disappears after metaphase I. In *Saprolegnia*, as in higher organisms (Smetana & Busch, 1974) there is often a close association between condensing chromatin and the contracted nucleolus (Ellzey & Huizar, 1977).

Conclusions

Ultrastructural studies of Phycomycete fungi over the past fifteen years have greatly improved understanding of their nuclear behaviour (see Robinow & Bakerspigel, 1965). In particular, it has shown that vegetative nuclear divisions are truly mitotic, involving the segregation of daughter chromatids along microtubular, often intranuclear, spindle apparatuses. The additional information obtainable with the electron microscope (e.g. structure of polar bodies, and behaviour of nuclear envelope) has been particularly useful in helping to trace possible phylogenetic relationships in this complex and varied group of organisms (see Fuller, 1976; Kubai, 1978). In contrast, studies of meiosis are sparse and, in particular, accounts of Peronosporalean and Zygomycete meiosis are still awaited. Now that the pioneering descriptive accounts of Phycomycete nuclei are nearing completion, it is to be hoped that in the future electron microscopists will increasingly collaborate with biochemists, cell biologists and geneticists to unravel the largely unknown functional secrets of Phycomycete nuclei.

References

Amerson, H. V. & Bland, C. E. (1973). The occurrence of polycomplexes in the nucleus of encysting spores of *Lagenidium callinectes*, a marine phycomycete. *Mycologia*, **65**, 966–9.

Beakes, G. W. (1976). The effects of streptomycin and divalent cations on the development and fine-structure of *Saprolegnia*. PhD Thesis, University of London.

Beakes, G. W. (1980). Electron microscopic study of oospore maturation and germination in an emasculate isolate of *Saprolegnia ferax*. 4. Nuclear cytology. *Canadian Journal of Botany*, **58**, 228–40.

Beakes, G. W. & Gay, J. L. (1977). Gametangial nuclear division and fertilization in *Saprolegnia furcata* as observed by light and electron microscopy. *Transactions of the British Mycological Society*, **69**, 459–71.

Beakes, G. W. & Gay, J. L. (1980). The effects of streptomycin on growth and sporulation of *Saprolegnia*. *Journal of General Microbiology*, **119**, 361–71.

Beckett, A., Heath, I. B. & McLaughlin, D. J. (1974). *An Atlas of Fungal Ultrastructure*. London: Longman.

Bland, C. E. & Lunney, C. Z. (1975). Mitotic apparatus of *Pilobus crystallinus*. *Cytobiologie*, **11**, 382–91.

Bouteille, M., Laval, M. & Dupoy-Coin, A. M. (1974). Localization of nuclear functions as revealed by ultrastructural autoradiography and cytochemistry. *The Cell Nucleus*, vol. 1, ed. H. Busch, pp. 3–71. New York & London: Academic Press.

Braselton, J. P. & Miller, C. E. (1973). Centrioles in *Sorosphaera*. *Mycologia*, **65**, 220–5.

Braselton, J. P., Miller, C. E. & Pechak, D. G. (1975). The ultrastructure of cruciform nuclear division in *Sorosphaera veronicae* (Plasmodiophoromycetes). *American Journal of Botany*, **62**, 349–58.

Bryant, T. R. & Howard, K. L. (1969). Meiosis in the oomycetes: 1. A microspectrophotometric analysis of nuclear deoxyribonucleic acid in *Saprolegnia terrestris*. *American Journal of Botany*, **56**, 1075–83.

Dylewski, D. P., Braselton, J. P. & Miller, C. E. (1978). Cruciform nuclear division in *Sorosphaera veronicae*. *American Journal of Botany*, **65**, 258–67.

Ellzey, J. T. (1974). Ultrastructural observations of meiosis within antheridia of *Achlya ambisexualis*. *Mycologia*, **66**, 32–47.

Ellzey, J. T. & Huizar, E. (1977). Synaptonemal complexes in antheridia of *Achlya ambisexualis* E 87. *Archives of Microbiology*, **112**, 311–13.

Ellzey, J. T., Huizar, E. & Yanez, D. (1976). Microfilament bundles in antheridial nuclei of *Achlya ambisexualis* E 87. *Archives of Microbiology*, **107**, 113–14.

Franke, W. W. & Reau, P. (1972). The mitotic apparatus of a zygomycete *Phycomyces blakesleeanus*. *Archiv für Mikrobiologie*, **90**, 121–9.

Franke, W. W. & Scheer, U. (1974). Structures and functions of the nuclear envelope. *The Cell Nucleus* vol. 1, ed. H. Busch, pp. 219–347. New York & London: Academic Press.

Fuller, M. S. (1976). Mitosis in fungi. *International Review of Cytology*, **45**, 113–53.

Garber, R. C. & Aist, J. R. (1979a). Mitosis in *Plasmodiophora brassicae* (Plasmodiophorales). *Journal of Cell Science*, **40**, 89–110.

Garber, R. C. & Aist, J. R. (1979b). The ultrastructure of meiosis in *Plasmodiophora brassicae* (Plasmodiophorales). *Canadian Journal of Botany*, **57**, 2509–18.

Gleason, F. H. (1973). Nuclear tubules in *Saprolegnia*. *Cytobios*, **8**, 185–7.

Gotelli, D. (1979). Synaptonemal complexes in the gametangia of *Sapromyces androgynus* (Leptoniatales, Oomycetes). *Mycotaxon*, **9**, 90–2.

Grove, S. N. & Bracker, C. E. (1978). Protoplasmic changes during zoospore eruption and cyst germination in *Pythium aphanider-matum*. *Experimental Mycology*, **2**, 51–98.

Gull, K. & Trinci, A. P. J. (1974a). Nuclear division in *Basidiobolus ranarum Transactions of British Mycological Society*, **63**, 457–60.

Gull, K. & Trinci, A. P. J. (1974b). Effects of griseofulvin on the mitotic cycle of the fungus *Basidiobolus ranarum*. *Archives of Microbiology* **95**, 57–65.

Heath, I. B. (1974a). Mitosis in the fungus *Thraustotheca clavata*. *Journal of Cell Biology*, **60**, 204–20.

Heath, I. B. (1974b). Centrioles and mitosis in some oomycetes. *Mycologia*, **66**, 354–9.

Heath, I. B. (1974c). Genome separation mechanisms in prokaryotes, algae, and fungi. *The Cell Nucleus*, vol. 2, ed. H. Busch, pp. 487–515. New York & London: Academic Press.

Heath, I. B. (1975a). The effect of antimicrotubule agents on the growth and ultrastructure of the fungi *Saprolegnia ferax* and their ineffectiveness in disrupting hyphal microtubules. *Protoplasma*, **85**, 147–76.

Heath, I. B. (1975b) Ultrastructure of freshwater phycomycetes. *Recent Advances in Aquatic Mycology*, ed. E. B. Gareth-Jones, pp. 603–50. London: Paul Elek Press.

Heath, I. B. (1975c). The possible significance of variations in the mitotic systems of the aquatic fungi (phycomycetes). *Biosystems*, **7**, 351–9.

Heath, I. B. (1978). Experimental studies of mitosis in the fungi. *Nuclear Division in the Fungi*, ed. I. B. Heath, pp. 89–176. New York & London: Academic Press.

Heath, I. B. & Greenwood, A. D. (1968). Electron microscope observations of dividing somatic nuclei in *Saprolegnia*. *Journal of General Microbiology*, **53**, 287–9.

Heath, I. B. & Greenwood, A. D. (1970). Centriole replication and nuclear division in *Saprolegnia*. *Journal of General Microbiology*, **62**, 139–48.

Heath, I. B. & Greenwood, A. D. (1971). Ultrastructural observations on the kinetosomes, and Golgi bodies during the asexual life cycle of *Saprolegnia*. *Zeitschrift für Zellforschung und Mikroscopische Anatomie*, **112**, 371–89.

Heath, I. B., Greenwood, A. D. & Griffiths, H. B. (1970). The origin of flimmer in *Saprolegnia, Dictyuchus, Synura*, and *Cryptomonas. Journal of Cell Biology*, **7**, 445–61.

Hemmes, D. E. & Bartnicki-Garcia, S. (1975). Electron microscopy of gametangial interaction and oospore development in *Phytophthora capsici*. *Archives of Microbiology* **103**, 91–112.

Hemmes, D. E. & Hohl, H. R. (1973). Mitosis and nuclear degeneration: simultaneous events during secondary sporangium formation in *Phytophthora palmivora*. *Canadian Journal of Botany* **51**, 1673–75.

Hickey, E. L. & Coffey, M. D. (1977). A fine structure study of the pea downy mildew fungus *Peronospora pisi* in its host *Pisum sativum*. *Canadian Journal of Botany*, **55**, 2845–58.

Hoch, H. C. & Mitchell, J. E. (1972). The ultrastructural of zoospores of *Aphanomyces enteiches* and of their encystment and subsequent germination. *Protoplasma*, **75**, 113–38.

Howard, K. L. & Moore, R. T. (1970). Ultrastructure of oogenesis in *Saprolegnia terrestris*. *Botanical Gazette*, **131**, 331–36.

Ichida, A. & Fuller, M. S. (1968). Ultrastructure of mitosis in the aquatic fungus *Catenaria anguillulae*. *Mycologia*, **60**, 141–55.

Jordan, E. G., Birkett, J. A. & Trinci, A. P. J. (1980). Effect of temperature on nucleolar size in *Basidiobolus ranarum*. *Transactions of the British Mycological Society*, **74**, 111–18.

Kazama, F. Y. (1974). The ultrastructure of nuclear division in *Thraustochytrium* sp. *Protoplasma*, **82**, 155–75.

Keskin, B. (1971). Beitrag zur protomitase bei *Polymyxa betae*. *Archiv für Mikrobiologie*, **77**, 334–48.

Khan, S. E. (1976). Electron microscope observations of dividing somatic nuclei of *Albugo*. *Canadian Journal of Botany*, **54**, 168–72.

Kubai, D. F. (1974). The evolution of the mitotic spindle. *International Review of Cytology*, **43**, 167–227.

Kubai, D. F. (1978). Mitosis and fungal phylogeny. In *Nuclear Division in the Fungi*, ed. I. B. Heath, pp. 177–228. New York & London: Academic Press.

Lessie, P. E. & Lovett, J. S. (1968). Ultrastructural changes during sporangium formation and zoospore differentiation in *Blastocladiella emersonii*. *American Journal of Botany*, **55**, 220–236.

Madelin, M. F. & Beckett, A. (1972). The production of planonts by thin-walled sporangia of the fungus *Coelomyces indicus*, a parasite of mosquitoes. *Journal of General Microbiology*, **72**, 185–200.

McCully, E. K. & Robinow, C. F. (1973). Mitosis in *Mucor hiemalis*. A comparative light and electron microscopic study. *Archiv für Mikrobiologie*, **94**, 133–48.

McNitt, R. (1973). Mitosis in *Phlyctochytrium irregulare*. *Canadian Journal of Botany*, **51**, 2065–74.

Mohberg, J. (1974). The nucleus of the plasmodial slime moulds. In *The Cell Nucleus*, vol. 1, ed. H. Busch, pp. 187–218. New York & London: Academic Press.

Moorman, G. W. (1976). Mitosis in *Ancyclistes*. *Mycologia*, **68**, 902–9.

Olson, L. W. (1972). Colchicine and the mitotic spindle of the aquatic phycomycete *Allomyces*. *Archiv für Mikrobiologie*, **84**, 327–38.

Olson, L. W. (1974a). Mitosis in the aquatic phycomycete *Allomyces neo-moniliformis*. *Comptes Rendus du Laboratoire Carlsberg*, **40**, 125–47.

Olson, L. W. (1974b). Meiosis in the aquatic phycomycete *Allomyces maerogynus*. *Comptes Rendus du Laboratoire Carlsberg*, **40**, 113–51.

Olson, L. W. & Reichle, R. (1978a). Meiosis and diploidization in the aquatic phycomycete *Catenaria anguillulae*. *Transactions of the British Mycological Society*, **70**, 423–37.

Olson, L. W. & Reichle, R. (1979b). Synaptonemal complex formation and meiosis in the resting sporangium of *Blastocladiella emersonii*. *Protoplasma*, **97**, 261–73.

Olson, L. W. & Rønne, M. (1975). Induction of abnormal gametes and androgenesis in the aquatic phycomycete *Allomyces*. *Protoplasma*, **84**, 327–44.

Perkins, F. O. (1970). Formation of centriole and centriole-like structures during meiosis and mitosis in *Labyrintha* sp. (Rhizopodea, Labyrinthulida). An electron-microscope study. *Journal of Cell Science*, **6**, 629–53.

Perkins, F. O. & Amon, J. P. (1969). Zoosporulation in *Labyrinthula* sp.; an electron microscope study. *Journal of Protozoology*, **16**, 235–57.

Pickett-Heaps, J. D. (1969). The evolution of the mitotic apparatus: an attempt at comparative ultrastructural cytology of dividing plant cells. *Cytobios*, **3**, 257–80.

Pickett-Heaps, J. D. (1972). Variation in mitosis and cytokinesis in plant cells: its significance in the phylogeny and evolution of ultrastructural systems. *Cytobios*, **5**, 59–77.

Pommerville, J. & Fuller, M. S. (1976). The cytology of gametes and fertilization of *Allomyces macrogynus*. *Archives of Microbiology* **109**, 21–30.

Powell, M. J. (1975). Ultrastructural changes in nuclear membranes and organelle associations during mitosis of the aquatic fungus *Entophlyctis sp*. *Canadian Journal of Botany*, **53**, 627–46.

Robinow, C. F. & Bakerspigel, A. (1965). Somatic nuclei and forms of mitosis in fungi. In *The Fungi an Advanced Treatise*, vol. 1, *The Fungal Cell*, ed. G. C. Ainsworth & A. S. Sussman, pp. 119–42. New York & London: Academic Press.

Sansome, E. (1980). Reciprocal translocation heterozygosity in heterothallic species of *Phytophthora* and its significance. *Transactions of the British Mycological Society*, **74**, 175–85.

Schnepf, E. & Deichgräber, G. (1978). Development and ultrastructure of the marine, parasitic oomycete, *Lagenisma coscinodisci* Drebes (Lagenidiales). *Archives of Microbiology*, **116**, 141–50.

Smetana, K. & Busch, H. (1974). The nucleolus and nucleolar DNA. *The Cell Nucleus*, vol. 1, (ed. H. Busch), pp. 73–147. New York & London: Academic Press.

Sun, N. C. & Bowen, C. C. (1972). Ultrastructural studies of nuclear division in *Basidiobolus ranarum* Eidam. *Caryologia*, **25**, 471–94.

Tanaka, K. (1970). Mitosis in the fungus *Basidiobolus ranarum* as revealed by electron microscopy. *Protoplasma*, **70**, 423–40.

Westergaard, M. & von Wettstein, D. (1972). The synaptonemal complex. *Annual Review of Genetics*, **6**, 71–109.

Whisler, H. C. & Travland, L. B. (1973). Mitosis in *Harpochytrium*. *Archiv für Protistenkunde*, **115**, 69–74.

Win-Tin & Dick, M. W. (1975). Cytology of oomycetes. *Archives of Microbiology*, **105**, 283–93.

Ultrastructure and behaviour of nuclei and associated structures within the meiotic cells of Euascomycetes

A. BECKETT

Department of Botany, Bristol University,
Bristol BS8 1UG, UK

Introduction

Since the yeast nucleus is receiving separate treatment in this volume (Zickler, pp. 63–83), I shall deal here with those fungi of the subclass Euascomycetidae *sensu* Alexopoulos (1962). However, since ultrastructural work is restricted to relatively few genera, my usage of the term Euascomycete will in practice encompass members of the Pyrenomycetes and Discomycetes only.

One of the earliest ultrastructural studies on the nuclear cytology of the meiotic cells in a Euascomycete fungus is that on the Discomycete *Neottiella rutilans* (Westergaard & von Wettstein, 1966). This and subsequent papers (Westergaard & von Wettstein, 1970*a, b*) provide a detailed account of this fungus.

Two points arising from the work on *Neottiella* are particularly relevant to the scope and emphasis of this chapter. One is that 'meiotic induction' in the form of DNA replication occurs in the crozier cells prior to karyogamy. Therefore, I shall consider here the stalk cell, penultimate cell and terminal cell of the crozier as well as the ascus itself. Secondly, the exceptionally large size of the nucleus in *Neottiella* has enabled excellent light microscope preparations to be made (Rossen & Westergaard, 1966). This illustrates the usefulness of a correlated light and electron microscope study. Such a correlation is essential for a clear understanding of most preparations at the ultrastructural level and in my view is a prerequisite for a study of nuclei and associated structures in fungi. Comprehensive reviews of meiotic events at the light microscope level have been provided by Olive (1953, 1965) and should be consulted for general background information. In this chapter I shall discuss certain aspects of meiotic nuclear behaviour at the light microscope level only when it is necessary to do so in order to clarify ultrastructural observations.

Prefusion nuclei and associated structures

The term prefusion nucleus (*sensu stricto*) presupposes that two such nuclei will, at some point in the nuclear cycle, fuse together to form a diploid fusion nucleus. In *Neottiella rutilans* for example this would include the nucleus of the terminal cell and that of the stalk cell but may exclude those of the penultimate cell (see below). Since information is lacking on ultrastructural differences between nuclei that do ultimately fuse and those that do not, I shall consider nuclei of the crozier cells under this one heading.

The nuclear envelope

Specific information on the structure of nuclear envelope in fungal nuclei is sparse. As in all eukaryotic organisms interphase nuclei are bounded by an envelope of two membranes. The inner and outer of these membranes are structurally continuous around the edges of a variable number of perforations or nuclear pores. Frequently the nuclear pores in the envelopes of prefusion nuclei are associated with electron-opaque deposits of granular material (Westergaard & von Wettstein, 1966; Wells, 1970; Zickler, 1977). Little is known of the frequency and distribution of pores in the nuclear envelope. In *Neottiella rutilans* large numbers of well developed pores occur in the envelopes of the prefusion stalk and terminal cell nuclei. Similar pores are found in the envelope of the diploid fusion nucleus. In contrast to this, nuclear pores in the penultimate cell nuclei are fewer in number and are differentiated to a lesser degree. These nuclei of the penultimate cell may not fuse in *N. rutilans* and it is possible that these structural differences may represent different nuclear envelope 'types'. It is tempting to speculate that such a typing system, involving structural differences at the nuclear envelope level and which could be mediated via the nuclear pores, might play a role in determining which nuclei fuse together and which do not. In *Neurospora crassa* (Singleton, 1953), *Podospora anserina* (Beckett & Wilson, 1968), *Sordaria humana* (Fig. 1*a*) and most other Euascomycetes (Olive, 1953; Reeves, 1967; Carroll, 1969; Wells, 1970, 1972), it is the nuclei of the penultimate cell which fuse and not those of the stalk and terminal cells.

Structural continuity between the outer membrane of the nuclear envelope and endoplasmic reticulum (ER) membranes is now well documented in eukaryotes (Franke, 1974). In *Neurospora crassa* and *Sordaria humana*, membrane continuities of this kind are common with the prefusion nuclei of the crozier cells. In *S. humana* the degree of membrane continuity is such that the nuclear envelope of the prefusion nucleus in the terminal or stalk cell may be structurally continuous, via the ER and a specialized membrane complex in the pores of the cross walls between these cells, with that of the fusion nucleus in the ascus (Fig. 1*a*, *c*). Here also the possibility exists for a nuclear control system, operating by way of membrane continuities, which might govern the rate of particular pathway of differentiation within certain cell types. Evidence for this is circumstantial but in *N. crassa* and *S. humana* these complex septal pore membrane structures, composed of specialized ER and which connect the nucleus of one cell with that of another, occur only in the crozier cells where the dikaryon is being established, and in the very young ascus initials where karyogamy has just taken place. In addition, membrane continuities can be seen between the individual septal pore apparatuses of, for example, the stalk and terminal cell and/or the stalk cell and that subtending it.

Structural relationships between these cytoplasmic membranes and the nuclear envelope inevitably mean that there is also a luminal continuity between the perinuclear space, the ER lumen, the spaces between the membranes of the septal pore apparatus and hence with corresponding intramembranous spaces in adjacent cells. This luminal continuity (Fig. 1*c*) is clearly demonstrated in sections of material which have been impregnated with a mixture of osmium tetroxide and zinc iodide solution (Marty, 1973; Harris, 1978). Numerous examples of continuities between the perinuclear space and cytoplasmic membrane lumena have been reported for a variety of eukaryotes (Franke, 1974), but this is the first reported instance of such a complex relationship in the fungi and will be discussed in more detail elsewhere (A. Beckett, *Protoplasma*, in press).

The inner membrane of the prefusion nuclear envelope serves as a site for the attachment and localization of condensed chromatin. This is particularly well illustrated in the nuclei of the stalk and terminal cells of the crozier of *Neottiella rutilans* (Westergaard & von Wettstein, 1966). Similar regions of electron-opaque, granular material in other organisms have been referred to as peripheral heterochromatin (Franke, 1974). However, Fuller (1976) suggests that in some cases the term heterochromatin may be inappropriate, since little is known of the molecular constitution of these granular areas.

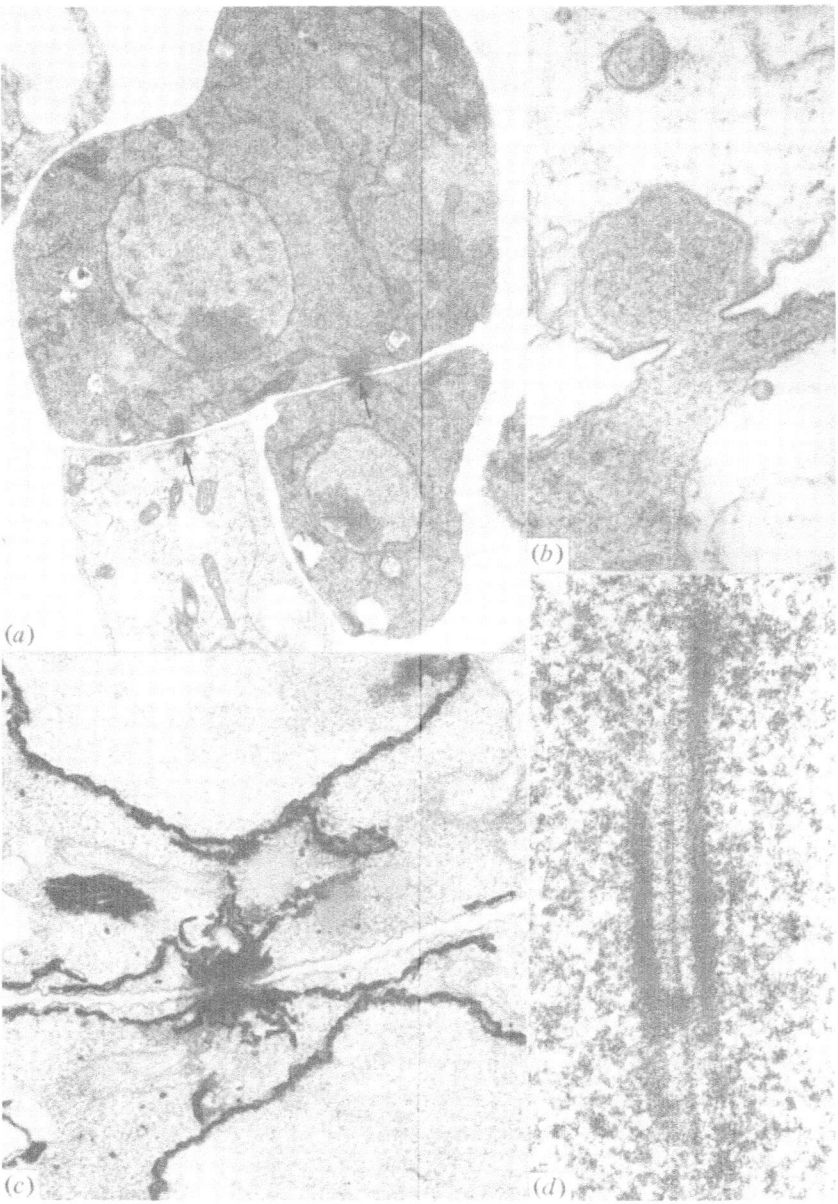

Fig. 1 (*a*) *Sordaria humana:* LS of crozier showing stalk cell, penulti-mate cell with fusion nucleus and terminal cell with prefusion nucleus. Note septal pore apparatuses in cross walls (arrows). × 8000. (*b*) *S. humana:* septal pore apparatus in the thick cross wall at the base of a mature ascus. × 40 000. (*c*) *S. humana:* LS through two cells of

Even denser clumps of chromatin are associated with the inner membrane of the envelope of nuclei in the penultimate cell. As already suggested, these are probably not prefusion nuclei. Condensed chromatin is also attached to the nuclear envelope in the prefusion nuclei of *Ascobolus stercorarius* (Wells, 1970; Zickler, 1970) and *Xylaria polymorpha* and *X. longipes* (A. Beckett, unpublished). The attachment of short synaptic structures, comprised of lateral and central components, to the inner membrane of the nuclear envelope of what are presumably prefusion nuclei in *Ascophanus carneus*, has been reported by Zickler (1973a). The attachment of these structures via the telomeres with the nuclear envelope of the diploid nucleus is normal and will be discussed later, but the occurrence of them in nuclei of ascogenous hyphae, as in *Ascophanus*, is apparently unique.

The nucleolus and chromatin

In most species of Euascomycetes the nucleolus is seen in sections of aldehyde- and osmium tetroxide-fixed specimens, as an electron-opaque, largely granular and usually circular body (Westergaard & von Wettstein, 1966; Beckett & Crawford, 1970; Wells, 1970; Zickler, 1970; Hung & Wells, 1971). In some cases (Westergaard & von Wettstein, 1966; Hung & Wells, 1971) internal differentiation of the nucleus is visible in the form of irregular, comparatively electron-transparent areas which may represent nucleolar vacuoles. In contrast to the situation in the fusion nucleus (see later), there is little published evidence for the presence of granular and fibrillar regions within the nucleolus of prefusion nuclei.

Chromatin in prefusion nuclei is usually in the form of electron-opaque, condensed material and is possibly heterochromatin. In some cases little or no chromatin seems to have been preserved, or at least, not in a visible state (Hung & Wells, 1971). Whether this reflects differences in nuclear physiology between those with and those apparently without visible chromatin, or whether it simply reflects differences in fixation technique, is not clear.

ascogenous hypha stained with osmium tetroxide/zinc iodide mixture. Membrane and luminal continuity can be seen between the nuclei of adjacent cells via the septal pore apparatus. × 20 000. (*d*) *Xylaria longipes:* LS through part of a synaptonemal complex showing the banded lateral components, the thin central component and a recombination nodule. Note transverse filaments traversing the central region between central and lateral components. × 45 000.

Spindle pole bodies and microtubule organizing centres

The term spindle pole body (SPB) has been widely used by mycologists and electron microscopists to describe the electron-opaque structure(s) which is associated with microtubules, both of the cytoplasm and of the nuclear spindle, and which is usually found adjacent to the nuclear envelope of both interphase and dividing nuclei in the non-flagellate fungi. The origin of this term, its synonyms and the connotations of its usage are discussed in detail by Fuller (1976). Microtubule organizing centre (MTOC) is the term coined by Pickett-Heaps (1969) for a morphologically similar structure, which he proposed was responsible for microtubule organization. It is usually associated with microtubules but may be located in positions other than at the poles of the spindle of dividing nuclei. Fuller (1976) has suggested that SPBs may be a special type of MTOC. I believe this is justifiable and I shall use the terms separately here giving my reasons where relevant.

In aldehyde-fixed Euascomycetes, electron-opaque structures are associated to varying extents with prefusion nuclei of the crozier cells. These structures fall into two distinct morphological categories. (1) An amorphous body of variable size and shape (0.4–1.0 μm wide; 1.0–1.6 μm long) lies outside the nuclear envelope in a zone of exclusion (i.e. devoid of ribosomes). Examples of this type are found in *Xylaria polymorpha* (Beckett & Crawford, 1970; Schrantz, 1970), *X. longipes* and *Sordaria humana* (A. Beckett, unpublished). This I shall refer to as an MTOC. (2) A plate-like disc of the material is closely associated with, or is even a specialized region of, the nuclear envelope. This form resembles that seen at the poles of dividing nuclei in asci. Examples of this type are found in *Ascobolus stercorarius* (Wells, 1970; Zickler, 1970) and *Pyronema domesticum* (Hung & Wells, 1971). Since at interphase this structure is not associated with a nuclear spindle, I believe the term MTOC is most suitable for it. It is not obvious whether this structure is identical to or one and the same, as that found at a later stage polarizing the division spindles in the same organisms. These structures will be dealt with in more detail later.

Fusion nuclei and meiotic prophase

Studies with the light microscope have shown that the so-called 'Neurospora-type' of meiosis (Westergaard, 1964; Rossen & Westergaard, 1966) in which the fusion nucleus immediately enters meiotic prophase and chromosomes pair while in a highly contracted state (Singleton, 1953), occurs in several Euascomycetes (Carr & Olive, 1958; Lu, 1967; Barry, 1969). The fact that the chromosomes contract shortly after karyogamy has led to difficulties in the recognition and interpretation of early prophase stages. Indeed it has been suggested that the leptotene stage is absent in the 'Neurospora-type' meiosis (Burnett, 1976). Singleton (1953), Carr & Olive (1958) and Barry (1969) used ascus length and morphology as an aid to the recognition of meiotic prophase stages in *Neurospora crassa, Sordaria fimicola* and *N. crassa* respectively. A relationship has also been shown between the sizes and volumes of nuclei and nucleoli and between the lengths of chromosomes, with specific prophase stages (Singleton, 1953; Carr & Olive, 1958; Lu, 1967). In general this is seen as a gradual increase in the diameter of the nucleus and nucleolus from early prophase until late pachytene, followed by a decrease during diplotene. Lu (1967) related changes in the size of the nucleolus to the physiological requirements of the ascus which increases in length dramatically during this period.

Correlated light and electron microscope studies of these stages are rare but in *Neottiella rutilans*, Westergaard & von Wettstein (1970a) have shown that changes in the ultrastructure of the nucleolus include the loss of nucleolar material into the ascus cytoplasm during late pachytene and diplotene. In *Sordaria macrospora* variations in the size, density and ultrastructure of the nucleolus have been correlated with different stages of prophase (Zickler, 1977). Recently, serial sectioning and subsequent reconstruction of whole nuclei have greatly improved our understanding of the ultrastructural events of meiotic prophase. In *N. rutilans* (Westergaard & von Wettstein, 1970b; von Wettstein, 1977), *S. macrospora* (Zickler, 1977) and *Neurospora crassa* (Gillies, 1979) it is now possible to recognize leptotene and zygotene as well as the more obvious pachytene and diplotene on the basis of precise ultrastructural features.

The nuclear envelope

Membranous and luminal continuity between the diploid nucleus and prefusion nuclei of stalk or terminal cells is a transient relationship, since once meiosis has taken place the septal pore apparatuses in the cross walls at the base of the ascus lack the elaborate membrane complex (Fig. 1b). The timing of these changes in relation to nuclear division and ascus development has not yet been precisely documented but it would seem to be an important aspect of the ultrastructural differentiation of asci, at least for *Sordaria* and *Neurospora*.

During meiotic prophase, the inner membrane of the envelope of the fusion nucleus serves as an attachment site for individual lateral components prior to pairing and subsequently for the fully formed synaptonemal complex (Schrantz, 1970; Westergaard & von Wettstein, 1970*b*; Gillies, 1972, 1979; Zickler, 1973*a*, 1977; von Wettstein, 1977). No obvious structural modifications of the membrane occur at these sites in fungi, although von Wettstein (1977) illustrated diagrammatically a thickening of the nuclear envelope at the points where the synaptonemal complex was attached. Thickenings or 'end plates' have been reported in insects and animals but not in plants (Gillies, 1975 and references cited).

The attachment sites are apparently distributed randomly around the nuclear envelope in *Neurospora crassa* (Gillies, 1979) although in *Sordaria macrospora* at least one end for each of bivalents 1 and 3–7 occupies a non-random site on the nuclear envelope, while the opposite ends are located randomly (Zickler, 1977). Zickler (1977) terms the non-random ends 'fixed' ends and suggests the 'fixed' end of bivalent 2 is anchored in the nucleolus.

The nuclear envelope has been implicated in the mediation of pairing of homologues during leptotene and zygotene (Gillies, 1975, 1979; Zickler, 1977). In *Sordaria macrospora* it has been postulated that the 'fixed' ends for one set of chromosomes may be established in the haploid nucleus prior to nuclear fusion, so that once karyogamy has taken place the seven chromosomes from one nucleus each have one 'fixed' end on the nuclear envelope. The seven chromosomes from the other nucleus presumably have their ends distributed randomly on the envelope. One of the original two MTOCs of the initial fusion nucleus is lost during leptotene. Bivalent 5 is apparently always attached to the nuclear envelope opposite to the retained MTOC and Zickler (1977) suggests that this point has been carried over, as it were, from the original haploid nucleus. She further postulates that the 'fixed' ends of the other six chromosomes may also have originated on the same nuclear envelope as that 'carrying' the retained MTOC. Thus it may be that the process of homologous pairing is in some way assisted by the early establishment of one 'fixed' end for one homologue of each bivalent.

An extension of this hypothetical mechanism might involve structural differences in the membrane(s) of the nuclear envelopes of two prefusion nuclei, and it may be that freeze-etch studies, which at present are lacking, might provide insight into this possibility. There is evidence to suggest that in some flowering plants there is a colchicine-sensitive protein either present in the nuclear envelope as in lily-meiocytes (Hotta & Shepard, 1973), or as a fibrillar attachment structure anchoring the chromosomes to the inner nuclear membrane as in wheat meiocytes (Bennett, Stern & Woodward, 1974). No similar findings have been reported for fungi, but one might speculate that alignment of attached ends of homologous chromosomes, as in *Sordaria macrospora*, could occur by the movement of specific proteinaceous structures through a fluid lipid system such as that described for the fluid mosaic model by Singer & Nicolson (1972). Observations on the meiotic-1 (*mei*-1) mutant of *Neurospora crassa* by Lu & Galeazzi (1978) indicate that the formation of lateral components during leptotene cannot itself bring about homologous pairing since although these structures are present, chromosome alignment fails to occur in over 90% of asci. It is not clear from this work however whether the lateral components, which appear to be of normal structure, are attached to the nuclear envelope. The nucleolus may also mediate chromosome pairing (Moens, 1973, see later).

In Euascomycetes the nuclear envelope remains intact during meiosis and mitosis until telophase. In *Neottiella rutilans* (Westergaard & von Wettstein, 1970a) the nucleolus is expelled from the nucleoplasm during late prophase but remains bounded by a portion of envelope which is derived from the nuclear envelope. The nucleolus remains in this state until telophase, at which time it is seen to be attached to the outer membrane of the envelope of one daughter nucleus. Here it subsequently aborts.

In *Pustularia cupularis* (Schrantz, 1970), the nuclear envelope constricts around each daughter nucleus at telophase in such a way as to leave a long narrow, membrane-bounded isthmus between the two. Remnants of the spindle microtubules and a portion of nucleoplasm from the former anaphase nucleus are trapped within the isthmus. This configuration is reminiscent of that found during mitosis in the Chytrids (Fuller, 1976). As with the Chytrids, nothing is known of the way(s) in which the envelope finally breaks at the constriction points and units around the separate daughter nuclei.

Work done with yeast by Severs, Jordan & Williamson (1976) suggests that freeze-etching may provide valuable information on the structure and behaviour of nuclear envelopes during the cell and nuclear cycle.

The intranuclear spindle during meiosis is essentially similar to that of mitotic nuclei and is composed of chromosomal and continuous microtubules. Possible interactions between chromosomes, microtubules and the nuclear envelope during nuclear division are discussed in Chapter 4.

The synaptonemal complex

The synaptonemal complex (SC) in all Euascomycetes consists of two, banded, electron-opaque *lateral components* which lie parallel to each other and between which is a single, electron-opaque, amorphous *central component*. The *central region*, between the lateral components and within which the central component lies, is generally composed of less electron-opaque material and is remarkably uniform in width throughout all genera and species studied.

The width of the central region is 90–120 nm, the diameter of the lateral component is 30–50 nm and that of the central component ±20 nm.

This uniformity of dimensions is found in a wide range of eukaryotic organisms and has been related to the universality of four-strand crossing-over at meiosis (Westergaard & von Wettstein, 1972).

The most detailed work on the development and structure of SCs is that on *Neottiella rutilans* (Westergaard & von Wettstein, 1970b; von Wettstein, 1977), *Sordaria macrospora* (Zickler, 1977) and on *Neurospora crassa* (Gillies, 1972, 1979). In addition much is known of the structure of SCs in *Pustularia cupularis*, *Galactinia plebeia* and *Xylaria polymorpha* (Schrantz, 1970), *Ascobolus immersus*, *A. stercorarius*, *S. fimicola*, *S. humana*, *Podospora anserina*, *P. setosa*, and *Ascophanus carneus* (Zickler, 1973a).

In all Ascomycetes the lateral components are seen to be regularly banded by a periodic configuration of thick and thin, electron-opaque bar-like structures (Figs. 1d, 2a). The precise periodic arrangement of these bands varies considerably between different genera and between different species of the same genus (Zickler, 1973a). Von Wettstein (1971) has suggested that variations in fine structure of lateral and central components between different organisms reflect differences in composition rather than of function. In some species, fine fibrils traverse the central region between lateral and central components (Westergaard & von Wettstein, 1970b; Fig. 1d).

Oval or spindle-shaped, electron-opaque thickenings of the central component (Fig. 1*d*) have been reported for most Euascomycetes (Schrantz, 1970; Gillies, 1972, 1979; Westergaard & von Wettstein, 1972; Zickler, 1973*a*, 1977). These structures, originally referred to as *nodes* by Gillies (1972) and subsequently termed *recombination nodules* by Carpenter (1975), are now considered to represent areas, along the central component, where genetic crossing-over takes place at pachytene. In *Sordaria macrospora* recombination nodules apparently form on the developing central components during mid to late zygotene and by pachytene one to four nodules are present on each bivalent. The location of at least one nodule seems to be constant while the others are more variably, though not randomly, positioned (Zickler, 1977). Gillies (1979) found that recombination nodules in *Neurospora crassa* tended to be localized distally while the centromeres were median. The numbers and distribution of nodules for each bivalent suggest a close correlation with observed cross-over frequency (Zickler, 1977; Gillies, 1979).

The formation of SCs has been studied from serial sections and completely reconstructed nuclei at various stages of meiotic prophase (Gillies, 1972, 1979; Westergaard & von Wettstein, 1970*b*; von Wettstein, 1971, 1977; Zickler, 1977).

In *Neottiella rutilans* (Westergaard & von Wettstein, 1970*b*) a single lateral component forms *de novo* between the sister chromatids of each, unpaired, homologous chromosome at leptotene. At the same time, amorphous granular material of the central region is assembled within the particulate zone of the nucleolus (see below). The electron-opaque central component structure is also formed in the nucleolus in conjunction with the central region material (Westergaard & von Wettstein, 1970*b*; von Wettstein, 1971, 1977). Recombination nodules have been seen on these pre-formed pieces of central component within the nucleolus. Although *N. rutilans* is so far unique in this respect, it would appear that these presumed genetic exchange sites are structurally established, in the nucleolus, prior to homologous pairing of chromosomes. This feature is not discussed by Westergaard & von Wettstein (1972) although their figure 15b illustrates the point.

The assembled central component must be in some way transported from the nucleolus to a position between the homologous chromosomes which, during late leptotene, achieve approximate alignment to within ±300 nm of each other. In *Neurospora crassa* and *Sordaria macrospora* (Gillies, 1972, 1979; Zickler, 1977), the central component forms *in situ* at this point, possibly involving an interdigitation of the transverse filaments, and is not associated with the nucleolus. In all ascomycetes the precise point-to-point pairing of chromosomes involving the completion of SC formation takes place during zygotene and is accomplished by pachytene. In *S. macrospora* several initiation sites for the formation of the central component may occur along the length of a bivalent. Neither the formation of these sites, nor precise pairing need be synchronous among the seven bivalents of a nucleus (Zickler, 1977). However, pairing usually occurs preferentially from the point of contact on the nuclear envelope.

Westergaard & von Wettstein (1970*b*) and von Wettstein (1971, 1977) have suggested that point-to-point pairing requires the synthesis of specific recognition sites to enable a portion of central component to bring together homologous portions of lateral components, each with their associated sister chromatids. They envisage that these sites form by a structural change in the opposing free surface of a central component 'block', once it has established contact on one side, with the lateral component of one homologue. This free surface is then specifically changed to fit only with the counterpart on the lateral component of the other homologue. This hypothesis is illustrated diagrammatically by von Wettstein (1971, 1977).

At pachytene the SC has been shown by serial sectioning to exist uninterrupted along the length of the bivalents. Homologous chromosomes are held uniformly together with a gap of 100–120 nm between them in which the SC lies. Both ends of each bivalent are anchored on the nuclear envelope with the exception of those bivalents that are associated with the nucleolus. In *Neurospora crassa* and *Sordaria macrospora* (Gillies, 1972; Zickler, 1977) and *Xylaria polymorpha* (Fig. 2*b*) the end of one bivalent terminates in the nucleolus. In *Neottiella rutilans* (Westergaard & von Wettstein, 1972) four bivalents have one end embedded in the nucleolus.

Moens (1973) suggested that since in *Neurospora crassa* one chromosome end was embedded in the nucleolus, then movement of this chromosome and presumably its homologue during pairing may be mediated by the nucleolus. In many ascomycetes nucleoli of prefusion nuclei themselves fuse after karyogamy (Moens, 1973) and this may influence pairing of the associated homologues. This might also be true for *Sordaria macrospora* (Zickler, 1977) and *Neottiella rutilans* (Westergaard & von Wettstein, 1970*a*, *b*). A role for the nucleolus in chromosome pairing was also inferred by Heywood & Magee (1976).

The structural relationship between chromatin and the two lateral components of a bivalent during meiotic prophase is not clearly understood for most ascomycetes or indeed for many fungi. This is because in most genera chromatin is not well preserved when prepared for electron microscopy (see for example the yeasts, Zickler, next chapter). However, *Xylaria polymorpha* (Fig. 2*a*), *Pustularia cupularis* (Schrantz, 1970) and particularly *Neottiella rutilans* (Westergaard & von Wettstein, 1970*b*) are exceptions to this tendency. This had enabled von Wettstein (1971, 1977) to formulate a hypothesis for the structural nature and behavioural characteristics of the chromatin with respect to the SC throughout meiotic prophase. During leptotene the two chromatids of each homologue lie on opposite sides of a single lateral component. Prior to the point-to-point pairing at zygotene, the sister chromatids become realigned side by side along one face of the lateral component. The arrival of the central component 'blocks' from the nucleolus and their assembly between the lateral components of the two homologues results in the progressive pairing at the 100–120 nm distance. Only a small portion of either of the sister chromatids is attached to the lateral component and von Wettstein (1971) suggests that in order to accommodate the four-strand crossing-over event, portions of only one chromatid may be represented in any one segment of the lateral component. Adjacent segments may however be linked with different sister chromatids. At pachytene, sister chromatids are not visible in thin sections as separate entities but calculations have been made which show that the ratio of the cross-section area of a pair of sister chromatids to that of the lateral component in *N. rutilans* is about 43 to 1 (Westergaard & von Wettstein, 1970*b*). Similarly in *Neurospora crassa* it has been calculated that the DNA in all of the pachytene SCs would account for only 0.3% of the total genome (Westergaard & von Wettstein, 1972).

Once crossing-over has been accomplished at pachytene, repulsion of chromosomes occurs at diplotene. At this stage thin sections show small portions of SC scattered throughout the nucleus. In *Sordaria macrospora* most of these portions contain a single recombination nodule (Zickler, 1977). In *Neottiella rutilans* (Westergaard & von Wettstein, 1970*b*) the repulsing homologues are held together only by short stretches of electron-opaque material, possibly derived from modified central component.

The lateral components, and subsequently the unmodified central component, are shed as fibrillar material into the nucleoplasm. During late diplotene and diakinesis the short stretches of modified SC are also

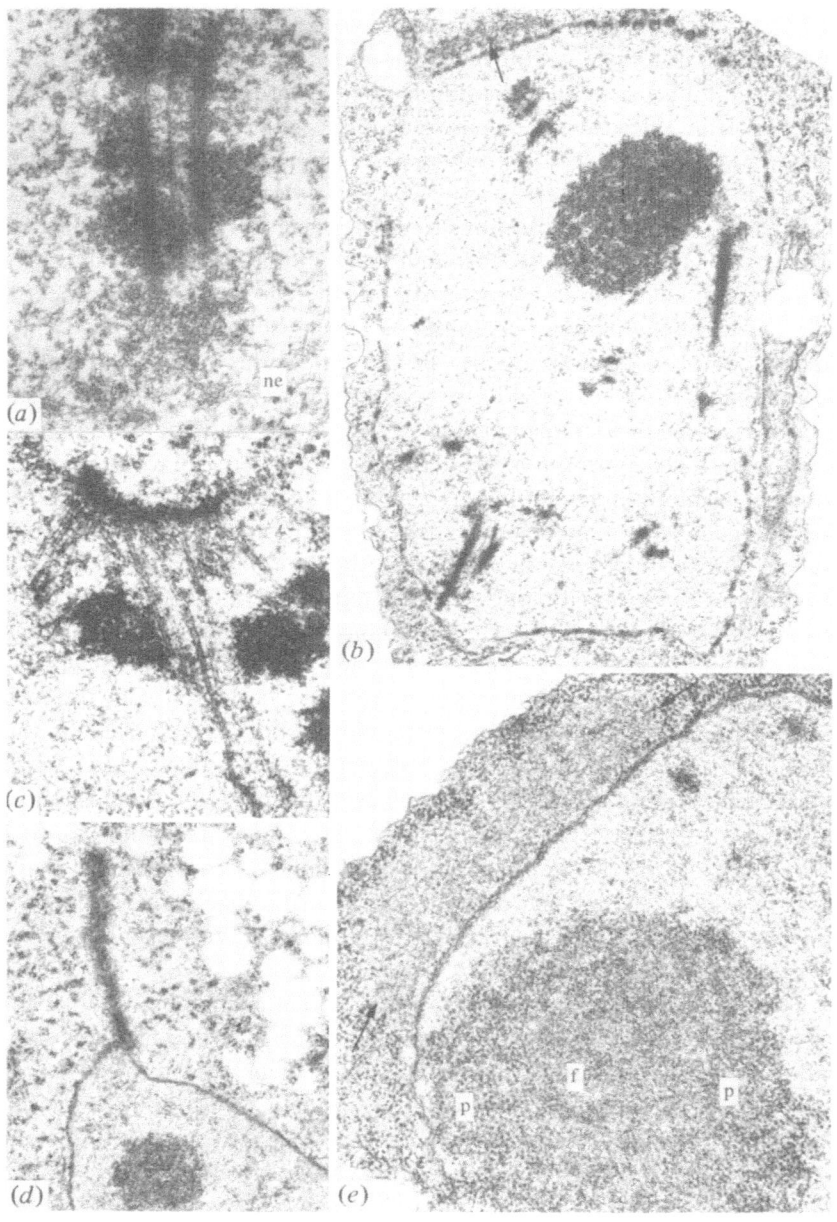

Fig. 2 (*a*) *Xylaria polymorpha:* LS through part of a synaptonemal complex. Note chromatin attached to the lateral component. A region of fibrillar material connects the ends of the lateral components with the nuclear envelope (ne). × 40 000. (*b*) *X. polymorpha:* LS through a removed and the chiasmata are represented by continuous chromatin bridges between the rejoined chromatids. This stage is marked by the presence of highly contracted ring- and rod-shaped bivalents (Westergaard & von Wettstein, 1970b).

The so-called 'diffuse diplotene' stage described in detail for *Neurospora crassa* from light microscope observations (Barry, 1969) cannot obviously be correlated with meiotic stages defined by electron microscopy. It has been suggested, however, that the diffuse appearance in the light microscope may be due to the shedding of the lateral components of the SC which may alter the affinity of the chromatin to light microscope stains (Zickler, 1977). This would therefore fit most closely with the early diplotene stages described for *Sordaria macrospora* (Zickler, 1977) and *Neottiella rutilans* (Westergaard & von Wettstein, 1970b).

The composition of the SC has been studied in *Neottiella rutilans* (Westergaard & von Wettstein, 1970a, b). They showed that the lateral components of the SC in formaldehyde-fixed cells react strongly with silver ions when stained at a pH below 10. This they attributed to the presence, in these components, of basic proteins. Digestion with trypsin of glutaraldehyde-fixed asci at the pachytene stage led to the removal of all components of the SC but did not affect the structure of the chromatin (Westergaard & von Wettstein, 1970b). The central region and central component was removed by RNase digestion while the banded lateral component was only partially digested. DNase digestion after formalin prefixation caused extensive dissolution of the bivalent chromatin but the components of the SC were unaffected. On the basis of these results Westergaard & von Wettstein (1970b) concluded that the lateral and central components together with the central region of the SC in *Neottiella* consist of ribonucleoproteins. The banded structure of the lateral component, being largely resistant to RNase, is probably composed of protein alone, or it may contain RNA in a form not

---

fusion nucleus at early pachytene. A lateral component of one bivalent is attached to the nucleolus. Note the dense pore complexes in the nuclear envelope and the MTOC material (arrow). × 17 000. (*c*) *X. polymorpha:* LS through one pole of an anaphase II spindle showing the plate-like SPB, spindle microtubules and dense chromatin. × 36 000. (*d*) *S. humana:* LS through part of an interphase nucleus after mitosis III. Note orientation of MTOC. × 17 000. (*e*) *S. humana:* part of leptotene nucleus showing fibrillar MTOC (between arrows) and fibrillar (f) and particulate (p) regions of nucleolus. × 20 000.

accessible to the RNase used. DNA is not a major component of the SC though the possible presence of small amounts cannot be excluded.

The nucleolus

In meiotic nuclei the nucleolus is seen as an electron-opaque body, comprising a coarse, granular region together with an amorphous or fibrillar region (Fig. 2e). In *Sordaria humana* the fibrillar region is surrounded by the coarse particulate region (Fig. 2e). This arrangement also occurs in *Neottiella rutilans* (Westergaard & von Wettstein, 1970a), *S. macrospora* (Zickler, 1977), *Xylaria polymorpha* and *Pustularia cupularis* (Schrantz, 1970).

Westergaard & von Wettstein (1970a) measured the nucleolar particles and found that although ribosome-like in appearance, they were significantly smaller than the cytoplasmic ribosomes. They also found that both the cytoplasmic ribosomes and the nucleolar particles were larger at metaphase and later stages than they were at prophase. The significance of this observation is unclear.

Nucleolar vacuoles have been seen in *Neurospora crassa* (Gillies, 1972; Lu & Galleazzi, 1978), *Ascobolus stercorarius* (Wells, 1970), *Pyronema domesticum* (Hung & Wells, 1971), *Galactinia plebeia* and *Pustularia cupularis* (Schrantz, 1970). In *Sordaria macrospora* the size of the nucleolus and the incidence and configuration of the nucleolar vacuole(s) varies throughout meiotic prophase (Zickler, 1977). Nucleolar diameter reaches a maximum at late zygotene and tends to decrease slightly at pachytene. During leptotene the nucleolus, which at first is homogeneous, develops several small cavities. These coalesce during zygotene to form a large vacuole which persists during early pachytene but later fragments.

In *Neottiella rutilans* what appear to be small pieces of the nucleolus, bounded by a membraneous envelope, have been seen in the ascus cytoplasm at diplotene adjacent to peripheral nucleolar vacuoles (Westergaard & von Wettstein, 1970a). It was suggested that these arose as a result of the transport of nucleolar material into the cytoplasm during the active growth phase of the ascus. At diakinesis separation of the nucleolus from the nucleus begins and culminates, at telophase, with the complete expulsion of the nucleolus into the ascus cytoplasm. This 'old' nucleolus remains bounded by a double membrane envelope which is separate from, but attached to the envelope of the telophase nucleus. Eventually the envelope breaks down and the particulate and fibrillar components of the 'old' nucleolus disperse into the cytoplasm. Fibrillar pronucleoli form at various sites on the chromatin of the telophase nuclei. These eventually fuse and the particulate component is then formed (Westergaard & von Wettstein, 1970*a*).

RNase and trypsin digest the particulate component but DNase has no apparent effect on the nucleolus. As with the lateral components of the SC, basic proteins of the nucleolus react strongly with ammoniacal silver ions particularly at a pH of 10 to 8 (Westergaard & von Wettstein, 1970*a*).

Microtubule organizing centres and spindle pole bodies

As suggested previously, a distinction can be made between MTOCs which vary in structure and SPBs which are always plate- or bar-like and which only occur at the poles of dividing nuclei. During meiotic prophase the MTOCs in *Sordaria humana* (Figs. 2*e*, 3*a*), *Xylaria polymorpha* (Beckett & Crawford, 1970; Fig. 2*b*) and *X. longipes* (Fig. 3*c*) consist of an aggregation of electron-opaque, fibrillar material closely associated with the nuclear envelope. Throughout leptotene and zygotene this material is extensive and in thin sections often surrounds the nuclear profile (Fig. 3*c*).

The nuclear envelope adjacent to the MTOC material is structurally distinct from that elsewhere around the nucleus. The pore complexes are either heavily stained or contain electron-opaque material within them (Figs. 3*c*, *d*) and in *Xylaria* the pores are clustered immediately beneath the MTOC (Fig. 3*b*). The MTOC material lies within a 'zone of exclusion' devoid of cytoplasmic ribosomes and other organelles (Figs. 2*e*, 3*c*, *d*). A narrow zone of cytoplasm separates the MTOC from the outer membrane of the nuclear envelope, but in *Xylaria* fibrillar material traverses this zone, forming connections between the nuclear pore complexes and the MTOC material (Fig. 3*d*). Microtubules have also been seen between the MTOC and the nuclear envelope (Beckett & Crawford, 1970). MTOCs sometimes lie within depressions of the nuclear envelope (Fig. 3*c*) or next to a flattened region of the envelope (Fig. 2*e*). Similar modifications of the nuclear envelope have been reported for other Ascomycetes (Schrantz, 1970; Martin, Gay & Jackson, 1976).

Schrantz (1970) adopted the term 'calotte polaire' or polar cap of Feldmann (1956) for MTOCs of the type described above and which he found in *Xylaria polymorpha* and *Galactinia plebeia*. They have also been seen with the light microscope in mutants of *Neurospora crassa* (Raju, 1978) and, according to Raju (1978), in *Sordaria brevicollis* by

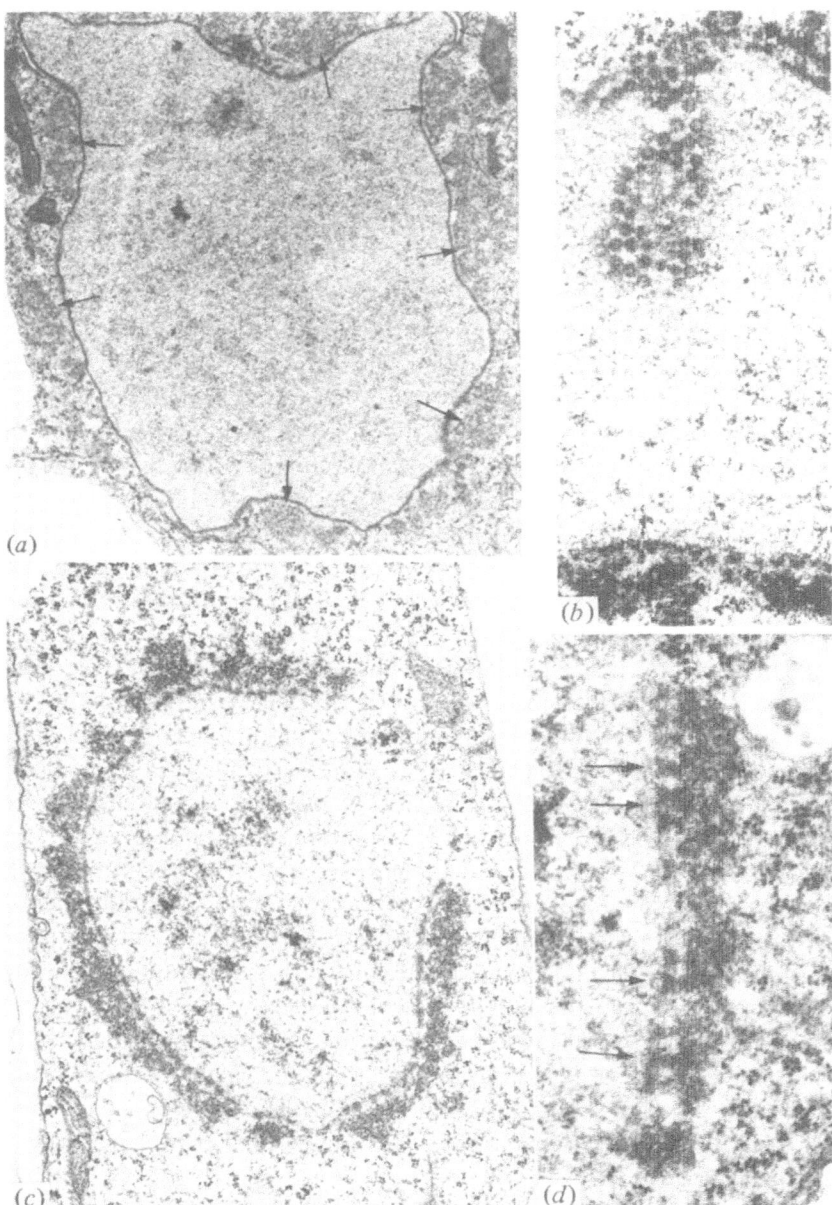

Fig. 3 (*a*) *Sordaria humana:* LS through early leptotene nucleus showing MTOC material surrounding the nuclear profile (arrows). × 10 000. (*b*) *Xylaria longipes:* part of a leptotene nucleus showing the dense aggregation of nuclear pores (seen in surface view) which lie beneath the MTOC material. × 20 000. (*c*) *X. longipes:* LS through

Mu'Azu (1973). MTOCs of the polar cap type are seen with (i) fusion nuclei in ascogenous hyphae and (ii) at the base of newly delimited ascospores (Beckett & Crawford, 1970). With the onset of meiotic metaphase I they disappear and the intranuclear spindle is seen to be polarized by the plate-like SPBs (Fig. 2c) which are more intimately associated with the nuclear envelope (Beckett & Crawford, 1970; Schrantz, 1970; Zickler, 1970; Wells, 1970; Simonet & Zickler, 1972). In those genera which do not have MTOCs of the polar cap type, a plate-like MTOC is found adjacent to interphase prefusion nuclei and prophase fusion nuclei. Here it is closely associated with the nuclear envelope, but since the spindle has not formed at these stages the term SPB is inappropriate. Examples of this type are seen in *Sordaria macrospora* (Zickler, 1977), *Pustularia cupularis* (Schrantz, 1970) and *Ascobolus stercorarius* (Wells, 1970; Zickler, 1970). The structure and behaviour of MTOCs and SPBs in Euascomycetes is in need of careful investigation and as yet systematic studies, comparable to that on Hemiascomycetes (Ashton & Moens, 1979), have not been made. However, information extracted from some of the publications cited above and from my own results reported here is summarized diagrammatically in Fig. 4. From this it appears that: (i) interphase prefusion nuclei and prophase fusion nuclei of some genera and species are associated with a fibrillar MTOC; (ii) other, often related species, have instead plate-like MTOCs closely pressed to the envelope of nuclei at these same stages; (iii) at metaphase, anaphase and telophase, once the spindle has formed, all Euascomycetes have plate-like SPBs at the spindle poles; (iv) normally these SPBs are flattened along their entire length against the (modified?) nuclear envelope; (v) in some ascomycetes, during post-meiotic mitosis, the MTOC or SPB is 'attached' by one end only to the nuclear envelope (Zickler, 1969, 1970; Beckett, Heath & McLaughlin, 1974; Sakai, 1974; see also Fig. 2d).

In *Ascobolus stercorarius* (Zickler, 1970), the SPB may consist of an extranuclear, dense, outer zone and an intranuclear, less dense, inner zone. The former is associated with astral ray microtubules, the latter with the spindle microtubules. This configuration is also seen in *Galactinia plebeia* (Schrantz, 1970) and *Neottiella rutilans* (Westergaard

leptotene/zygotene nucleus showing extensive MTOC material. × 15 000. (*d*) *X. longipes:* part of the prophase nuclear envelope shows dense pore complexes and associated extranuclear MTOC material. Note material connecting pore complexes with MTOC (arrows). × 30 000.

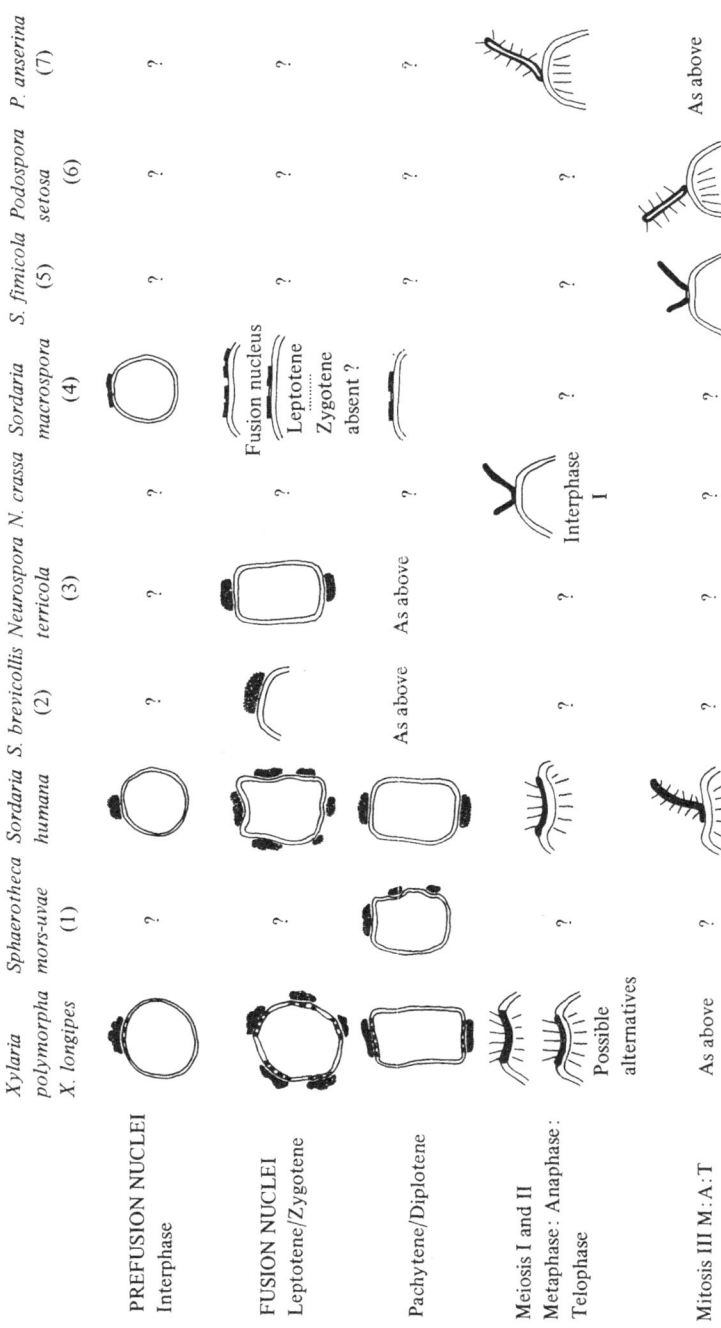

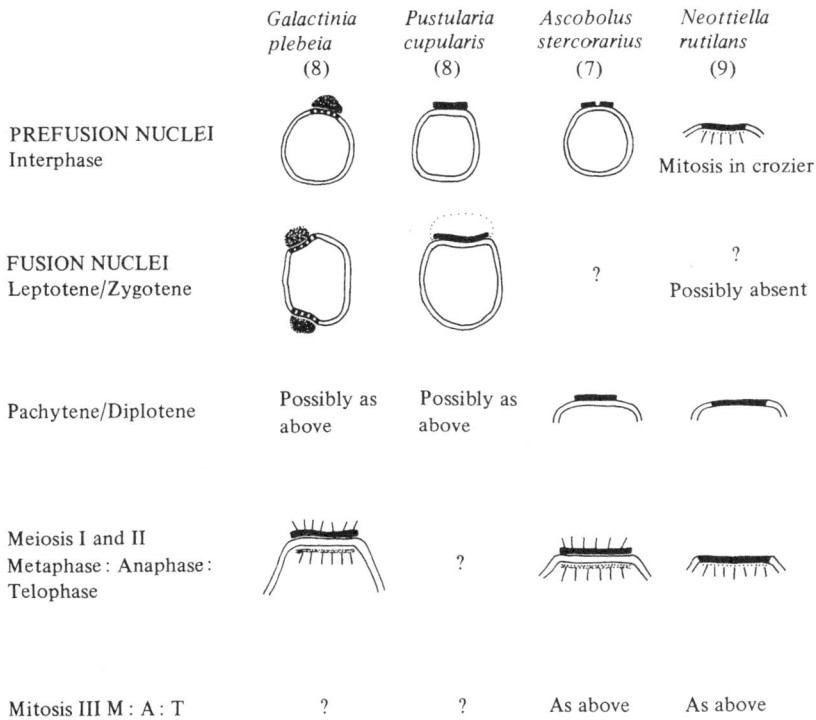

Fig. 4. Diagrammatic summary of the morphology and behaviour of MTOCs and SPBs during meiosis and post-meiotic mitosis in Euascomycetes. (*Left opposite*) Pyrenomycetes; (*above*) Discomycetes.?, not known or information not published. Sources: (1) Martin *et al.* (1976); (2) Mu'Azu (1973); (3) Raju (1978); (4) Zickler (1977); (5) Beckett *et al.* (1974); (6) Zickler (1969); (7) Zickler (1970); (8) Schrantz (1970); (9) Westergaard & von Wettstein (1970*a*).

& von Wettstein, 1970*a*). It may be significant that all of these are Discomycetes.

Little is known of the composition of MTOCs or SPBs. Schrantz (1970) suggested that the fibrillar, polar cap type MTOCs of *Xylaria polymorpha* and *Galactinia plebeia* were composed of RNA fibrils. The fibrillar connections between the MTOC and the nuclear pore complexes in *Xylaria* (Fig. 3*d*) may represent a stage of formation. There is some evidence that the SPBs in *Ascobolus stercorarius* contain proteins that are digested by pronase and trypsin, and that they may also contain DNA. The presence or absence of RNA has not been determined conclusively (Zickler, 1973*b*).

The mechanics of MTOC or SPB division are not understood, but structures have been seen in thin sections which resemble a dividing SPB at telophase III in *Sordaria fimicola* (Beckett *et al.*, 1974) and an MTOC at interphase I in *Neurospora crassa* (Fig. 5). Division of the

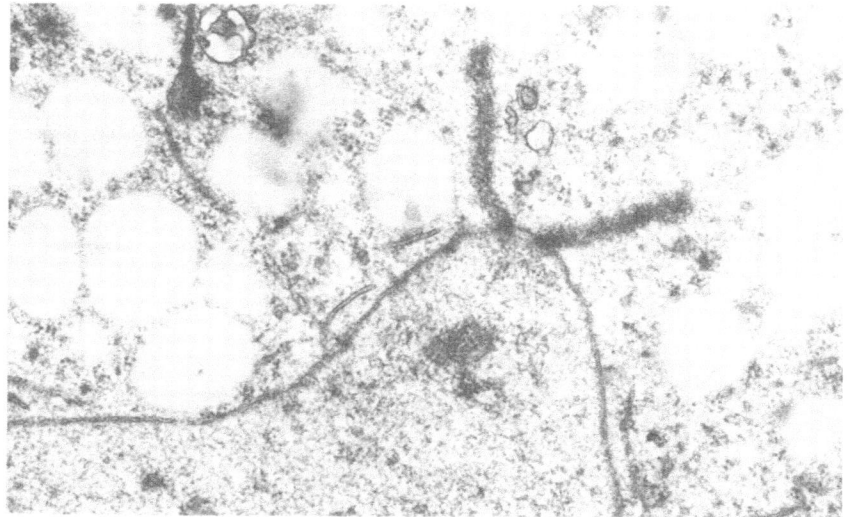

Fig. 5. *Neurospora crassa*. Part of an interphase I nucleus showing what is presumably a divided MTOC. × 32 000.

MTOC material in *Xylaria* may occur by a gradual polarization over the nuclear envelope during zygotene and pachytene, so that by diplotene the MTOCs come to occupy opposite ends of the nucleus, i.e. the positions where the SPB will subsequently lie at metaphase. Thus in *Xylaria, Galactinia, S. humana* and other species which have polar cap type MTOCs, the SPB may form as a result of a concentration and modification of the fibrillar MTOC material, following its division and polarization during zygotene and pachytene. If this is the case then the SPB is a specialized form of MTOC as Fuller (1976) suggested. Furthermore the nuclear envelope may mediate the process of division and polarization of the MTOC material.

Conclusions
(1) Ultrastructural studies based on serial sections of nuclei are limited to relatively few genera. There is therefore a need for similar studies on other genera and species.

(2) Little is known of the mechanics of karyogamy, homologous chromosome pairing, cell differentiation within the crozier, or the division and migration of MTOCs. It is possible that the nuclear envelope could mediate all or any one of these processes.

(3) There is a need for detailed freeze-etch studies on meiotic fungal nuclei.

(4) Lateral components of the synaptonemal complex are structurally distinct in ascomycetes but the dimensions of the SC are uniform and similar to those of other eukaryotes.

(5) The ultrastructural relationship between chromatin and the lateral and central components of the SC is not clear, since chromatin is rarely well-preserved in electron microscope preparations. Only a small proportion of the genome DNA is associated with SCs.

(6) A systematic study of the ontogeny and behaviour of MTOCs and SPBs in ascomycete fungi is necessary. SPBs might be a special type of MTOC.

References

Alexopoulos, C. J. (1962). *Introductory Mycology*, 2nd edn. New York & London: John Wiley & Sons.

Ashton, M. L. & Moens, P. B. (1979). Ultrastructure of sporulation in the hemiascomycetes *Ascoidea corymbosa, A. rubescens, Cephaloascus fragrans,* and *Saccharomycopsis capsularis. Canadian Journal of Botany,* **57,** 1259–84.

Barry, E. G. (1969). The diffuse diplotene stage of meiotic prophase in *Neurospora. Chromosoma, Berlin,* **26,** 119–29.

Beckett, A. & Wilson, I. M. (1968). Ascus cytology of *Podospora anserina. Journal of General Microbiology,* **53,** 81–7.

Beckett, A. & Crawford, R. M. (1970). Nuclear behaviour and ascospore delimitation in *Xylosphaera polymorpha. Journal of General Microbiology,* **63,** 269–80.

Beckett, A., Heath, I. B. & McLaughlin, D. J. (1974). *An Atlas of Fungal Ultrastructure.* London: Longman.

Bennett, M. D., Stern, H. & Woodward, M. (1974). Chromatin attachment to nuclear membrane of wheat pollen mother cells. *Nature, London,* **252,** 395–96.

Burnett, J. H. (1976). *Fundamentals of Mycology,* 2nd edn. London: Edward Arnold.

Carpenter, A. T. C. (1975). Electron microscopy of meiosis in *Drosophila melanogaster* females. II: the recombination nodule – a recombination-associated structure at pachytene? *Proceedings of the National Academy of Sciences, USA,* **72,** 3186–9.

Carr, A. J. H. & Olive, L. S. (1958). Genetics of *Sordaria fimicola.* III. Cytology. *American Journal of Botany,* **45,** 142–50.

Carroll, G. C. (1969). A study of the fine structure of ascosporogenesis in *Saccobolus kerverni. Archiv für Mikrobiologie,* **66,** 321–39.

Feldmann. G. (1956) Développement d'une Plasmodiophorale marine: *Plasmodiophora bicaudata* J. Feldm. parasite du *Zostera nana* Roth. *Revue Générale de Botanique*, **63**, 390–420.

Franke, W. W. (1974). Structure biochemistry and functions of the nuclear envelope. *International Review of Cytology*, Supp. 4, ed. G. H. Bourne & J. F. Danielli, pp. 71–236. New York, San Francisco & London: Academic Press.

Fuller, M. S. (1976). Mitosis in fungi. *International Review of Cytology*, vol. 45, ed. G. H. Bourne & J. F. Danielli, pp. 113–53. New York & London: Academic Press.

Gillies, C. B. (1972). Reconstruction of the *Neurospora crassa* pachytene karyotype from serial sections of synaptonemal complexes. *Chromosoma, Berlin*, **36**, 119–30.

Gillies, C. B. (1975). Synaptonemal complex and chromosome structure. *Annual Review of Genetics*, **9**, 91–109.

Gillies, C. B. (1979). The relationship between synaptonemal complexes, recombination nodules and crossing over in *Neurospora crassa* bivalents and translocation quadrivalents. *Genetics*, **91**, 1–17.

Harris, N. (1978). Nuclear pore distribution and relation to adjacent cytoplasmic organelles in cotyledon cells of developing *Vicia faba*. *Planta*, **141**, 121–8.

Heywood, P. & Magee, P. T. (1976). Meiosis in protists. Some structural and physiological aspects of meiosis in algae, fungi and protozoa. *Bacteriological Reviews*, **40**, 190–240.

Hotta, Y. & Shepard, J. (1973). Biochemical aspects of colchicine action on meiotic cells. *Molecular and General Genetics*, **122**, 243–60.

Hung, C-Y. & Wells, K. (1971). Light and electron microscopic studies of crozier development in *Pyronema domesticum*. *Journal of General Microbiology*, **66**, 15–27.

Lu, B. C. (1967). The course of meiosis and centriole behaviour during the ascus development of the ascomycete *Gelasinospora calospora*. *Chromosoma, Berlin*, **22**, 210–26.

Lu, B. C. & Galeazzi, D. R. (1978). Light and electron microscope observations of a meiotic mutant of *Neurospora crassa*. *Canadian Journal of Botany*, **56**, 2694–706.

Martin, M., Gay, J. L. & Jackson, G. V. H. (1976). Electron microscopic study of developing and mature cleistothecia of *Sphaerotheca mors-uvae*. *Transactions of the British Mycological Society*, **66**, 473–87.

Marty, M. F. (1973). Sites réactifs à l'iodure de zinc-tetroxyde d'osmium dans les cellules de la racine d'*Euphorbia characius*. *Compte Rendu Hebdomadaire des Séances de l'Académie des Sciences*, **227**, 1317–20.

Moens, P. B. (1973). Mechanisms of chromosome synapsis at meiotic prophase. *International Review of Cytology*, vol. 35, ed. G. H. Bourne & J. F. Danielli, pp. 117–34. New York & London: Academic Press.

Mu'Azu, S. (1973). Cytogenetic studies on *Sordaria brevicollis*. PhD Thesis, University of Cambridge, UK.

Olive, L. S. (1953). The structure and behaviour of the fungus nuclei. *Botanical Reviews*, **19**, 439–586.

Olive, L. S. (1965). Nuclear behaviour during meiosis. In *The Fungi; an Advanced Treatise*, vol. 1, ed. G. C. Ainsworth & A. S. Sussman, pp. 143–61. New York & London: Academic Press.

Pickett-Heaps, J. D. (1969). The evolution of the mitotic apparatus: an attempt at comparative ultrastructural cytology in dividing plant cells. *Cytobios*, **3**, 257–80.

Raju, N. B. (1978). Meiotic nuclear behaviour and ascospore formation in five homothallic species of *Neurospora*. *Canadian Journal of Botany*, **56**, 754–63.

Reeves, F. Jr (1967). The fine structure of ascospore formation in *Pyronema domesticum*. *Mycologia*, **59**, 1018–33.

Rossen, J. M. & Westergaard, M. (1966). Studies on the mechanism of crossing over. II. Meiosis and the time of meiotic chromosome replication in the ascomycete *Neottiella rutilans* (Fr.) Dennis. *Comptes Rendus des Travaux du Laboratoire Carlsberg*, **35**, 233–60.

Sakai, A. (1974). Centriolar plaque and spindle microtubules in the ascomycete *Sordaria humana*. *Botanical Magazine, Tokyo*, **87**, 341–45.

Schrantz, J-P. (1970). Etude cytologique en microscopie optique et électronique de quelques ascomycètes. I. Le noyau. *Revue de Cytologie et de Biologie Végétale*, **33**, 1–100.

Severs, N. J., Jordan, E. G. & Williamson, D. H. (1976). Nuclear pore absence from areas of close association between nucleus and vacuole in synchronous yeast cultures. *Journal of Ultrastructure Research*, **54**, 374–87.

Simonet, J. M. & Zickler, D. (1972). Mutations affecting meiosis in *Podospora anserina*. I. Cytological studies. *Chromosoma, Berlin*, **37**, 327–51.

Singer, S. J. & Nicolson, G. L. (1972). The fluid mosaic model of the structure of cell membranes. *Science*, **175**, 720–31.

Singleton, J. R. (1953). Chromosome morphology and the chromosome cycle in the ascus of *Neurospora crassa*. *American Journal of Botany*, **40**, 124–44.

Wells, K. (1970). Light and electron microscope studies of *Ascobolus stercorarius*. I. Nuclear divisions in the ascus. *Mycologia*, **62**, 761–90.

Wells, K. (1972). Light and electron microscopic studies of *Ascobolus stercorarius*. II. Ascus and ascospore ontogeny. *University of California Publications in Botany*, **62**, 1–33.

Westergaard, M. (1964). Studies on the mechanism of crossing over. I. Theoretical considerations. *Comptes Rendus des Travaux du Laboratoire Carlsberg*, **34**, 359–405.

Westergaard, M. & von Wettstein, D. (1966). Studies on the mechanism of crossing over. III. On the ultrastructure of the chromosomes in *Neottiella rutilans* (Fr.) Dennis. *Comptes Rendus des Travaux du Laboratoire Carlsberg*, **35**, 261–86.

Westergaard, M. & von Wettstein, D. (1970a). The nucleolar cycle in an ascomycete. *Comptes Rendus des Travaux du Laboratoire Carlsberg*, **37**, 195–237.

Westergaard, M. & von Wettstein, D. (1970b). Studies on the mechanism of crossing over. IV. The molecular organization of the synaptonemal complex in *Neottiella* (Cooke) Saccardo (Ascomycetes). *Comptes Rendus des Travaux du Laboratoire Carlsberg*, **37**, 239–68.

Westergaard, M. & von Wettstein, D. (1972). The synaptonemal complex. *Annual Review of Genetics*, **6**, 71–110.

von Wettstein, D. (1971). The synaptonemal complex and four-strand crossing over. *Proceedings of the National Academy of Sciences, USA*, **68**, 851–5.

von Wettstein, D. (1977). The assembly of the synaptonemal complex. *Philosophical Transactions of the Royal Society of London, Series B*, **277**, 235–243.

Zickler, D. (1969). Sur l'appareil de quelques Ascomycètes. *Comptes Rendus Hebdomadaires des Séances de l'Académie des Sciences*, **268**, 3040–2.

Zickler, D. (1970). Division spindle and centrosomal plaque during mitosis and meiosis in some ascomycetes. *Chromosoma, Berlin*, **30**, 287–304.

Zickler, D. (1973a). Fine structure of chromosome pairing in ten ascomycetes: meiotic and premeiotic (mitotic) synaptonemal complexes. *Chromosoma, Berlin*, **40**, 401–16.

Zickler, D. (1973b). Evidence for the presence of DNA in the centrosomal plaques of *Ascobolus*. *Histochemie*, **34**, 227–38.

Zickler, D. (1977). Development of the synaptonemal complex and the 'recombination nodules' during meiotic prophase in the seven bivalents of the fungus *Sordaria macrospora* Auersw. *Chromosoma, Berlin*, **61**, 289–316.

Ultrastructure of the yeast nucleus

DENISE ZICKLER

Laboratoire de Génétique, Bât 400, Université de Paris-Sud,
91405 Orsay, France

Introduction

Yeasts have been successfully employed in about all phases of genetics and biochemistry while their cytology has long been a controversial subject, mainly because of the small size of the nuclei. In addition the cell wall has also proved to be a barrier to fixation, and its partial removal by enzymatic treatment (Peterson, Gray & Ris, 1972) not only improved the quality of ultrastructural studies, but permitted demonstration that the mitotic (Byers & Goetsch, 1974, 1975a; Peterson & Ris, 1976) and meiotic divisions (Zickler & Olson, 1975) of *Saccharomyces cerevisiae* proceed in an essentially orthodox manner. The chromosome number was found (Byers & Goetsch, 1975b) to be consistent with genetic evidence (Sherman & Lawrence, 1974) for 17 bivalents per diploid cell. Elegant demonstration has been given that the spindle pole bodies, first described in mitosis by Robinow & Marak (1966), and in meiosis by Moens & Rapport (1971a), serve indeed as foci for the initiation of microtubules (Borisy *et al.*, 1975; Hyams & Borisy, 1978; Byers, Shriver & Goetsch, 1978). Furthermore yeasts offer uniquely advantageous systems for isolating mutants defective in specific steps of the cell cycle (see reviews by Hartwell, 1974; Simchen, 1978; Esposito, M. S. & Esposito, R. E., 1978). The mutants are deficient in functions required for cell progression and therefore make it possible to define developmental steps in the cycle and to recognize the characteristics of their cytological fine structure (Moens, 1974; Moens, Esposito & Esposito, 1974; Byers & Goetsch, 1974; Moens *et al.*, 1977; Schild & Byers, 1978; Horesh, Simchen & Friedmann, 1979).

The data presented here do not represent a catalogue of all information available on the respective topics. Rather they are intended to show that progress in our understanding of the cell cycle is often obtained by a clear knowledge of what is happening in the nucleus. Recent reviews have surveyed the ultrastructural (Fuller, 1976; Heath, 1974, 1978) as well as the phylogenic aspects (Kubai, 1975, 1978) of mitosis in fungi. Therefore more attention will be paid to meiosis and sporulation of *Saccharomyces cerevisiae* and related species. In essence, yeasts provide a good illustration of the fact that our understanding of the observed cell components and elucidation of their functions will come from a combination of genetic, cytological and biochemical approaches.

Generalities

The obvious common feature concerning yeast nuclei is their lack of chromosome 'condensation' on electron-microscopic (EM) pictures. The mitotic and meiotic division sequences have therefore been described with respect to microtubule and spindle pole body development. Only during meiotic prophase can the bivalent behaviour be studied with the aid of the synaptonemal complex (Zickler & Olson, 1975; Byers & Goetsch, 1975b; Horesh *et al.*, 1979).

Another general observation is the fact that the nuclear envelope remains intact during mitosis and meiosis. Moreover, as first noted by Moens & Rapport (1971a), the two spindles of the second meiotic division are formed in a single nucleus. Numerous EM studies conclude that the nuclear envelope has the same fundamental fine structure as in other eukaryote nuclei: inner and outer membranes, a perinuclear space, pores (especially clear in freeze-etch preparations; Moor & Mühlethaler, 1963; Guth, Hashimoto & Conti, 1972), and annular material.

The nucleolus persists during divisions, but further evidence is needed to ascertain whether it always becomes constricted into two at telophase as was shown by McCully & Robinow (1971) in *Schizosaccharomyces pombe*, or whether it is extruded into the cytoplasm by the nuclear envelope constriction.

It should be emphasized that gross nuclear development as well as parts of nucleolus and spindle behaviour can be followed successfully during mitosis and meiosis in living cells under phase contrast. Moreover, light microscopy of stained yeast provides evidence that there are chromosomes even if they are not usually resolved into countable numbers (Matile, Moor & Robinow, 1969; McCully & Robinow, 1971; Robinow, 1975, 1977).

Spindle pole body and microtubules during the cell cycle
Mitosis, budding and conjugation

Fortunately for yeast cytologists, a yeast nucleus always shows a spindle pole body (SPB) appressed to the nuclear envelope, even during interphase. The SPB ultrastructure of *Saccharomyces cerevisiae* has been studied during mitosis by Robinow & Marak (1966); Matile *et al.* (1969); Moens & Rapport (1971*a*); Byers & Goetsch (1974, 1975*a*) and Peterson & Ris (1976). Mitosis in *Schizosaccharomyces pombe* has been followed by McCully & Robinow (1971).

The single SPB seen in unbudded cells and ascospores is a small (about 0.15 μm), disc-like, dense structure embedded in the nuclear envelope. It always bears a bundle of intranuclear microtubules which never end in the dense layer of the SPB, but in an adjacent inner zone of lower electron density. Symmetrical, amorphous electron-opaque material is seen at the cytoplasmic side of the SPB.

At the time of bud emergence, the chromosomes replicate (Williamson, 1965) and the SPB duplicates. However, this duplication seems independent of DNA replication as shown by the behaviour of three conditional 'cell division cycle' (*cdc*) mutants isolated by Hartwell (1974). Mutant strains defective either in initiation (*cdc*7) or elongation (*cdc*8 and *cdc*21) of DNA replication, duplicate their SPBs and form spindles. On the contrary, bud emergence and SPB duplication seem to be obligatorily coupled events. A lack of budding leads, in strain *cdc*28, to a failure of SPB duplication, and in *cdc*4, a series of buds are formed in cells which are mononucleate but have duplicated SPBs (Byers & Goetsch, 1974). As clearly illustrated by the same authors, during later G1 dense amorphous material is seen on the cytoplasmic side of the single SPB and connected to it by a half-bridge. True double SPBs (Fig. 1*a*), connected by a bridge and each bearing a bundle of intranuclear microtubules, are only found when bud emergence has started. At this stage, the nucleus is in close proximity to the budding site, and several extranuclear microtubules (Fig. 3.1*b*), emanating from both the SPBs and the bridge, enter the bud (Byers & Goetsch, 1975*a*). The bridge breaks and the two SPBs separate each retaining one half of the bridge. The formed spindle (approximately 1 μm long) remains in the mother cell whereas the nucleus migrates into the bud neck. When bud and mother-cell have reached the same size, the spindle elongates rapidly up to 6–8 μm (Byers & Goetsch, 1974). The two SPBs are then opposite in the two cells and the nucleus shows a characteristic dumb-bell configuration. Nuclear separation occurs by constriction of the formed isthmus.

A complete cytological analysis of microtubule behaviour during spindle formation has been undertaken by Peterson & Ris (1976). Initially, after SPB duplication, the two bundles of intranuclear microtubules have the same average length. Afterwards, short and long microtubules elongate from each SPB. When these latter separate, the long microtubules (1 μm) span through the nucleus from one SPB to the other, and form continuous microtubules whereas the short ones (500 nm), overlapping in the middle of the spindle, probably represent the 'chromosomal microtubules'. Unfortunately neither the chromosomes nor the kinetochores can be distinguished in the conventional EM preparations. However, two convincing arguments for their assignment as chromosomal microtubules are given:(a) while the continuous microtubules elongate, the discontinuous ones shorten more and more; (b) in thick sections, dense chromatin-like material seen in the middle of the spindle just after its formation first separates into two layers, when the discontinuous microtubules shorten, and then disappears. Serial cross-sections permitted the determination of an average number of one discontinuous microtubule per chromosome (14 in haploid, 29 in diploid cells) for every 7 (n) or 9 (2n) continuous microtubules per spindle.

Using isolated SPBs, Hyams & Borisy (1978) elegantly demonstrated that SPBs from exponentially growing cell cultures initiate only 5 to 12 microtubules in haploid and diploid cells, whereas SPBs from stationary phase cells (in G1) nucleate 5 to 22 (haploid) and 14 to 42 (diploid) microtubules. These results not only confirm the data of Peterson & Ris (1976), but show clearly that the ability of the SPBs to nucleate microtubules is cycle-dependent.

Cell separation by septum formation (Cabib & Farkas, 1971) leaves permanent scars on both cell surfaces (Beran, 1968). In the fission yeast, *Schizosaccharomyces pombe*, the septum growth begins at the end of the long dumb-bell stage (McCully & Robinow, 1971). Although septum formation requires mitosis to be completed as shown by the behaviour of ten *cdc* mutants (Nurse, Thuriaux & Nasmyth, 1976), the regulation of their number is separately controlled (Minet *et al.*, 1979).

During the first stages of conjugation in *Saccharomyces cerevisiae* the SPB of both nuclei is single with a satellite attached to the half-bridge (Byers & Goetsch, 1975a). When the nuclei fuse, the two SPBs fuse by the half-bridge and the satellites–SPB fusion is followed by SPB duplication and the bud made from the zygote is formed near this double SPB.

Meiosis

The basic outlines of the spindle structure changes associated with meiosis in *Saccharomyces cerevisiae* have so far been investigated by Rapport (1971), Moens & Rapport (1971*a*), Guth *et al.* (1972), Peterson, Gray & Ris (1972), Zickler & Olson (1975) and Horesh *et al.* (1979). Although major features of spindle behaviour are shared by meiotic and mitotic divisions, several modifications in the detailed ultrastructure of the SPBs occur, especially during the second division, presumably in connection with ascospore wall formation.

As for mitosis, in the absence of chromosome morphology the different stages are defined by SPB development examined in serial sections.

When stationary growth phase cells are transferred to sporulation medium, they show a single SPB, consisting of a dense central zone with fibrillar layers on both the cytoplasmic and the nuclear side (Fig. 1*c*). The inner zone bears a bundle of microtubules which terminate in the middle of the nucleus. Microtubules emanating from the outer layer are rare.

The SPBs probably duplicate during, or prior to (Horesh *et al.*, 1979) the premeiotic S phase. Analysis of several mutants suggests that premeiotic DNA synthesis is not required for SPB duplication. Moens *et al.* (1974) found that homozygous *spo*1-1 strains exhibit premeiotic DNA synthesis and genetic recombination at restrictive temperatures without SPB duplication, whereas *spo*11 (Moens *et al.*, 1977) with nearly blocked DNA synthesis, shows SPB duplication and spindle formation. Likewise *cdc*8-1 and *cdc*21-1 mutants, which fail to undergo premeiotic DNA synthesis at the restrictive temperature, and wild-type cells in which the synthesis has been inhibited by hydroxyurea, show duplicated SPBs (Schild & Byers, 1978).

The duplicated SPBs remain in a typical side-by-side position (Fig. 1*d*) until the end of prophase. In cross sections, they appear almost round, connected by a rectangular bridge, each SPB bearing intranuclear microtubules which often branch off widely. At metaphase I, when the two SPBs are found to face each other, they both show a clear outer layer (Fig. 1*e*). Anaphase I, compared with metaphase, is characterized by spindle and nucleus elongation and by a reduced number of microtubules (14–17) always maintained in parallel array. The nucleolus, located in a protrusion of the nuclear envelope, often appears double. At the same time, the cytoplasmic layer of the SPB increases in size and density. In some nuclei the partial or complete

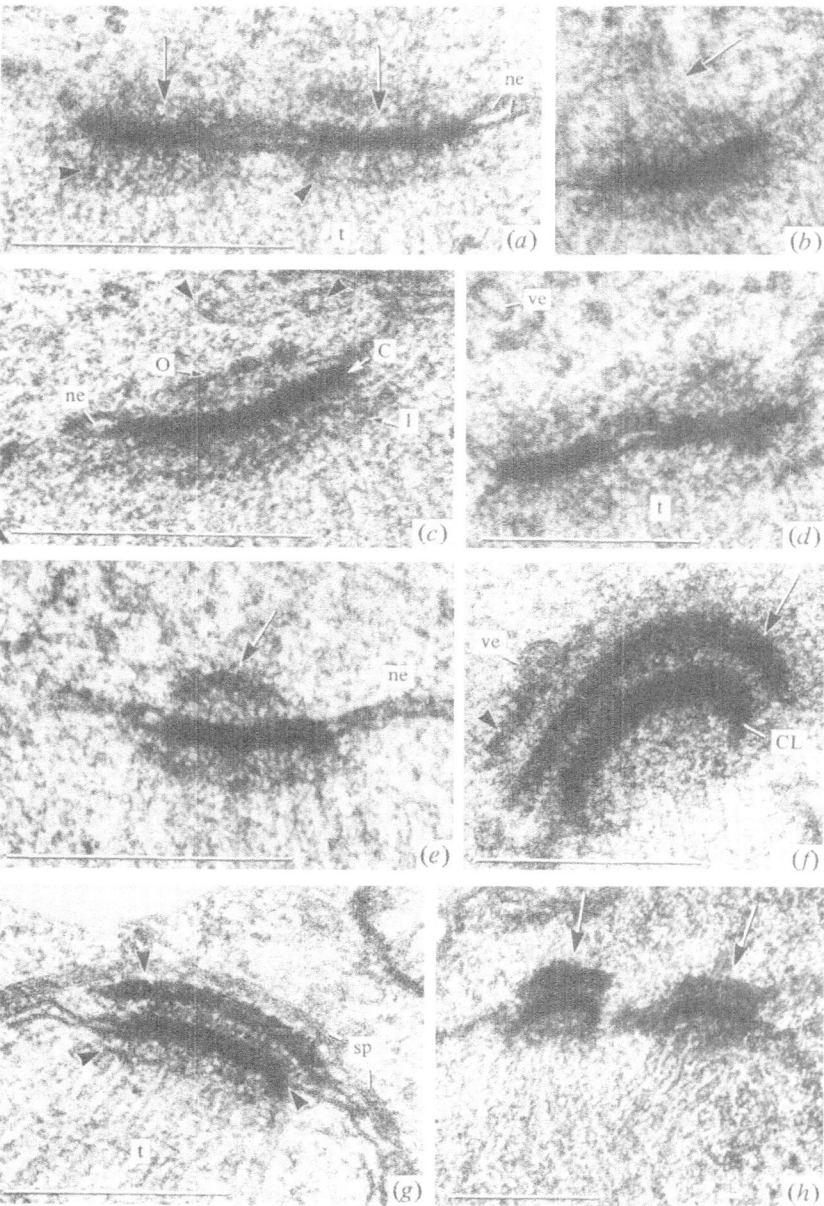

Fig. 1 (*a*, *b*) Mitosis; (*c–h*) meiosis. (*a*) Cross section through nuclear envelope (ne) with two SPBs (arrows). The intranuclear tubules (t) end in the fibrillar layer (arrowhead) of the SPBs. (*b*) Single SPB at metaphase with a bundle of cytoplasmic microtubules (arrow). (*c*) Single SPB from cell after 1 h in sporulation medium. It is composed of third dense layers observed on top of the normal SPB have been interpreted as probably representing stages of SPB doubling (Zickler & Olson, 1975).

During the second division, the SPBs develop prominent central and outer plates (Fig. 1*f*). The outer zone, often double and connected by fibrillar material to the middle layer, increases notably in diameter and density. In profile, it appears crescent-shaped over the middle-layer (Figs. 1*g*; 2*c*). At prophase II (Fig. 1*h*), the nuclei display two sets of duplicated SPBs connected by a bridge. The four bundles of microtubules slightly overlap in the middle of the nucleus. Unlike meiosis I, the SPBs are situated on a lobe of the nuclear envelope. They separate to form new spindles, and thus each nucleus contains two spindles. Initially short (Fig. 2*a*), the spindles elongate significantly while the nucleus develops several lobes. A large SPB covered by the ascospore wall primordium is located on each of four of these lobes. Nuclear constriction occurs coincidently with ascospore delimitation.

This particular 'uninuclear' (Moens & Rapport, 1971*a*) meiosis can be disturbed by mutation in a single gene: in homozygous *spo2–1* strains, the nucleus divides after each division thus leading to four nucleated cells (Moens *et al.*, 1974). It shows also that meiosis divisions I and II are under the same control.

Few observations of meiotic divisions in other Hemiascomycetes are available, especially because older data based on permanganate fixation preclude not only chromosome but also spindle and SPB visualization. The shape of the *Hansenula wingei* nucleus during sporulation (Black & Gorman, 1971) suggests a *Saccharomyces cerevisiae* type of meiosis. Clearly *Wickerhamia fluorescens* has a 'uninuclear' meiosis with similar SPB development (Roony & Moens, 1973*a*). In fission yeast, the SPB,

---

an inner layer (I), a dense central zone (C) and an outer layer (O). Circular vesicles (arrowheads) are often found near the SPB. (*d*) Duplicated SPBs from pachytene nucleus. ve, vesicle. (*e*) Cross section through a single SPB at metaphase I. It shows a dense layer on the cytoplasmic side (arrow). (*f*) Oblique section through a meiosis II SPB. The outer layer (arrow) is noticeably larger and denser than in metaphase I (*e*). A round vesicle (ve) is appressed to a third layer of the SPB (arrowhead). CL, central layer of the SPB. (*g*) The three components (arrowheads) of meiosis II SPB are most clearly visible: the inner layer bears a bundle of intranuclear microtubules (t); the double dense outer layer is connected by small tubules to the central one. The profile of spore wall primordium (sp) is seen on top of the outer layer. (*h*) Prophase II; duplicated SPBs (arrows). Bar scales: 0·3 μm.

as in mitosis (McCully & Robinow, 1971; Heath, 1978), is slightly different (Olson *et al.*, 1978); it is located outside the nucleus near a denser part of the nuclear envelope and a cytoplasmic bundle of microtubules runs along the nucleus during prophase. No indications are given of later stages.

Ascospore formation

Our knowledge of ascospore development is based mainly on studies of *Saccharomyces cerevisiae* (Lynn & Magee, 1970; Moens, 1971; Guth *et al.*, 1972; Beckett, Illingworth & Rose, 1973; Zickler & Olson, 1975).

As seen above, the membrane system which will give rise to the ascospore wall is formed on top of the outer layer of the SPB. First, when the two spindles are still short (presumably metaphase II), round vesicles consisting of two membranes are seen close to or on top of each SPB (Fig. 2*b*). They coalesce and flatten when the spindles elongate. The four ascospore walls have thus separate origins and display separate development. At the same time, mitochondria and large lipid vesicles aggregate around the nucleus. The noticeable lipid accumulation at this stage, first noted by Lynn & Magee (1970), is probably related to the four-fold increase of lipid content described by Illingworth, Rose & Beckett (1973) during sporulation.

When the wall primordia extend over the dense outer layer of the SPBs, they consist of two membranes enclosing an amorphous lumen, the density of which increases during development. Until joining, the primordia remain like flattened vesicles, the two round extremities of which are always clearly seen in longitudinal sections (Fig. 2*c*). They expand along the four prominent nuclear lobes, thus dividing the single nucleus by constriction. Similarly, mitochondria edge between the primordia and the nuclear envelope (Fig. 3*a*).

The SPBs remain associated with the primordia until complete ascospore delimitation. Then the outer layer loses its structure and remains as amorphous material appressed to the spore wall, whereas the middle layer separates from the wall and decreases in size and density, but is still associated with the nuclear envelope. Mature ascospores exhibit a single SPB with a bundle of intranuclear and few astral microtubules.

A certain amount of wall precursors is deposited between the two membranes before complete spore demarcation. In mature ascospores, the inner double membrane is now the spore plasma membrane whereas

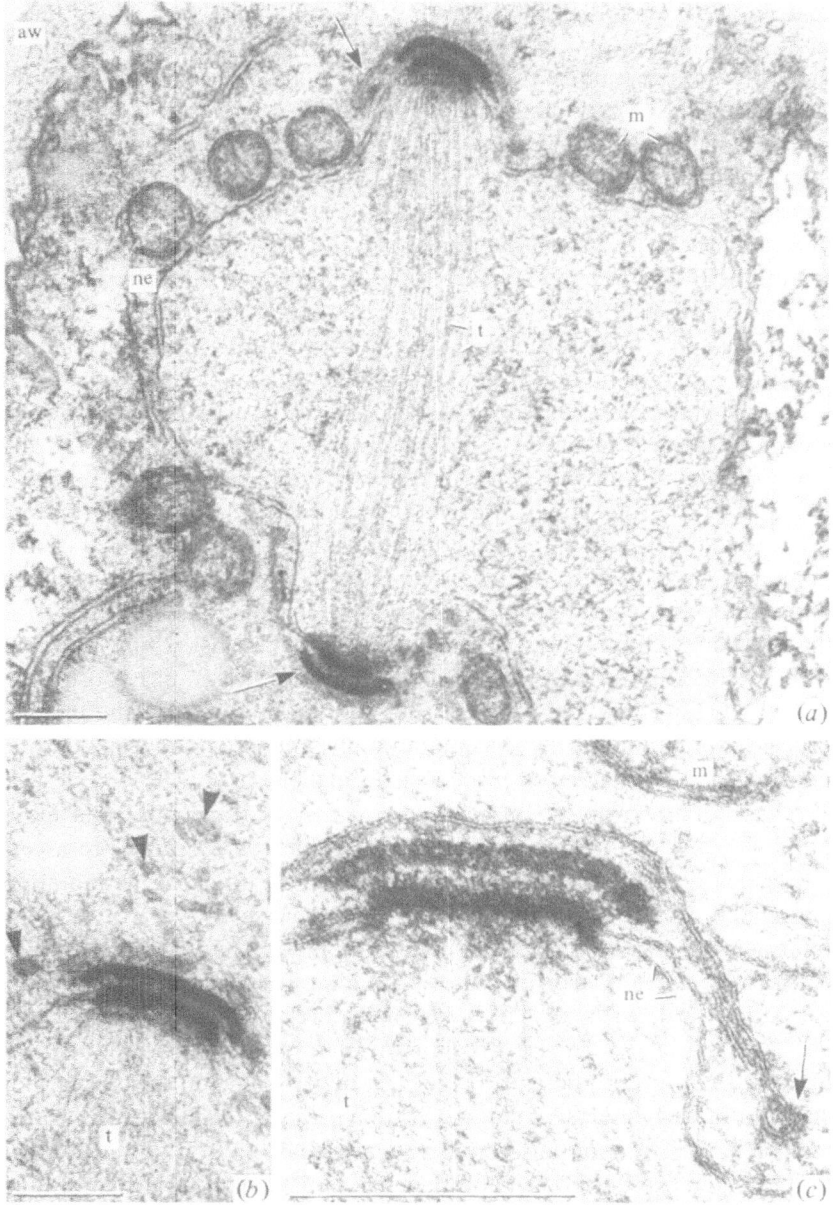

Fig. 2 (a) Metaphase II–anaphase II spindle extended between the two
SPBs which on the cytoplasmic side are capped by the spore wall
primordia (arrows). The mitochondria (m) aggregate around the
nucleus. aw, ascus wall; ne, nuclear envelope; t, microtubule. (b)
Numerous vesicles (arrowheads), believed to be involved in the
formation of the spore wall, can always be observed around the early
metaphase II SPBs. (c) Detail of the association between a metaphase
II – anaphase II SPB and ascospore wall primordium. The arrow
points to the round extremity of the vesicle. Bar scales: 0·3 μm.

the outer dense one is separated from it by the several layers of ascospore wall. A large amount of cytoplasm, vacuoles and even a few mitochondria (or a single plurilobular one) are left behind between the four ascospores and the cell wall (Fig. 3.3*a*).

Ascospores of *Hansenula* appear formed in the same way (Black & Gorman, 1971). In *Wickerhamia fluorescens*, the wall also develops on top of the SPB although the course of events is slightly different because mainly one ascospore matures (Rooney & Moens, 1973*b*).

Ascomycete, and particularly yeast, sporulation represents an interesting case of cellular differentiation: in each ascus, the nuclei around which the spores are organized correspond to the four products of one meiosis and an observable temporal sequence of events leads to the development of four independent cellular forms inside a single cell. In order to probe the sequence of events leading to ascospore formation and to examine their dependency relationships with meiosis, a systematic search for sporulation-deficient and recombination mutants has been undertaken by several groups, using different systems and criteria for the recovery of both recessive and dominant mutations (review in Baker *et al.*, 1976; Heywood & Magee, 1976; Esposito, M. S. & Esposito, R. E., 1978). Mitotic functions were shown to be required for the meiotic process, since among the *cdc* mutants all strains defective in mitotic division were also arrested in meiosis (Simchen, 1974, 1978).

It should, however, be emphasized that most of the meiotic and spore mutants show pleiotropic phenotypes. It is therefore difficult to ascertain which defect is the primary and/or the single effect of the mutation and which are the secondary effects. However, several features of the interplay of sporulation development and meiosis have been found. In contrast to what is seen in mitosis (Hartwell, 1974), meiotic spindle formation (at least in *spo*11 strain) seems independent of DNA synthesis (Moens *et al.*, 1977). Prospore wall formation is dependent on meiosis II differentiation of SPBs; it is never found in mutant strains arrested in early meiosis I stages. Moreover, in apomictic strains which make a single division between meiotic prophase (based on recombination rate and presence of synaptonemal complexes) and sporulation, the two unique SPBs formed exhibit meiosis II morphology (Moens, 1974; Moens *et al.*, 1977). Co-ordination of the four prospore walls' growth requires the presence of the single lobed nucleus: normal capsulation is never found in the *spo*2–1 strains which form four separated nuclei after meiosis. However, spore delimitation is possible without entry of nuclear material (Moens *et al.*, 1974).

Several mutants blocked in meiosis and sporulation have been isolated in *Schizosaccharomyces pombe* (Bresch, Müller & Egel, 1968), but only one, blocked after DNA synthesis and commitment to meiosis, has been checked so far for ultrastructure (Olson *et al.*, 1978).

Spindle pole body functions

As seen above, the SPB is always present in the yeast cell. This observation suggests that the SPB plays an active role not only during nuclear division but also in the control of several events of the cell cycle.

The most evident function is related to spindle formation. All intranuclear and, if any, all cytoplasmic microtubules converge to the SPB. The number of microtubules comprised in the spindle is a function of the ploidy (Peterson & Ris, 1976) and is correlated with the size of the SPB. Isolated mitotic and meiotic SPBs deprived of native microtubules (Borisy *et al.*, 1975; Byers *et al.*, 1978; Hyams & Borisy, 1978) can indeed nucleate the *in vitro* assembly of microtubules. The number of microtubules obtained is not only limited (Byers *et al.*, 1978), but also ploidy-dependent (Hyams & Borisy, 1978) as described *in vivo* by Peterson & Ris (1976). Moreover, the ability to nucleate microtubules depends on the physiology of the isolated SPBs (Hyams & Borisy, 1978). Thus in *Saccharomyces cerevisiae*, SPBs are real 'microtubule organizing centres' (Pickett-Heaps, 1969).

A second function of the SPBs appears during budding. Oriented toward the bud emergence site, they are always associated with the bud cytoplasm by long microtubules formed on their outer layer (Byers & Goetsch, 1974, 1975a). Also, as seen above, the study of *cdc* mutants (Byers & Goetsch, 1974) indicates a temporal relationship between SPB duplication and budding, and suggests that the duplicated SPBs control bud emergence.

The third apparent function is a leader role during conjugation (Byers & Goetsch, 1975a). When cells begin to fuse, the SPBs of both nuclei are oriented toward the isthmus. Again cytoplasmic microtubules are formed and enter the opposite cell. Moreover, some microtubules interconnect the SPB of the two nuclei which will first meet at the level of their SPBs. The specificity of their role in leading the two nuclei to meet is enhanced by the observation that during mitosis they never play an active part in the nuclear migration into the bud. This migration starts on a nuclear lobe whereas the spindle, with variable orientations, remains in the mother cell.

Another interesting role of the SPB appears during ascospore wall formation. Although the origin of the spore-delimiting membranes remains obscure, they indisputably form and elongate on top of the modified meiosis II SPBs (Moens, 1971; Peterson *et al.*, 1972; Zickler & Olson, 1975). The prospore wall always grows out from the SPB, even if later it can separate and form anucleated ascospores (Moens *et al.*, 1974). This role seems to be a specific feature of *Saccharomyces cerevisiae* and closely related species (Rooney & Moens, 1973b). It is 'lost' in Euascomycetes where the ascospore delimiting membranes are first like a big bag (vesicle) formed along the ascus wall, far from the SPBs (Carroll, 1967; Zickler, 1970; Stiers, 1976). This single vesicle is built up even if the nuclei abort (Simonet & Zickler, 1972). The observed increase of the SPBs and of the aster's microtubules during sporulation probably helps to provide a guidance for ascus vesicle invaginations around the nuclei, but spores can be isolated in total absence of the SPBs (Simonet & Zickler, 1972; D. Zickler, unpublished).

Finally, an aspect of the SPBs' role, this time not related to cell control, is their possible usefulness in assessing relationships among yeasts. The SPB structure, combined with nuclear envelope integrity and uninuclear meiosis II division, can be used as a good marker for taxonomic research (McCully & Robinow, 1972) and for clarifying certain relationships between Hemiascomycetes and Euascomycetes (Ashton & Moens, 1979).

Meiotic prophase, genetic recombination and synaptonemal complex formation

Although synaptonemal complexes (SCs) had been accepted as being universally present during meiotic prophase (Westergaard & von Wettstein, 1972), yeast long remained a puzzling problem because only short stretches of SCs and polycomplexes could be found (Engels & Croes, 1968; Moens & Rapport, 1971b).

Zickler & Olson (1975), using a new fixation method according to Peterson *et al.* (1972), revealed the presence of standard SC development in *Saccharomyces cerevisiae*. Since then another method using glass beads instead of enzyme has been used with success by Olson *et al.* (1978). In the absence of contrasted chromatin, like in other Ascomycetes (Gillies, 1972; Zickler, 1977), bivalent behaviour was followed with the aid of the SC, and a typical meiotic prophase with discernible leptotene, zygotene, pachytene and diplotene stages was described

(Zickler & Olson, 1975). Recently, Horesh *et al.* (1979), using different strains, provided confirmation of these previous observations. However, as SCs often remain fuzzy in tangential sections, karyotyping was shown to be easy only in a homozygous *cdc4* strain with prolonged pachytene stage (Byers & Goetsch, 1975*b*). Reconstruction of SCs from diploid and tetraploid nuclei permitted the assignment of an average number of respectively 17 and 34 chromosome pairs. Quadrivalent pairing was also found in all tetraploid nuclei.

After shifting to sporulation medium, almost all nuclei show a spherical dense body, which may or may not be associated with the nucleolus, and is probably present before premeiotic DNA replication (Horesh *et al.*, 1979) (Fig. 3.3*b*). The latter authors also found that the following stage, i.e. a dense body 'cut' by pieces of central region-like structures, is coincident with the premeiotic S phase (Fig. 3*c*), since in their synchronization procedure the SCs are seen later (toward the rise of recombination commitment), the authors conclude that the SC central elements are formed in the dense body before transfer between the lateral elements. It must be said that timing of the different stages in hours is dependent on the employed synchrony system and is therefore not used here but can be found in Simchen, Pinòn & Salts (1972); Petersen, Olson & Zickler (1978), and Horesh *et al.* (1979).

During the first hours in sporulation medium, a few cells show dense unpaired lateral elements, whereas some others correspond to zygotene pictures. After that, most cells contain typical SCs (Figs. 3*d*, 4*f*) with two lateral elements paired at a distance of approximately 10–11 nm and a central element. In transverse sections, the central elements exhibit dense bodies (Fig. 4*a*) similar to the 'recombination nodules' found in other species (review in Carpenter, 1979). Most SCs are found attached to the nuclear envelope except the 'nucleolar organizers' which end in the nucleolus.

Parallel to the decrease of complete SC, pieces of SC of variable lengths, with one or two thickened lateral elements, appear in most nuclei (Fig. 4*b*, *c*). Aggregates of up to seven SCs called polycomplexes (Fig. 4*d*, *e*) are also found before complete release of the SCs and are thus attributed to diplotene nuclei (Zickler & Olson, 1975).

As long as SCs or polycomplexes are present, the nucleus always shows a typical duplicated SPB with two bundles of microtubules never found to be connected to the SCs (Fig. 4*f*).

The meiotic prophase in other yeasts and especially in *Schizosaccharomyces pombe* remains puzzling. Olson *et al.* (1978) have described

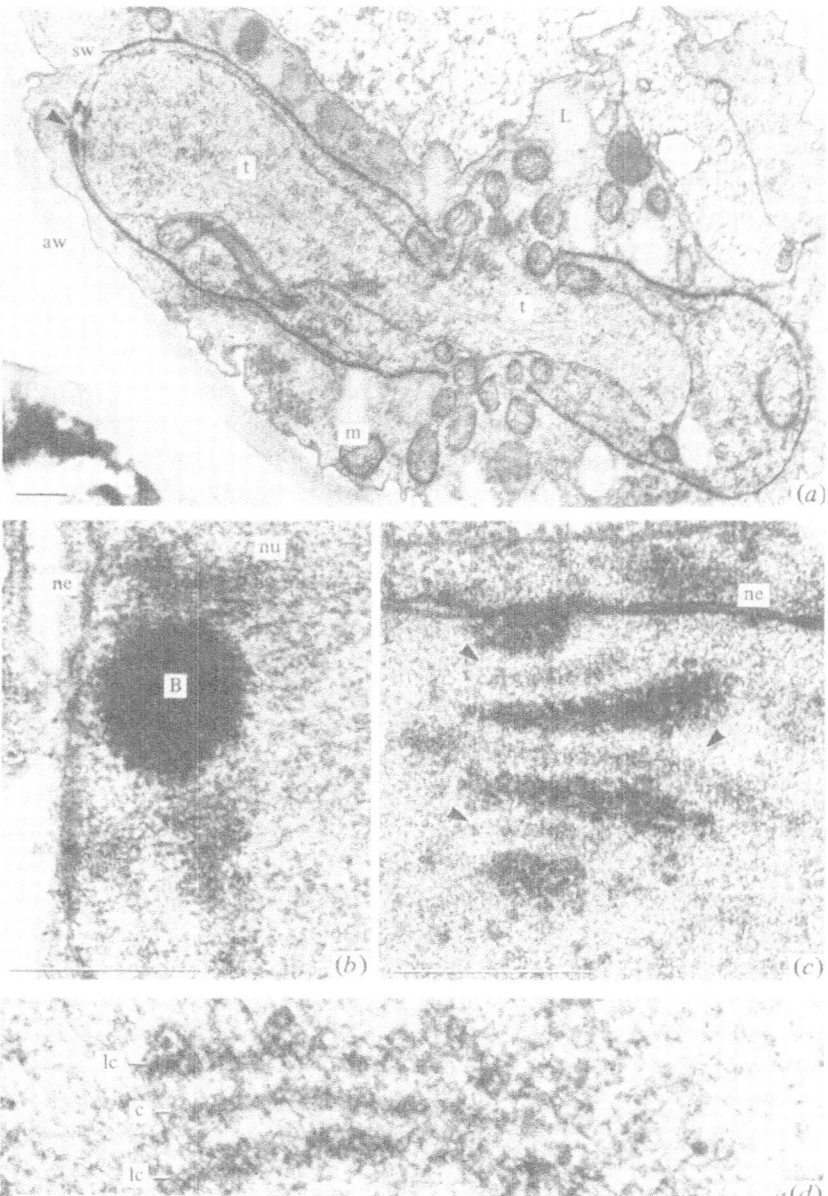

Fig. 3 (*a*) Advanced stage of ascospore formation with aggregation of
mitochondria (m). aw, ascus wall; t, microtubule; sw, spore wall; L,
lipid vesicle; arrowhead, SPB. (*b*) Nucleolus (nu) with dense body (B)
from cell after one hour in sporulation medium. ne, nuclear envelope.
(*c*) Dense body, or dense polycomplex, with three pieces of central
region-like structures (arrowheads). (*d*) Longitudinal section of SC at
pachytene. lc, lateral component; c, central component. Bar scales: 0·3
μm.

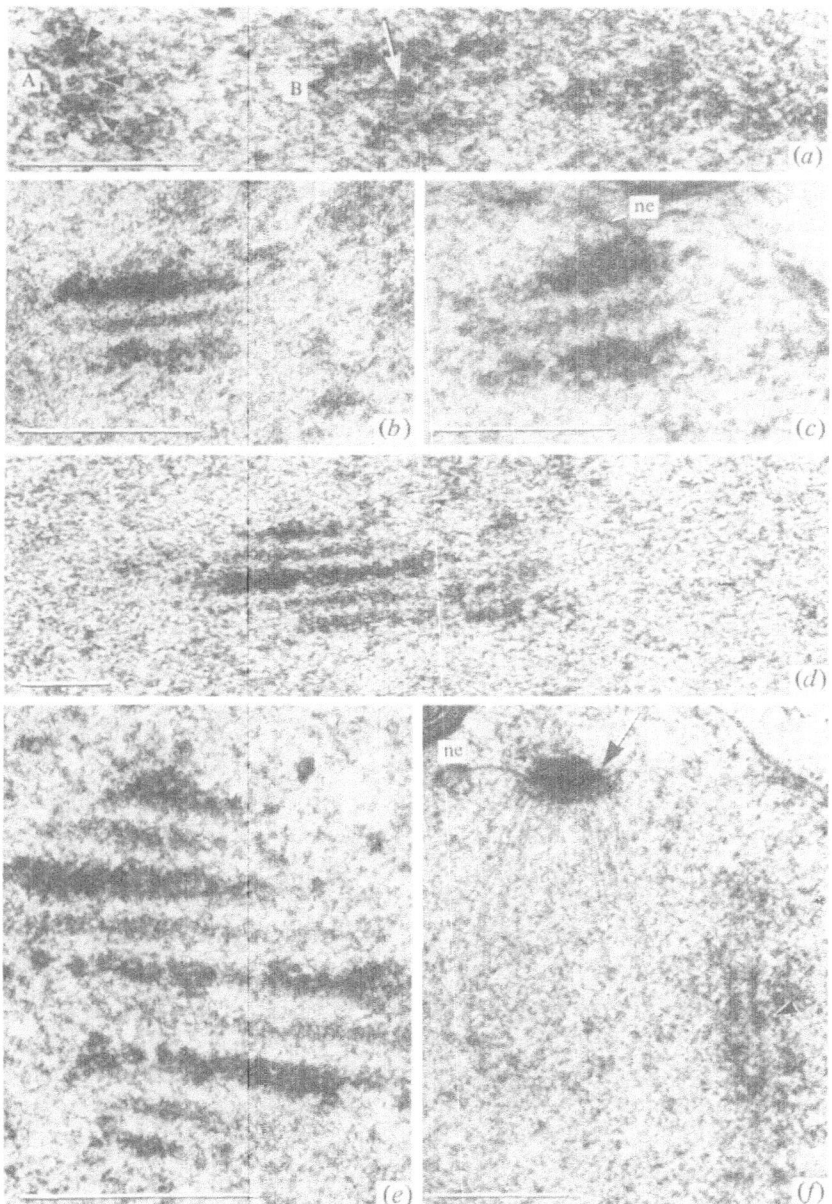

Fig. 4 (*a*) Aspects of SC in cross (A) and longitudinal (B) sections. The three arrowheads point to the lateral and central components. Note a clear 'recombination nodule' (arrow) on the central component of the SC. (*b*, *c*) Two SCs with locally thickened lateral elements at late pachytene. (*d*, *e*) Polycomplex of diplotene nuclei. Cells contain 2(*d*), 4(*e*) or more stacked SCs, mainly located in the middle of the nucleus. (*f*) Pachytene nucleus with SC (arrowhead points to the nodule) and one of the duplicated SPBs (arrow) with two bundles of microtubules Bar scales: 0·3 μm.

dense 'linear elements' which resemble the lateral elements, but not true SCs.

The study of SCs in yeasts is especially interesting because they offer a unique situation: SCs are present during pairing in almost all meiotic cells (Westergaard & von Wettstein, 1972) and therefore might be related to recombination (Carpenter, 1975, 1979; Zickler, 1977). But analysis of recombination is usually only possible after the completion of meiosis, thus making a temporal study between SC occurrence and recombination rate difficult. In *Saccharomyces cerevisiae*, cells can be reverted to mitotic behaviour after meiotic DNA synthesis and before commitment to sporulation (Simchen *et al.*, 1972). By removal of the cells from meiosis-inducing conditions to vegetative growth medium, genetic recombination can be estimated in wild-type and in sporulation-defective mutants: if the strain used is heteroallelic for an auxotrophic marker, all prototrophic recombinants can easily be screened. When removed at an appropriate time, the reverted cells display intergenic as well as intragenic recombination with meiosis-like frequencies and this is before cells are committed to complete meiosis (Esposito, R. E. & Esposito, M. S., 1974).

The first interesting evidence from these observations is that commitment to genetic exchange at the meiotic level is not sufficient to commit cells to meiotic disjunction. There is also a functional separation between premeiotic DNA synthesis and commitment to recombination as shown by mutants that inhibit premeiotic DNA synthesis (Roth & Fogel, 1971), by DNA replication inhibitor (Silva-Lopez, Zamb & Roth, 1975) and by sporulation-defective *cdc* mutants (Schild & Byers, 1978).

Using appropriate markers, Olson & Zimmerman (1978) found (although an overlap resulting from insufficient synchrony is a critical point) that there is first a sequential commitment to gene conversion, and later to crossing-over. As no SCs were found during the same interval of time, they concluded that gene conversion and a limited amount of crossing-over can be completed before SC formation.

Initiation of lateral elements and the polycomplexes are not dependent on DNA synthesis: homozygous *spo*11 strains with nearly blocked DNA synthesis show both structures (Moens *et al.*, 1977). Similar results are given by *cdc*4 strains (Simchen, cited in Moens *et al.*, 1977; Horesh *et al.*, 1979) which show, at restrictive temperature, a high recombination rate combined with limited DNA synthesis. They exhibit normal SCs. Thus, even SC formation seems independent of the completion of premeiotic DNA synthesis, at least in these two mutants.

Confirmation of the association between SC and genetic recombination seen in the *cdc4* strain is also clearly given by another *cdc* mutant (Schild & Byers, 1978). Homozygous *cdc7–1* and *cdc7–4* strains synthesize DNA, but fail to undergo commitment to recombination at restrictive temperature. The arrested cells become committed to recombination when shifted to permissive temperature. While no SCs are detected in the first experiment, most cells contained SCs and nodules at the permissive temperature.

Further characterization of *cdc* and *spo* mutants in *Saccharomyces cerevisiae* and *Schizosaccharomyces pombe* will hopefully provide additional insights into the relationships of DNA replication, genetic recombination and SC formation.

Concluding remarks

The purpose of this review was to outline the more important features of nuclear behaviour during the yeast life cycle. It appears clearly that we need to fill some major gaps in our knowledge about yeasts other than *Saccharomyces cerevisiae*, especially *Schizosaccharomyces pombe* in which meiosis remains almost unstudied. Several studies with *Saccharomyces cerevisiae* have not been motivated primarily by an interest in cellular ultrastructure or function: the widespread use of yeast for genetic research has prompted the need for cytological data which has important genetic consequences. On the other hand, related cellular processes have been clarified with the aid of mutants because yeast has a very well-studied genetic system. Furthermore *S. cerevisiae* has a mitotic and meiotic cell cycle which allows both biochemical and physiological studies.

Thus, reviewing ultrastructure appears frustrating for several reasons: for example, although the morphology of the SPBs is well documented, almost nothing is known about their duplication, movements and ascospore wall association. Important questions, like how the structure and functions of the SPBs are inherited, remain unsolved. However, the most striking conclusion for an ultrastructural review of the nucleus still remains that although synaptonemal complexes display the presence of bivalents during meiotic prophase, nobody has yet been able to see chromosomes in yeast!

Acknowledgements. My own work with yeast presented in this chapter was performed in collaboration with L. W. Olson in Copenhagen. I am deeply grateful to Dr Olson for his friendship and patience during the course of this project. I wish to express my gratitude to Professor D. von Wettstein of the Carlsberg Laboratory, Copenhagen, whose interest and hospitality in his laboratory provided the stimulus for this work. I especially thank N. Randsholt for help in the English translation and M. Dahuron for typing the manuscript.

References

Ashton, M. L. & Moens, P. B. (1979). Ultrastructure of sporulation in the Hemiasco-mycetes *Ascoidea corymbosa, A. rubescens, Cephaloascus fragrans*, and *Saccharomyces capsularis. Canadian Journal of Botany*, **57**, 1259–84.

Baker, B. S., Carpenter, A. T. C., Esposito, M. S., Esposito, R. E. & Sandler, L. (1976). The genetic control of meiosis. *Annual Review of Genetics*, **10**, 53–134.

Beckett, A., Illingworth, R. F. & Rose, A. H. (1973). Ascospore wall development in *Saccharomyces cerevisiae. Journal of Bacteriology*, **113**, 1054–7.

Beran, K. (1968). Budding of yeast cells, their scars and ageing. *Advances in Microbial Physiology*, **2**, 143–71.

Black, S. H. & Gorman, C. (1971). The cytology of *Hansenula*. III. Nuclear segregation and envelopment during ascosporogenesis in *Hansenula wingei. Archiv für Mikrobiologie*, **79**, 231–48.

Borisy, G. G., Peterson, J. B., Hyams, J. S. & Ris, H. (1975). Polymerization of microtubules onto the spindle pole body of yeast. *Journal of Cell Biology*, **67**, 38a.

Bresch, C., Müller, G. & Egel, R. (1968). Genes involved in meiosis and sporulation of a yeast. *Molecular and General Genetics*, **102**, 301–6.

Byers, B. & Goetsch, L. (1974). Duplication of spindle plaques and integration of the yeast cell cycle. *Cold Spring Harbor Symposia in Quantitative Biology*, **38**, 123–31.

Byers, B. & Goetsch, L. (1975a). Behavior of spindles and spindle plaques in the cell cycle and conjugation of *Saccharomyces cerevisiae. Journal of Bacteriology*, **124**, 511–23.

Byers, B. & Goetsch, L. (1975b). Electron microscopic observations on the meiotic karyotype of diploid and tetraploid *Saccharomyces cerevisiae. Proceedings of the National Academy of Sciences, USA*, **72**, 5056–60.

Byers, B., Shriver, K. & Goetsch, L. (1978). The role of spindle pole bodies and modified microtubule ends in the initiation of microtubule assembly in *Saccharomyces cerevisiae. Journal of Cell Science*, **30**, 331–52.

Cabib, E. & Farkas, V. (1971). The control of morphogenesis: an enzymatic mechanism for the initiation of septum formation in yeast. *Proceedings of the National Academy of Sciences, USA*, **68**, 2052–56.

Carpenter, A. T. C. (1975). Electron microscopy of meiosis in *Drosophila melanogaster* females. II. The recombination nodule – a recombination associated structure at pachytene? *Proceedings of the National Academy of Sciences, USA*, **72**, 3186–9.

Carpenter, A. T. C. (1979). Recombination nodules and synaptonemal complex in recombination-defective females of *Drosophila melanogaster. Chromosoma, Berlin*, **75**, 259–92.

Carroll, G. C. (1967). The ultrastructure of ascospore delimitation in *Saccobolus kerverni*. *Journal of Cell Biology*, **33**, 218–24.

Engels, F. M. & Croes, A. F. (1968). The synaptoinemal complex in yeast. *Chromosoma, Berlin*, **25**, 104–6.

Esposito, R. E. & Esposito, M. S. (1974). Genetic recombination and commitment to meiosis in *Saccharomyces*. *Proceedings of the National Academy of Sciences, USA*, **71**, 3172–6.

Esposito, M. S. & Esposito, R. E. (1978). Aspects of the genetic control of meiosis and ascospore development inferred from the study of *spo* (sporulation-deficient) mutants of *Saccharomyces cerevisiae*. *Biologie cellulaire*, **33**, 93–101.

Fuller, M. S. (1976). Mitosis in fungi. *International Review of Cytology*, **45**, 113–53.

Gillies, C. B. (1972). Reconstruction of the *Neurospora crassa* pachytene karyotype from serial sections of synaptonemal complexes. *Chromosoma*, Berlin, **36**, 119–30.

Guth, E., Hashimoto, T. & Conti, S. F. (1972). Morphogenesis of ascospores in *Saccharomyces cerevisiae*. *Journal of Bacteriology*, **109**, 869–80.

Hartwell, L. H. (1974). *Saccharomyces cerevisiae* cell cycle. *Bacteriological Reviews*, **38**, 164–98.

Heath, I. B. (1974). Genome separation mechanisms in Prokaryotes, Algae and Fungi. In *The Cell Nucleus*, ed. H. Busch, vol. 2, pp. 487–515. New York & London: Academic Press.

Heath, I. B. (1978). Experimental studies of mitosis in the fungi. In *Nuclear division in the Fungi*, ed. I. B. Heath, pp. 89–176. London & New York: Academic Press.

Heywood, P. & Magee, P. T. (1976). Meiosis in Protists. Some structural and physiological aspects of meiosis in Algae, Fungi and Protozoa. *Bacteriological Reviews*, **40**, 190–240.

Horesh, O., Simchen, G. & Friedmann, A. (1979). Morphogenesis in the synapton during yeast meiosis. *Chromosoma, Berlin*, **75**, 101–15.

Hyams, J. S. & Borisy, G. G. (1978). Nucleation of microtubules *in vitro* by isolated spindle pole bodies of the yeast *Saccharomyces cerevisiae*. *Journal of Cell Biology*, **78**, 401–14.

Illingworth, R. F., Rose, A. H. & Beckett, A. (1973). Changes in the lipid composition and fine structure of *Saccharomyces cerevisiae* during ascospore formation. *Journal of Bacteriology*, **113**, 373–86.

Kubai, D. F. (1975). The evolution of the mitotic spindle. *International Review of Cytology*, **43**, 167–227.

Kubai, D. F. (1978). Mitosis and fungal phylogeny. In *Nuclear Division in the Fungi*, ed. I. B. Heath, pp. 177–229. New York & London: Academic Press.

Lynn, R. R. & Magee, P. T. (1970). Development of the spore wall during ascospore formation in *Saccharomyces cerevisiae*. *Journal of Cell Biology*, **44**, 688–92.

Matile, P., Moor, H. & Robinow, C. F. (1969). Yeast cytology. In *The Yeasts*, vol. 1, ed. A. H. Rose & J. S. Harrison, pp. 219–302. London & New York: Academic Press.

McCully, E. K. & Robinow, C. F. (1971). Mitosis in the fission yeast *Schizosaccharomyces pombe*: a comparative study with light and electron microscopy. *Journal of Cell Science*, **9**, 475–507.

McCully, E. K. & Robinow, C. F. (1972). Mitosis in heterobasidiomycetous yeasts. I. *Leucosporidium scottii*. *Journal of Cell Science*, **10**, 857–81.

Minet, M., Nurse, P., Thuriaux, P. & Mitchison, J. M. (1979). Uncontrolled septation in a cell division cycle mutant of the fission yeast *Schizosaccharomyces pombe*. *Journal of Bacteriology*, **137**, 440–6.

Moens, P. B. (1971). Fine structure of ascospore development in the yeast *Saccharomyces cerevisiae*. *Canadian Journal of Microbiology*, **17**, 507–10.

Moens, P. B. (1974). Modification of sporulation in yeast strains with two-spored asci (*Saccharomyces*, Ascomycetes). *Journal of Cell Science*, **16**, 519–27.

Moens, P. B., Esposito, R. E. & Esposito, M. S. (1974). Aberrant nuclear behavior at meiosis and anucleate spore formation by sporulation-deficient (*spo*) mutants of *Saccharomyces cerevisiae*. *Experimental Cell Research*, **83**, 166–74.

Moens, P. B., Mowat, M., Esposito, M. S. & Esposito, R. E. (1977). Meiosis in a temperature-sensitive DNA-synthesis mutant and in an apomictic strain (*Saccharomyces cerevisiae*). *Philosophical Transactions of the Royal Society of London. Series B*, **277**, 351–58.

Moens, P. B. & Rapport, E. (1971a). Spindles, spindle plaques, and meiosis in the yeast *Saccharomyces cerevisiae* (Hansen). *Journal of Cell Biology*, **50**, 344–61.

Moens, P. B. & Rapport, E. (1971b). Synaptic structures in the nuclei of sporulating yeast, *Saccharomyces cerevisiae* (Hansen). *Journal of Cell Science*, **9**, 665–77.

Moor, H. & Mühlethaler, K. (1963). Fine structure in frozen-etched yeast cells. *Journal of Cell Biology*, **17**, 609–28.

Nurse, P., Thuriaux, P. & Nasmyth, K. (1976). Genetic control of the cell division cycle in the fission yeast *Schizosaccharomyces pombe*. *Molecular and General Genetics*, **146**, 167–78.

Olson, L. W., Eden, U., Egel-Mitani, M. & Egel, R. (1978). Asynaptic meiosis in fission yeast. *Hereditas*, **89**, 189–99.

Olson, L. W. & Zimmermann, F. K. (1978). Meiotic recombination and synaptonemal complexes in *Saccharomyces cerevisiae*. *Molecular and General Genetics*, **166**, 151–59.

Peterson, J. B., Gray, R. H. & Ris, H. (1972). Meiotic spindle plaques in *Saccharomyces cerevisiae*. *Journal of Cell Biology*, **53**, 837–41.

Peterson, J. B. & Ris, H. (1976). Electron microscopic study of the spindle and chromosome movement in the yeast *Saccharomyces cerevisiae*. *Journal of Cell Science*, **22**, 219–42.

Petersen, J. G. L., Olson, L. W. & Zickler, D. (1978). Synchronous sporulation of *Saccharomyces cerevisiae* at high cell concentrations. *Carlsberg Research Communications*, **43**, 241–53.

Pickett-Heaps, J. D. (1969). The evolution of the mitotic apparatus: an attempt at comparative ultrastructural cytology in dividing plant cells. *Cytobios*, **3**, 257–80.

Rapport, E. (1971). Some fine structure features of meiosis in the yeast *Saccharomyces cerevisiae*. *Canadian Journal of Genetics and Cytology*, **13**, 55–62.

Robinow, C. F. (1975). The preparation of yeasts for light microscopy. In *Methods of Cell Biology*, vol. II, *Yeast Cells*, ed. D. M. Prescott, pp. 1–22. New York & London: Academic Press.

Robinow, C. F. (1977). The number of chromosomes in *Schizosaccharomyces pombe*: light microscopy of stained preparations. *Genetics*, **87**, 491–97.

Robinow, C. F. & Marak, J. (1966). A fiber apparatus in the nucleus of the yeast cell. *Journal of Cell Biology*, **29**, 129–51.

Rooney, L. & Moens, P. B. (1973a). Nuclear divisions at meiosis in the ascomycetous yeast *Wickerhamia fluorescens*. *Canadian Journal of Microbiology*, **19**, 1383–7.

Rooney, L. & Moens, P. B. (1973b). The fine structure of ascospore delimitation in the yeast *Wickerhamia fluorescens*. *Canadian Journal of Microbiology*, **19**, 1389–92.

Roth, R. & Fogel, S. (1971). A selective system for yeast mutants deficient in meiotic recombination. *Molecular and General Genetics*, **112**, 295–305.

Schild, D. & Byers, B. (1978). Meiotic effects of DNA-defective cell division cycle mutations of *Saccharomyces cerevisiae*. *Chromosoma, Berlin*, **70**, 109–30.

Sherman, F. & Lawrence, C. W. (1974). *Saccharomyces*. In *Handbook of Genetics*, ed. R. C. King, pp. 359–93. New York: Plenum Press.

Silva-Lopez, E., Zamb, T. J. & Roth, R. (1975). Role of premeiotic replication in gene conversion. *Nature, London*, **253**, 212–14.

Simchen, G. (1974). Are mitotic functions required in meiosis? *Genetics*, **76**, 745–53.

Simchen, G. (1978). Cell cycle mutants. *Annual Review of Genetics*, **12**, 161–91.

Simchen, G., Pinon, R. & Salts, Y. (1972). Sporulation in *Saccharomyces cerevisiae*: premeiotic DNA synthesis, readiness and commitment. *Experimental Cell Research*, **75**, 207–18.

Simonet, J. M. & Zickler, D. (1972). Mutations affecting meiosis in *Podospora anserina*. I. Cytological studies. *Chromosoma, Berlin*, **37**, 327–51.

Stiers, D. L. (1976). The fine structure of ascospore formation in *Ceratocystis fimbriata*. *Canadian Journal of Botany*, **54**, 1714–22.

Westergaard, M. & von Wettstein, D. (1972). The synaptonemal complex. *Annual Review of Genetics*, **6**, 71–110.

Williamson, D. H. (1965). The timing of deoxyribonucleic acid synthesis in the cell cycle of *Saccharomyces cerevisiae*. *Journal of Cell Biology*, **25**, 517–28.

Zickler, D. (1970). Division spindle and centrosomal plaques during mitosis and meiosis in some Ascomycetes. *Chromosoma, Berlin*, **30**, 287–304.

Zickler, D. (1977). Development of the synaptonemal complex and the 'recombination nodules' during meiotic prophase in the seven bivalents of the fungus *Sordaria macrospora* Auersw. *Chromosoma, Berlin*, **61**, 289–316.

Zickler, D. & Olson, L. W. (1975). The synaptonemal complex and the spindle plaque during meiosis in yeast. *Chromosoma, Berlin*, **50**, 1–23.

Mechanisms of nuclear division in fungi

I. B. HEATH

Biology Department, York University,
4700 Keele Street, Downsview,
Toronto, Ontario M3J 1P3,
Canada

Introduction

The way in which fungal nuclei undergo mitosis has remained a controversial question for many years (e.g. Olive, 1953; Robinow & Bakerspigel, 1965; Heath, 1974a, 1978; Fuller, 1976; Girbardt, 1978). It is now clear that in many major features, the mitotic apparatus of most fungi is not greatly dissimilar from that of other eukaryotes. However, there are numerous variations of detail which may help explain the fundamental mechanisms of mitosis. The purpose of this chapter is to examine these variations as they apply to the force-requiring events of mitosis. As in previous works on this topic (Heath, 1974a, 1975, 1978, 1980a, 1981), I shall select organisms on the basis of their interest to mycologists rather than the logic of current taxonomy.

There are four distinct mitotic processes which require the operation of some force-generating mechanism. (1) At an early stage in mitosis, the structures which form the polar foci of the mitotic spindle (hereafter referred to as nucleus-associated organelles (NAOs) following the arguments of Girbardt & Hädrich, 1975; Girbardt, 1978 and Heath, 1978, 1980a, 1981. N.B. centrioles are considered to be a type of NAO) move apart so that there is an inter-NAO region in which the spindle develops. This movement is referred to as *NAO migration.* (2) During mitotic prophase the chromatin typically undergoes condensation, then often moves to the approximate equator of the spindle to assume its metaphase configuration. This movement is referred to as *prometaphase movement.* The prophase movements characteristic of meiosis are not considered here. (3) Following metaphase the chromosomes move to the poles of the spindle during the well-known *anaphase movements.* (4) During anaphase and telophase the nuclei may undergo extensive elongation which will be termed *telophase elongation.*

Microtubules are the best described feature of the force-generating mechanisms of mitosis and therefore it is important to introduce briefly the terminology currently applied to them. Microtubules are composed of protein subunits consisting of two or more related species of tubulin which polymerize to produce the intact microtubule. A variety of agents can alter the equilibrium between tubulin dimers and the polymeric microtubule. In addition to tubulin, microtubules are composed of a number of 'microtubule associated proteins' (MAPs) some of which may form cross-bridges between adjacent microtubules. Details of this level of microtubule organization are considered further by Gull in the next chapter. Within the cell there are various classes of microtubules whose names reflect their cellular location. *Cytoplasmic microtubules* will not be further subdivided in the present context but the term will be used to exclude flagellar axonemes and flagellar root systems. Intranuclear or mitotic spindle microtubules may be divided into *kinetochore* (or chromosomal) *microtubules* and *non-kinetochore microtubules*. The latter category can be further subdivided into *continuous microtubules* (which run from one pole of the spindle to the other); *interdigitating microtubules* (which run from one pole of the spindle for some distance beyond the equator of the spindle but not to the opposite pole); *polar microtubules* (which are comparable to interdigitating microtubules but do not run as far as the equator of the spindle); and *free microtubules* (which lie in various parts of the spindle with no evident connection to either spindle poles or kinetochore). This terminology is derived from Heath (1974*b*).

Migration of nuclear associated organelles

Among the fungi there is substantial variability in details of NAO morphology and replication behaviour (Heath, 1980*a*, 1981). However, in all cases, replication is followed by separation of the duplicated NAOs so that they come to lie at the poles of the mitotic spindle. The timing of this migration varies with genera such as *Phlyctochytrium* (McNitt, 1973), *Sorosphaera* (Braselton, Miller & Pechak, 1975), *Plasmodiophora* (Garber & Aist, 1979), *Sapromyces* (Heath, 1975*a*), and *Catenaria* (Ichida & Fuller, 1968), showing migration prior to spindle formation and other genera having concomitant migration and spindle formation (e.g. *Fusarium*, Aist & Williams, 1972; *Pilobolus*, Bland & Lunney, 1975; *Saprolegnia*, Heath & Greenwood, 1968, 1970; *Thraustotheca*, Heath, 1974*b*). In the latter situation it is difficult to analyse the mechanism responsible for NAO migration since migration and spindle formation are so tightly co-ordinated; but the very earliest stages of migration in these genera, and the other patterns of migration, are more informative.

In most fungi, NAO migration occurs in the presence of a closely appressed intact nuclear envelope. The exceptions to this rule are the various Basidiomycetes where the NAOs enter the nucleus and migrate during spindle formation (e.g. *Trametes*, Girbardt, 1968; *Coprinus*, Thielke, 1974; *Rhodosporidium*, McCully & Robinow, 1972). In this latter case it seems most likely that spindle development provides the force necessary for NAO migration but in the other system this is less likely. McCully & Robinow (1972) suggested that appropriately localized membrane synthesis could be responsible for NAO migration. Whilst it is clear that nuclear envelope synthesis must occur at some stage during the nuclear cycle, there is little evidence to support the required timing and localization. There is also no reason to believe that such hypothesized synthesis would effect NAO migration. However, this criticism ignores recent information on the nature of the nuclear envelope. Whilst membranes *per se* behave as fluids which are unlikely to act in any way other than to permit the NAOs to move laterally through the phospholipid bilayer, it is now clear that the nuclear envelope contains, or is associated with, a system of filaments (Fig. 1) which interconnect the nuclear pores and presumably play a role in their disposition over the surface of the nucleus (reviewed in Franke & Scheer, 1974; see also Comings & Okada, 1976; Berezney & Coffey, 1977; Schatten & Thoman, 1978; Gerace & Blobel, 1980). The chemical identity, and even existence of such filaments, in the fungi remains unknown but their location and nature (i.e. a network) make them likely candidates for a role in moving the membrane-associated NAOs.

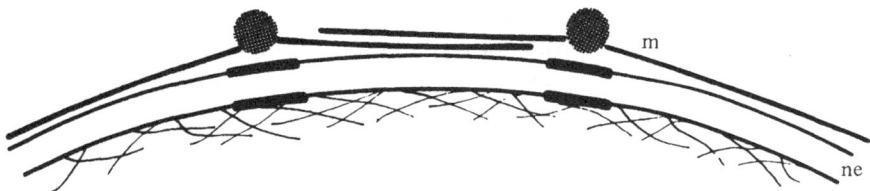

Fig. 1. Diagram of the observed structures that could be involved in NAO migration. Attached to the inside of the nuclear envelope (ne) is the fibrous 'nuclear envelope lamina' or 'residual envelope'. The NAOs (stippled) are associated with differentiated regions of the nuclear envelope (thickened lines) and form the foci of cytoplasmic microtubules (m). The microtubules run along the outer surface of the nuclear envelope, both ahead of and between the migrating NAOs.

Clearly, if membrane-associated filaments were involved, it is most likely that they would move the NAO-associated region of the nuclear envelope and the NAOs would be moved by attachment to this region. Such attachments are known to occur (e.g. Heath, 1978). The use of a nuclear envelope-associated network to effect NAO migration would suggest economy of cellular systems because some comparable system is probably involved in chromosome synapsis during meiotic prophase (Moens, 1973).

Whilst a nuclear envelope-associated network could be causally involved in NAO migration (certainly it would be rearranged during migration even if it were not active in the process) there is also evidence of a role for cytoplasmic microtubules. In a number of genera (e.g. *Thraustotheca*, Heath, 1974*b*; *Thraustochytrium*, Kazama, 1974; *Plasmodiophora*, Garber & Aist, 1979; *Sorosphaera*, Braselton *et al.*, 1975; and *Aspergillus*, B. R. Oakley, personal communication) there is typically a small population of cytoplasmic microtubules which radiate along the nuclear envelope from the pre-migration NAOs (Fig. 1). During migration these microtubules are rearranged (or perhaps, more accurately, broken down and reformed) to form an array emanating from each migrating NAO. This rearrangement clearly results in the production of a few (two or three) microtubules running between the migrating NAOs (Heath, 1974*b*) (Fig. 1). These inter-NAO microtubules form early on in the migration process, before the intranuclear spindle develops in the case of *Thraustotheca* (Heath, 1974*b*), and are excellent candidates for force generation via elongation (polymerization) or intermicrotubule sliding (with concomitant polymerization, since they clearly elongate). These cytoplasmic microtubule arrays persist throughout the migration process. However, as is obvious in the reports of Kazama (1974), McNitt (1973) and Garber & Aist (1979), the geometry of the nucleus and the migrating NAOs is such as to preclude a continuation of force production by simple 'pushing' of the inter-NAO cytoplasmic microtubules. In all reports the cytoplasmic microtubules are laterally appressed to the nuclear envelope. It is tempting to suggest that shear-force generation between the nuclear envelope (stabilized by the underlying fibrillar network mentioned above) and the microtubules propels the microtubules and their associated NAOs over the surface of the nucleus. There are no data to support, or refute, this suggestion, but if it were true, clearly the physical characteristics of the nuclear envelope would have to be varied locally to provide anchorage in one area and permit movement of the NAO in a closely adjacent area. An important implicit part of this suggestion is that microtubules with opposite polarity (discussed by Borisy, 1978 and Bergen & Borisy, 1980), i.e. those preceding and trailing each migrating NAO, would be working to generate force in the same direction. This would entail one set pushing and one set pulling, and would be analogous to bidirectional force generation in a flagellar axoneme or a muscle sarcomere, neither of which occur. Clearly, if this system is operating, its characteristics and controls will be complex. However, it could be that the microtubules play a more passive role in NAO migration. They could act as static linkages mediating an effective increase in the area of nuclear envelope to which the NAO is bound in a manner analogous to the tentacles of an octopus holding its body on a surface. In this hypothesis the force generation would be produced by the nuclear envelope-associated network discussed above. This postulated static linkage would be analogous to that operating in the interaction between flagellar roots and fungal zoospore nuclei (Heath, 1976).

Support for a role of microtubules in NAO migration does come from the observations of inhibition of the process by anti-microtubule agents such as colchicine, nocodazole (Heath, 1978) and methylbenzimidazol-2-yl carbamate (Künkel & Hädrich, 1977). However, since both of these reports deal with fungi in which NAO migration is synchronous with spindle formation, it is hard to evaluate which set of microtubules was involved. Künkel & Hädrich (1977) reported that after prolonged treatment with MBC some unspecified number of NAOs could undergo migration, but it was not possible to determine whether this did indeed occur in the absence of cytoplasmic microtubules (recovery and re-formation of microtubules is a known phenomenon; Wunderlich & Speth, 1970). Even if it did not involve microtubules, it is unclear whether the mechanism of movement was comparable to that normally encountered. Brownian motion could have sufficed, given the time available and the possibility of a drug-induced breakdown of the normal cellular cytoskeleton.

Analysis of NAO migration is complicated by Aist's (1969) observation that during interphase the NAO of *Ceratocystis* often exhibits erratic, rapid movements along the nuclear surface. These movements ceased at the onset of mitosis and it was implied that NAO migration occurred in a slower and more regular manner. These observations suggest the possibility of two different force-generating systems operating on the NAO, but their ultrastructural basis is not yet described.

As discussed later, it is sometimes possible to obtain an indication of the nature of the force-generating system by analysis of its rate of movement. This is rarely possible for NAO migration although one can deduce rates of 0.1 μm min^{-1} for *Saprolegnia* (Heath & Rethoret, 1980), 3.0 μm min^{-1} for *Fusarium* (Aist & Williams, 1972), and 5.0 μm min^{-1} for *Thielaviopsis* (Huang & Patrick, 1972). These rates are typically lower than the rate of anaphase movement for each species but this may not be due to different force-generating systems because the nuclear envelope may constitute an effective 'damper' on the motor. Thus one can draw few conclusions from the dynamic information currently available.

As a final point, it should be acknowledged that in some genera (e.g. *Uromyces*, Heath, I.B. & Heath, M.C., 1976) cytoplasmic microtubules appear to be absent during NAO migration. It is unlikely that this represents an artefactual loss during fixation so that a universal involvement of cytoplasmic microtubules is in doubt. Thus one must conclude that, in all but some Basidiomycetes, fungal NAO migration may involve a nuclear envelope component with or without an additional input from cytoplasmic microtubules. Clearly, when migration and spindle formation are concurrent events, polymerization of the spindle microtubules may be the primary 'motor' powering NAO migration.

Prometaphase movements

In plants, animals and those fungi with clear metaphase plates, prometaphase is simply defined as the movement of the dispersed prophase chromosomes onto the metaphase plate. However, in most fungi a metaphase plate is not established; instead the chromosomes are spread over much of the length of the spindle at all times prior to their anaphase separation (see listings in Heath, 1978, 1980*a*). Thus it is difficult to determine when metaphase has been attained. Furthermore, the prometaphase events evidently occur rapidly since they are rarely described in detail in studies of fungal mitoses – contrary to the elegant analyses of Rickards (1975), Roos (1976) and Wise (1978) in animal cells. Nevertheless, available reports suggest that there may be two patterns of behaviour, depending on the mode of spindle formation (Fig. 2). It will be argued that they both utilize a comparable mechanism to achieve the metaphase configuration. In the best-studied organisms in which spindle formation and NAO migration are simultaneous events, it is clear that the kinetochores are attached to the spindle poles from the earliest detectable stages of spindle formation (e.g. Heath & Greenwood, 1970; Roos, 1975; Moens, 1976; Heath, 1978, 1980*b* and

Fig. 2). This connection is mediated via one to three microtubules per kinetochore. During spindle development, through prometaphase and metaphase and until anaphase, these microtubules elongate to varying degrees with the consequent positioning of the kinetochores in a broad equatorial region of the spindle. It is clear that the kinetochores move relative to one another during this process (as opposed to passively retaining their prophase configuration), because at prophase they are clustered in an area with a diameter of about 0.6 μm (Moens, 1976; Heath, 1978, 1980*b*) whereas by metaphase they are spread over a central 1 μm length of spindle. Although comparable data are lacking from other species, it is likely that a similar situation is widespread, certainly the widely dispersed metaphase kinetochores occur in many species (see listings in Heath, 1978, 1980*a*).

Because the above prometaphase movements are only detected with the electron microscope one can only estimate their dynamic characteristics. For *Saprolegnia* one can see that the mean minimum rate of kinetochore movement at this stage must be about 0.01 μm min^{-1} (based on a dispersal from 0.6 μm to 1.0 μm during a time of 18 min, Heath & Rethoret, 1980). This rate assumes that the kinetochores do not oscillate, for which this is scant evidence, although Aist (1969) noted

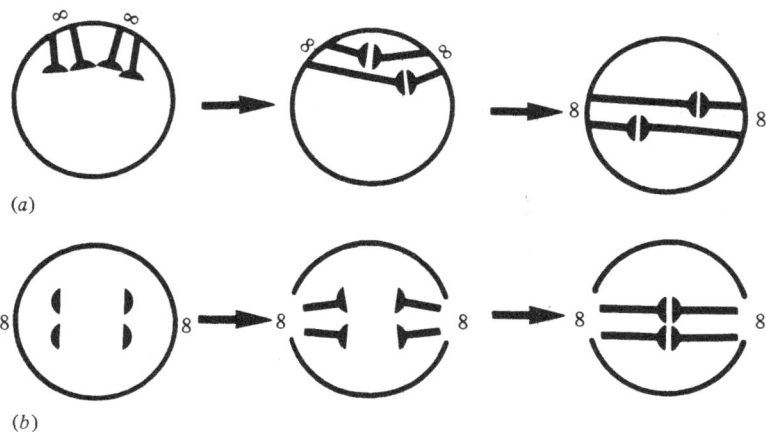

(*a*)

(*b*)

Fig. 2. Diagrammatic representation of the alternative forms of prometaphase behaviour. In (*a*) the kinetochore microtubules are initially found side by side, they reorientate, pairs are formed, and they position the chromatin along a large part of the spindle. In (*b*) the chromatin (semicircles) is initially dispersed in the nucleus, then kinetochore microtubules develop between it and the spindle poles as the chromatin comes to lie at the equator of the metaphase spindle.

that by metaphase the *Ceratocystis* chromosomes behave as if they were static with respect to the spindle. Whilst clearly a crude estimate, the above rate dramatically differs from the rates of about 1 μm min^{-1} observed during prometaphase in insects (Wise, 1978). This difference suggests a different mechanism; indeed with such slow movements one may question the existence of any mechanism other than Brownian movement. However, because the kinetochore movements clearly involve microtubule length changes, it seems most likely that the cell actively controls these changes, and that they are responsible for kinetochore movement. With the observed differences in length of the various kinetochore microtubules during prometaphase and metaphase, it seems likely that control of the events is mediated at the level of the individual microtubule rather than on a spindle-wide basis.

The above discussion relates only to fungi in which NAO migration and spindle formation are simultaneous events. However, a similar mechanism probably operates in the alternative system whereby the spindle invades the nucleus from previously separated NAOs (Fig. 2). McNitt (1973) has shown, in *Phlyctochytrium*, that when the chromatin condenses during prophase, it is dispersed throughout the nucleoplasm. During spindle formation it appears to be moved to the metaphase plate position as the spindle develops from each pole. A similar situation seems evident in *Plasmodiophora* (Garber & Aist, 1979). In both cases, the ultrastructural observations suggest that simple polymerization of the kinetochore microtubules effects the movement of the chromatin to the metaphase plate. An important unresolved question in these studies is the site of polymerization of the kinetochore microtubules. As most recently discussed by Borisy (1978), the site of polymerization (i.e. from kinetochore towards the poles or vice versa) has very important consequences for the mode of operation of the spindle and thus deserves more investigation.

Thus one may conclude that irrespective of the mode of spindle formation the most likely mechanism by which the kinetochores are translocated during prometaphase is microtubule polymerization. This does not rule out other contributing factors, but current data do not require the postulation of additional force-generating systems at this stage of mitosis.

Anaphase movements

In most mitotic systems, including the fungi, there are two movements which occur during anaphase, i.e. spindle elongation and

Table 1. *Rates of diverse cellular movements*

Type of movement	Rate (μm min^{-1})		Reference
	Mean	Range	
Striated muscle contraction	420	–	Wolpert, 1965
Inter-microtubule sliding in			
eukaryotic flagella	390	–	Brockaw, 1975
Saltation	300	–	Rebhun, 1972
Amoeba cytoplasm	< 1800	–	Allen & Taylor, 1975
Mycoplasma motility	22	14–31	Maniloff, 1979
Vorticellid spasmonemes	1.4×10^6	–	Amos, 1975
Fungal hyphal growth	20	0.2–100	Burnett, 1968,
			Heath & Heath, 1978
Fungal NAO migration	≈ 1	0.1–5	See text
Fungal prometaphase	$\approx 0.01?$	–	See text
Fungal anaphase	≈ 5	0.3–10	Heath, 1980*a*
Angiosperm and Metazoan			
anaphase	≈ 1	0.2–3	Heath, 1980*a*
Fungal spindle elongation	11	4–30	Heath, 1980*a*,
			Heath & Rethoret,
			1980
Angiosperm and Metazoan			
elongation	1.6	0.9–2.1	Heath, 1980*a*
Fungal organelle migrations	–	3–588	Heath & Heath, 1978

chromosome-to-pole migration. The temporal relationship between these two processes is variable between species and one or the other may be lacking entirely. This supports the experimental evidence (e.g. Oppenheim, Hauschka & McIntosh, 1973), suggesting that the two processes are essentially independent (although clearly co-ordinated) and may be due to different force-generating systems. In the fungi, most of the spindle elongation process occurs during telophase and will be discussed separately later. This discussion will deal exclusively with the possible mechanisms effecting chromosome-to-pole movements.

Chromosome-to-pole movements have three characteristics which are important in relation to force-generating systems. (1) They are very slow. In fungi they are typically faster than in other cells but they are still approximately two orders of magnitude slower than many other cell movements and are even at the low end of the range reported for hyphal growth (Table 1). (2) They traverse a short distance; for example, Heath & Rethoret (1980) report 1 – 2 μm, which is close to the value characteristic of many fungi. (3) They require very little force generation or energy expenditure. This point is not as clearly measurable because of uncertainty over the nature of the medium through which the chromosomes move, and their interaction (i.e. friction) with that medium. However, a reasonable estimate indicates that the hydrolysis of 30 ATP molecules could move a large insect chromosome all the way to the spindle pole (Forer, 1969). The force required to move a single large chromosome is in the order of 0.1 piconewtons (Forer, 1974) which is small compared with the 0.35 piconewtons which can be generated by a single dynein cross-bridge in a flagellum and the 3–4 piconewtons produced by a single myosin head working with actin in striated muscle (Brokaw, 1975). Extrapolating these figures to the fungi is difficult because, whilst fungal chromosomes are typically smaller, they may not always be condensed (e.g. Peterson & Ris, 1976; Gordon, 1977; Heath, 1980c), thus they might have a higher coefficient of friction relative to the nuclear matrix. However, it is clear that very little force is needed and thus the force-generating system may be small and therefore hard to identify.

The obvious structure which is specifically involved in anaphase movement is the kinetochore microtubule. The only fungi which have consistently failed to show convincing evidence for the presence of kinetochore microtubules are the mucoralean Zygomycetes. The dubious validity of these negative reports has been discussed elsewhere (Heath, 1978, 1980a), but an interesting explanation of a possible mode of mitosis in their absence has been presented (McCully & Robinow, 1973). This possible anomaly will not be considered further.

In all fungi that have been critically and adequately examined, the kinetochore microtubules clearly run the entire distance from the kinetochores to the spindle poles throughout metaphase and anaphase. Clearly then, they depolymerize as the kinetochores approach the poles during anaphase. These data support the concept, first most clearly proposed by Inoué & Sato (1967) (see also Margolis, Wilson & Kiefer, 1978), that microtubule depolymerization generates the necessary force (Fig. 3b). The fact that the rate of anaphase movement is comparable with rates of microtubule depolymerization in other fungal cell processes, e.g. zoospore encystment where Holloway & Heath (1977) reported $1.2 - 4$ μm min^{-1}, is consistent with the hypothesized role of depolymerization. Furthermore, Inoué & Ritter (1975) have calculated that microtubule depolymerization could generate enough force for anaphase movement. However, the difficult, and unresolved problem is differentiating between depolymerization *causing* the movement, versus its *permitting* movement generated by another force-producing system. There are no convincing data to show the existence of a system capable of excising tubulin subunits whilst the microtubule maintains tension between the kinetochore and the spindle pole. The necessity of such a system is most obvious when, as is typical for the fungi, one only has one microtubule per kinetochore. When there are multiple microtubules per kinetochore one can invoke the 'excision-hold model' of Hanzely & Schjeide (1973). The current ignorance of an adequate excision-hold system in a single microtubule is inadequate cause to discard a hypothesis, but it should be noted that, if microtubule depolymerization is the cause of anaphase movement, it will be the only known system in which such a motor is working. Since anaphase is so much slower than other cellular movements this may support the hypothesis. However, the case for microtubule depolymerization is not sufficiently strong to discourage analysis of other models.

Because the distances moved are so short, and the rates of movement are so slow, it is worth reconsidering the concept of directed Brownian motion as the source of anaphase movement. Using calculations and viscosity data of the type shown by Taylor (1965), Rebhun (1972) and Maniloff (1979), one can show that a hypothetical 0.1 μm diameter chromosome could indeed move the required distance (~ 1 μm) in the time of a typical fungal anaphase movement (~ 30 s, Aist & Williams, 1972; Heath & Rethoret, 1980). However, there are a number of points which argue against such a hypothesis. (1) The nucleoplasm is undoubtedly more structured than the assumed Newtonian fluid (see following discussion of the nuclear matrix). This structural complexity is likely to dampen Brownian motion. (2) The apparently uncondensed chromosomes of some fungi (Peterson & Ris, 1976; Gordon, 1977; Heath, 1980*c*) and the large, seemingly fused chromatin masses of others (McNitt, 1973; Braselton *et al.*, 1975; Garber & Aist, 1979) must impose more drag than an assumed 0.1 μm diameter spherical chromosome. (3) One must still identify a 'director' whose properties would presumably include a ratchet-like activity which has not yet been identified. Thus, it seems likely that another source of force generation should be sought.

There is evidence for five distinct hypothetical systems of force generation in anaphase movements (in addition to depolymerization as discussed above).

Inter-microtubule sliding

The essential feature of this hypothesis, as postulated by McIntosh, Hepler & Van Wie (1969) and in modified form by Nicklas

(1971), is that the kinetochore microtubules are propelled towards the spindle poles by dynein-like cross-bridges, acting against the non-kinetochore microtubules (Fig. 3c). A clear prediction of this hypothesis is an intermicrotubule spacing of approximately 20 nm as found in flagellar axonemes (e.g. Brokaw, 1975). Such a spacing may occur in the rather coherently arranged spindles of genera such as *Uromyces* (Heath, I. B. & Heath, M. C., 1976) but clearly does not occur over most of the length of the kinetochore microtubules in other genera, e.g. *Polysphon-dylium* (Roos, 1975), *Thraustotheca* (Heath, 1974b) and *Physarum* (Ryser, 1970). It has been suggested that the divergent kinetochore microtubules of the latter group argue against inter-microtubule sliding, but in all cases examined the spindles are convergent at the poles where the kinetochore microtubules come close to non-kinetochore microtubules. The spacing in this zone is adequate to permit dynein-like cross-bridges (Fig. 3c). As noted above, one may only need one or two such linkages to effect chromosome movement. These could be located only in the polar regions and be quite effective. Thus the original concept of inter-microtubule sliding, with cross-bridges present along the entire length of the kinetochore microtubules, may not be applicable to some fungi, but a modified version is consistent with current observations. This suggested low number of polar cross-bridges may explain why cross-bridges are not abundantly detected in fungal spindles, although fixation artefacts must always be considered.

A further point of interest with respect to the inter-microtubule sliding hypothesis concerns the polarity of the interacting microtubules. The original hypotheses postulated sliding between antiparallel micro-tubules. Recent data suggest the kinetochore microtubules are anti-parallel with the non-kinetochore microtubules of each half spindle (Borisy, 1978; Bergen & Borisy, 1980) as predicted. However, the important point of comparison is that in the best known cellular motor involving dynein cross-bridges, namely the eukaryotic flagellum, sliding occurs between parallel, not anti-parallel, microtubules (Haimo, Telzer & Rosenbaum, 1979). It thus seems likely that, if dynein is involved in anaphase movements, the way in which it interacts with adjacent microtubules differs from that found in the flagellum. Alternatively, one of the cross-bridge-forming microtubule-associated proteins (Dentler, Granett & Rosenbaum, 1975; Kim, Binder & Rosenbaum, 1979) may be the spindle equivalent of dynein (Connolly *et al.*, 1978; Sherline & Schiavone, 1978) with an intrinsic capacity to function between anti-parallel microtubules.

One may conclude that there is no compelling reason to exclude inter-microtubule sliding of some form as the motor for fungal anaphase movements. However, the comparison of rates of movement shows that if dynein is involved it is working differently relative to the flagellum or that the rate of anaphase movement is moderated by another factor such as microtubule depolymerization.

Inter-microtubule zipping

According to Bajer (1973), non-sliding lateral 'zipping' between adjacent microtubules could generate anaphase movements according to Fig. 3(*a*). This hypothesis explicitly predicts at least two divergent microtubules per kinetochore. It is difficult to explain fungal spindles on this hypothesis because many only have one microtubule per kinetochore (see Heath, 1978, 1980*a* for listings), and in all cases all spindle microtubules, including the kinetochore microtubules, are convergent towards the poles. It may be possible to reformulate the hypothesis to accommodate these facts, but there is little reason to believe that zipping generates the forces of anaphase movement in either fungal spindles or any other well-characterized cellular motile system.

Actin–myosin systems

According to Forer (1974, 1978) anaphase movements may be generated by actin–myosin interactions. These could take the form of an actin–myosin 'fibre' that is functionally independent of the microtubules (Fig. 3*e*) or the actin and myosin may interact with the microtubules as outlined by Oakley & Heath (1978). In either case, as emphasized by Forer (1974), the rate of movement would be moderated by microtubule depolymerization. This would explain the difference in rates between other actin–myosin systems such as saltations, muscle contraction, streaming and anaphase movements (Table 1). This hypothesis, like the postulated use of dynein, is attractive because it implies cellular economy in utilizing a standard motor for diverse processes. Its primary support comes from the observations of both actin and myosin in many spindles (recently critically reviewed by Forer, 1978). At present there is no evidence for actin or myosin in any fungal spindle although both intranuclear actin and myosin are reported in *Physarum* (Jockusch, Ryser & Behnke, 1973; Hauser *et al.*, 1975) and actin is found in *Dictyostelium* nuclei (Fukui, 1978). Heath (1975*b*) could find no evidence for actin in *Thraustotheca* spindles but the localization technique used does not give reliable negative results; only a positive identification is valid. However, it is important to note that Heath (1975*b*) did use serial sections, which are essential since the force considerations discussed earlier suggest that one may only need very few actin and myosin filaments for anaphase movements. Thus one must conclude that actin and myosin could be involved in fungal anaphase movement but there is, as yet, no positive evidence to support the hypothesis.

Microtubule–matrix interactions

Subirana (1968) and Heath (1975*a*) have suggested that the forces for anaphase movements are generated by a sliding interaction

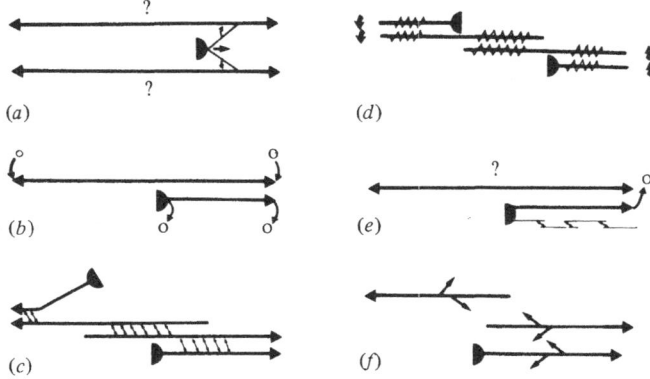

Fig. 3. Diagrams summarizing the salient features of the currently most viable models for anaphase movement and telophase elongation. In all cases the semicircles represent chromosomes and the broad arrow-heads show the direction of movement of the microtubules. In the zipper hypothesis (*a*) (Bajer, 1973), lateral zipping (curved arrows) between kinetochore and non-kinetochore microtubules moves the chromosome towards the spindle pole. The mode of spindle elongation is not specified. According to the polymerization–depolymerization model (*b*) (Inoué & Sato, 1967), appropriate addition and loss of tubulin subunits (circles) generates the necessary forces and length changes. The inter-microtubule sliding hypothesis (*c*) (McIntosh *et al.*, 1969) utilizes dynein-like cross-bridges (double-head arrows) between adjacent microtubules to propel them in the appropriate directions. The kinetochore microtubule on the left illustrates a derived concept in which sliding only occurs in the tip region of the kinetochore microtubule. According to the worm-gear mechanism (*d*) (Filner & Yadav, 1979), rotating (directions indicated by curved arrows) microtubules interact via helical 'worm gears' as indicated. Forer (1974) postulates an actin-involving fibre (overlapping fine lines) for force generation, with depolymerization as the 'governor', but does not specify the mode of spindle elongation (*e*). According to the matrix hypothesis (*f*) (Subirana, 1968; Heath, 1975*a*), microtubules are able to slide relative to the nuclear matrix to achieve the appropriate movements.

between the surface of the kinetochore microtubules and the nucleo-plasm, or, to use more current terminology, the nuclear matrix (Fig. 3*b*). The widely separated kinetochore microtubules of some fungi, already discussed in the context of inter-microtubule sliding, argue for this hypothesis since force generation could occur along the entire length of the microtubule. However, the nature of the matrix compo-nent is totally unidentified at present. Since the case for a microtubule–matrix interaction is more compelling with respect to telophase elonga-tion, further discussion of this hypothesis will be deferred to that section.

Worm gear mechanism

Recently two papers have suggested that the data indicating inter-microtubule sliding are equally consistent with force generation via a worm gear mechanism (Filner & Yadav, 1979; Schulz & Jarosch, 1980). This hypothesis invokes rotational movement of the microtubules being translated into longitudinal sliding between adjacent microtubules by means of an interposed helical 'gear' (Fig. 3*d*). Clearly this hypoth-esis necessitates close lateral interaction between co-operating micro-tubules, so that the inter-microtubule distance comments previously made with respect to the sliding hypothesis also apply to this concept. Whilst Behnke (1975) has shown a helical component surrounding microtubules *in vitro*, and the concept of biological rotary motors is well established for *bacterial* flagella (e.g. Berg, 1975), there is no direct evidence for, nor convincingly against, this hypothesis working in fungi.

In summarizing the above discussion, it is obvious that there are six basic hypothetical models that could account for anaphase movements. Fungal data are consistent with all except inter-microtubule zipping.

Telophase elongation

There are three important points that must be remembered when discussing mechanisms of telophase elongation in fungi. (1) Based on the limited available data, the rate of spindle elongation is compara-ble to the rate of anaphase movement (Table 1). The variability indicated in Table 1 is less when the two types of movement are compared in a single species. This similarity in rate suggests that there may be similarity in either the motor or the rate-limiting component of the mechanism. (2) Within fungal nuclei there appears to be a dicho-tomy in the arrangement of the non-kinetochore microtubules. In *Uromyces* (Heath, I. B. & Heath, M. C., 1976), the Mucorales

(McCully & Robinow, 1973; Franke & Reau, 1973; Bland & Lunney, 1975) and others listed in Heath (1978), they form a coherent parallel bundle in the centre of the spindle with the kinetochore microtubules and chromatin arranged around their periphery. Within the central bundle the microtubules do appear to run continuously from one pole to the other. In contrast, in *Saprolegnia* (Heath, 1975c, 1978), *Thraustotheca* (Heath, 1974b), *Sorosphaera* (Braselton *et al.*, 1975), *Catenaria* (Ichida & Fuller, 1968), *Ascobolus* (Zickler, 1970) and probably many other genera, they form a much more dispersed array, often intermingled with the chromosomes, and predominantly do *not* run from one pole to the other. Rather, they form two interdigitating arrays. Such arrays may also occur when the microtubules form a tight bundle as shown in *Dictyostelium* (Moens, 1976). This dichotomy between true pole-to-pole, predominantly bunched versus interdigitating, predominantly dispersed microtubules may be indicative of different force-generating systems as discussed below. However, when analysing spindles for this feature it is important to remember that a true pole-to-pole array of microtubules present early in division can become an interdigitating system later on (e.g. Tippit & Pickett-Heaps, 1977). (3) In most genera examined there is a population of cytoplasmic microtubules which radiate in various ways from the spindle poles. These microtubules may be present throughout mitosis (see p. 88 for a consideration of their potential role in NAO migration) or they may form specifically during anaphase–telophase (e.g. Heath, I. B. & Heath, M. C., 1976). In either case, but most obviously in the latter, it seems likely that these cytoplasmic microtubules are formed in response to an active role in telophase elongation. This point will be returned to when considering the cytoplasmic microtubules, but it clearly indicates the potential for there being two force-generating systems operating to effect telophase elongation.

Considering the above points, can one reach any conclusions concerning possible modes of force generation during telophase elongation? Clearly, when true pole-to-pole microtubules are present during elongation, microtubule polymerization could, in principle, generate the necessary force as suggested by Inoué & Sato (1967). Equally, when one has an interdigitating population, this model seems inappropriate because there is no obvious structure against which the non-polar ends of the microtubules could 'push'. This type of spindle argues for some form of sliding between adjacent microtubules or between microtubules and the matrix. Unfortunately, however, both types of spindle can be explained on both polymerization and sliding models if certain plausible assumptions are made. Thus, when the microtubules run from pole to pole, force generation could be by inter-microtubule sliding with concomitant polymerization maintaining the appropriate lengths of the microtubules. Functional advantages of this polymerization process could include (*a*) formation of increasing lengths of interacting micro-tubules in order to generate increasing force (as opposed to the decrease in force predicted by sliding apart of non-elongating microtubules), and (*b*) maintenance of strength throughout the spindle by maintaining microtubule numbers (as opposed to a reduction in numbers per half spindle, which would accompany sliding in the absence of polymeriza-tion). The clearly observed length changes occurring simultaneously in adjacent kinetochore and non-kinetochore microtubules during ana-phase (e.g. Heath, 1974*b*, 1978) show that the necessary fine level of polymerization control has evolved. Conversely, when one has an interdigitating type of spindle, static (as opposed to sliding) interactions between adjacent microtubules or between the microtubules and the matrix would provide a system against which microtubule polymeriza-tion could 'push' to move the poles apart. No fungal data permit unequivocal differentiation between these various possibilities. The best studied system in which intermicrotubule sliding seems to occur is the diatom spindle. In this system McDonald, Edwards & McIntosh (1979) have recently presented evidence for specific interactions between antiparallel, interdigitating, non-kinetochore microtubules. If the pat-tern that they demonstrate is essential to an inter-microtubule sliding system, the apparent lack of such a pattern in *Uromyces* (Heath, I. B. & Heath, M. C., 1976) may argue for the polymerization hypothesis, but the available data are too tenuous for a firm conclusion at present. The only other point which should be made here is that one can calculate a lateral movement of 40 nm per hydrolysis of 1 ATP molecule in the dynein-mediated sliding of flagellar axonemes (Brokaw, 1975), com-pared with elongation of about 1.2 nm per GTP molecule during polymerization (Heath, 1980*a*). The inherently greater efficiency of sliding argues for that hypothesis.

Whilst one cannot adequately differentiate between sliding and polymerization as force generators for telophase elongation, one can present arguments for the hypothesized role of the nuclear matrix in this process. These arguments take two forms. (1) At a simple level one can observe that the formation of both the typical dumbbell shape, and, perhaps more convincingly, the alternative doubly constricted shape of telophase nuclei is inconsistent with force application at the spindle poles of a homogeneous, deformable, fluid nucleus. Clear examples of the former shape are seen in McCully & Robinow (1973), Roos (1975), Moens (1976), Heath (1978) and Heath & Greenwood (1968), and of the latter in Ichida & Fuller (1968) and McNitt (1973). In order to explain these shapes one must postulate specific shape changes generated either by the nucleoplasm (nuclear matrix) itself, or by a nuclear envelope-associated system analogous to the cortical layer of animal cells. Evidence for both systems exists. Proteinaceous nuclear matrices have been described in a number of cells (following selective extraction of much of the nucleoplasm from isolated nuclei) by Comings & Okada (1976), Berezney & Coffey (1977), Wunderlich & Herlan (1977), Berezney *et al.* (1979) and Wunderlich, Giese & Bucherer (1978). These matrices retain the morphology of *in vivo* nuclei and some can be induced to undergo reversible contraction by changes in the ionic composition of the medium (Wunderlich & Herlan, 1977). A further characteristic of all reported matrices is a so-called 'residual envelope' (or 'nuclear envelope lamina'; Gerace & Blobel, 1980), which covers the surface of the matrix, contains residues of the nuclear pores and appears to be a tightly organized, fibrous layer. A comparable layer has been seen adjacent to the nuclear envelope of unextracted nuclei (Schatten & Thoman, 1978). This layer was referred to earlier as a candidate for force generation during NAO migration and equally is potentially capable of selectively effecting nuclear shape changes by localized expansions (relaxations) and contractions. An equatorial contraction as occurs during animal cell cytokinesis seems likely and would explain the observed nuclear morphology and bunching of the telophase spindle microtubules as shown by Heath (1978). Because the peripheral, residual envelope is intimately associated with the whole matrix, it is currently impossible to determine which, if either, is causing the relevant shape changes. However, the important point is that both represent testable candidates, independent of the spindle, for generating both the dumbbell shape and perhaps also nuclear elongation during telophase. The latter suggestion would leave the non-kinetochore microtubules with no function, yet they clearly elongate during nuclear elongation and must surely play a role. It is this role which may, in turn, also involve the nuclear matrix, along the second line of argument mentioned earlier. (2) When the non-kinetochore microtubules are closely spaced, they have the morphological arrangement that permits them to exert a pushing force on the spindle poles by one of the

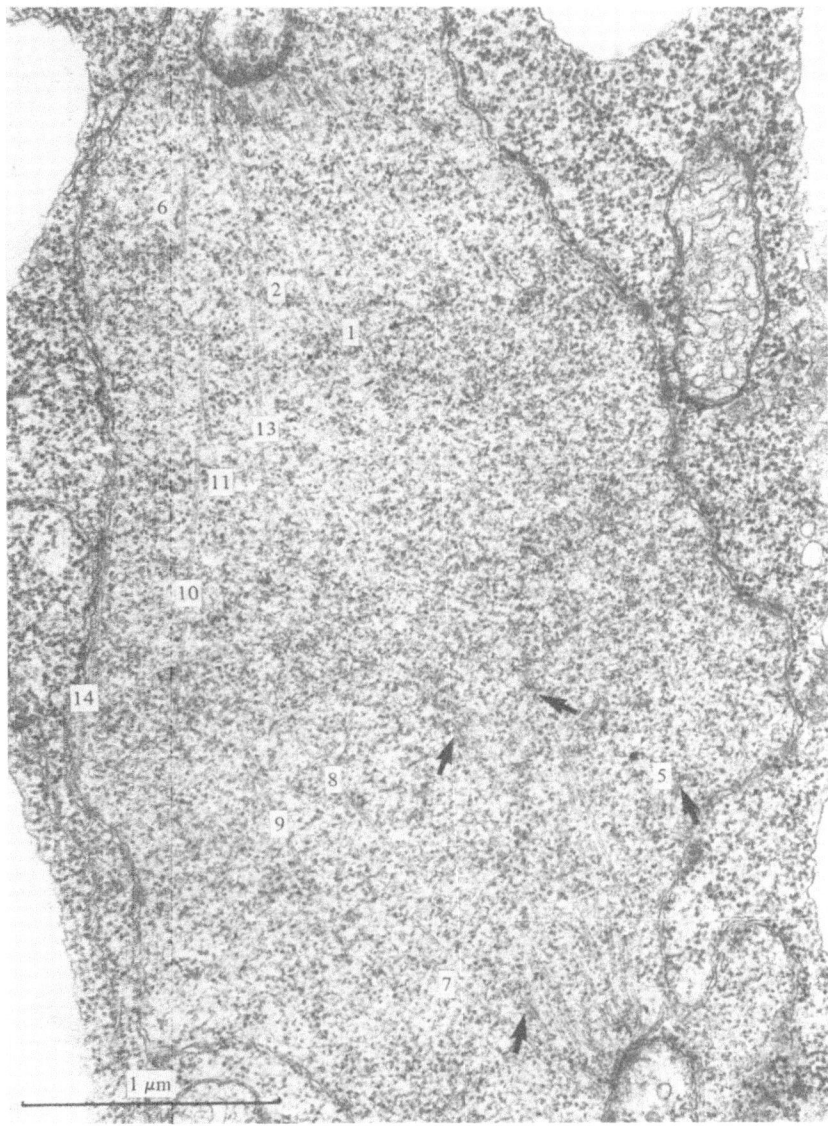

Fig. 4. Electron micrograph of the anaphase spindle of *Saprolegnia ferax* from which Figs. 5–7 were derived. Selected microtubules are numbered and can be cross-referenced to Figs. 5–7. Kinetochores are shown by arrows. This was one of a series of 33 sections through the spindle.

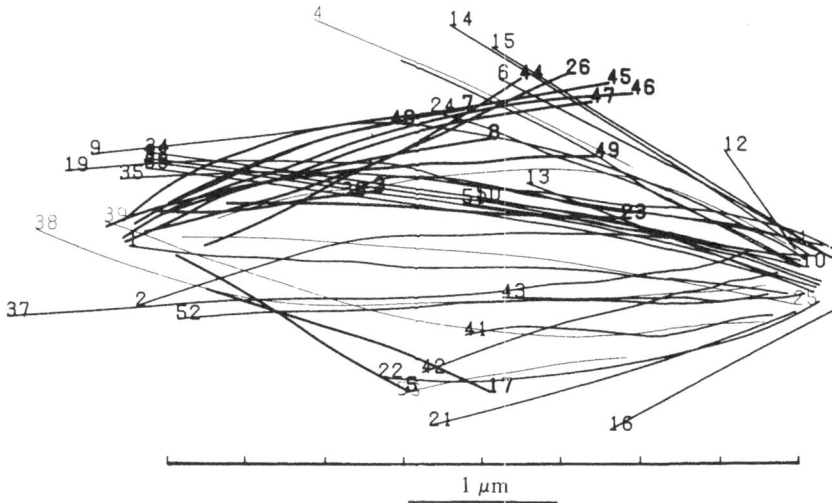

Fig. 5. Reconstruction of the non-kinetochore population of microtubules from the spindle shown in Fig. 4. Microtubules emanating from the left pole are shown by heavy lines, those from the right by medium weight lines and those which are either free or continuous are shown by fine lines. This is the view obtained in the same plane as the plane of the sections. The relative sparsity of microtubules from the left-hand pole is due to a partial deficiency in the series whereby about 10 microtubules which were highly divergent and running down below the plane of the paper were not plotted. The gradations along the bottom indicate the intervals at which cross sections were reconstructed.

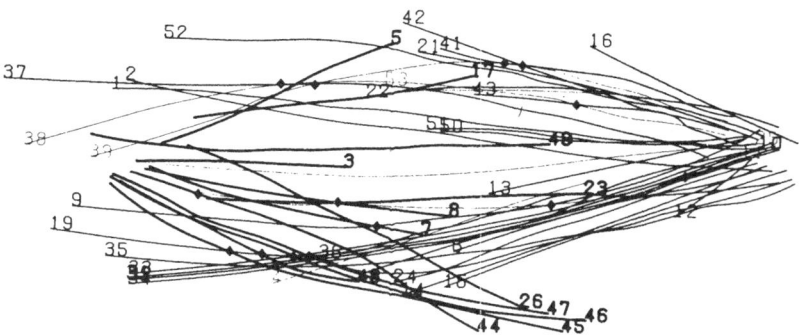

Fig. 6. The same spindle as Fig. 5 but rotated 140°. The diamonds indicate points at which microtubules from opposite poles come within approximately 50 nm of each other, or come within a similar distance of free or continuous microtubules, as determined from the reconstructed cross sections. Other details as for Fig. 5.

mechanisms discussed above. However, when they are widely spaced, they seem to lack this potential. Analysis of the non-kinetochore microtubules in two, presumably elongating, anaphase spindles of *Saprolegnia* both showed the same dispersal pattern, as illustrated in Figs. 4–7. A number of features of this spindle should be emphasized. Clearly true pole-to-pole microtubules are so rare as to be insignificant in force generation. Close lateral interactions between microtubules from opposite poles are rare; instances of such microtubules coming within 50 nm of each other are indicated in Figs. 6 and 7. These potential interactions are spread rather randomly throughout the spindle and rarely take the form of extensive regions of closely spaced antiparallel microtubules comparable to those reported in other spindles such as those of *Uromyces* (Heath, I. B. & Heath, M. C., 1976) or the diatoms (reviewed by Pickett-Heaps & Tippit, 1978). The frequency

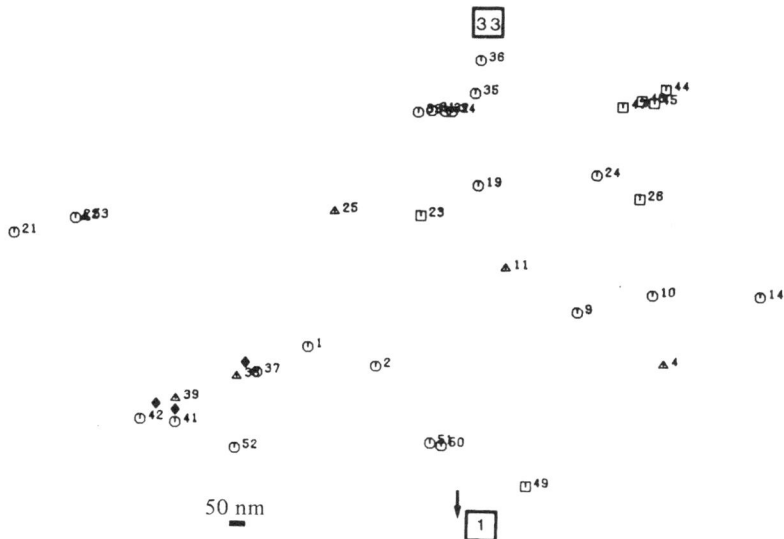

Fig. 7. Reconstructed cross section from the equator of the spindle shown in Fig. 5. Microtubules emanating from the left pole are drawn as squares, those from the right as octagons and the free and continuous ones as triangles. Numbers are for cross-reference to Figs. 4–6. Diamonds indicate intermicrotubule spacings of the type marked on Fig. 6. The numbers in squares indicate the position of the first and last sections in the series. The plane of sectioning was horizontal relative to this reconstruction. The few microtubules not plotted as indicated in the legend to Fig. 5 diverged in the direction of the arrow and if plotted would thus have appeared on this plot somewhere below the level of the figure '1'.

and distribution of these potential interaction sites seems to correspond to that which would be expected by chance in two intermingled, divergent arrays of microtubules, although such an assertion is difficult to prove in a statistically meaningful way. It would seem that the only way in which these microtubule arrays could contribute to the force needed for spindle elongation is by sliding relative to the nuclear matrix as suggested previously (Subirana, 1968; Dietz, 1972; Heath, 1975). The nature of the matrix component with which they may be interacting remains unknown. However, further evidence for such an interaction is seen even in spindles in which the microtubules are not widely dispersed. For example Roos (1975) shows clearly differentiated nucleoplasm specifically adjacent to the microtubules in *Polysphondylium*, so that the hypothesized microtubule–matrix interaction may eventually be extrapolated to all spindle types.

One can conclude that the spindle may be involved in telophase elongation in a number of different hypothetical force-generating systems, none of which can be clearly proven nor denied in most described spindle types. Only in the widely dispersed and carefully analysed *Saprolegnia* spindle is it possible to cast serious doubts on both polymerization and inter-microtubule sliding as force generators. Whether these spindles use a different force-generating system from other cells or represent a different arrangement of a universal system remains to be determined.

In addition to the intranuclear, spindle microtubules there is frequently a population of cytoplasmic microtubules that is present at the poles of mitotic nuclei. Typically, during spindle elongation, the separating spindle poles are preceded by an array of NAO-based microtubules which run for an undetermined distance into the cytoplasm. Clear examples of such microtubules are seen in the reports of Girbardt (1968), Aist & Williams (1972), Heath (1974b, 1978), Heath & Greenwood (1968), I. B. Heath & M. C. Heath (1976), Roos (1975), Powell (1975), McNitt (1973) and Ichida & Fuller (1968). They are clearly a widespread, possibly universal, feature of fungal mitoses. The formation of these microtubules, often specifically at anaphase–telophase (e.g. Aist & Williams, 1972; Girbardt, 1968; Heath & Heath, 1976), suggests that they do indeed aid in nuclear elongation. However, the mechanism by which they may achieve this is uncertain. Their arrangement suggests that they may exert a pull on the spindle poles, and if so the most obvious way would be by moving themselves through the cytoplasmic matrix (Wolosewick & Porter, 1979) in a manner analogous to the suggested spindle microtubule/nuclear matrix system discussed previously. This concept has been discussed in more detail by I. B. Heath & M. C. Heath (1978) and needs no further comment since the molecular details remain unknown. However, it is worth noting that cytoplasmic changes associated with mitosis, and consistent with this idea, have been reported by Girbardt (1968) and Robinow (1963). Finally, the polar interactions between nuclear envelope and cytoplasmic microtubules described by Heath & Greenwood (1968, 1970) and Heath (1974*b*) are consistent with this concept too, in that they could represent static interactions which serve to anchor the nuclear envelope (and thus the nucleus) to the moving microtubules. This is contrary to the author's original suggestion of a sliding interaction between the nuclear envelope and the microtubules; current data cannot differentiate between the alternative hypotheses.

In addition to the probable role in telophase elongation, the cytoplasmic microtubules are likely to be involved in the diverse types of nuclear movements reported for fungal nuclei. Direct experimental evidence for such a role is shown by Oakley & Morris (1980). This work supports the morphological evidence which was recently reviewed in detail by Heath (1978, 1981) and will not be considered further here. The only important point to be made is to emphasize the likelihood of the fungi utilizing a single force-generating system for both mitosis- and non-mitosis-associated events.

Conclusions

Fungal mitoses exhibit numerous force-requiring processes which may have common mechanisms and all of which can be explained by more than one current hypothesis. This unfortunate lack of unambiguous information is not restricted to the fungi; the force-generating systems operating during mitosis in any cell type are still unknown. However, the nature of the fungal spindle in many ways makes it uniquely suitable for study since (*a*) it is typically and clearly delineated from the cytoplasm (by the nuclear envelope), and thus may be isolated intact with relatively little cytoplasmic contamination, (*b*) it is small and thus relatively easily analysed by electron microscope techniques, (*c*) in at least some fungi (Myxomycetes especially), mitosis can be highly synchronous, thus yielding a large population at a suitable stage for analysis, and (*d*) many fungi are very amenable to genetic analysis (e.g. the work of Hartwell, 1974 and Morris, 1976). Work is only just beginning on fungal mitosis, but detailed analysis exploiting any of these characteristics must be rewarding both in the overall attack on the mechanisms of mitosis and in the intrinsic fascination of understanding better the fungi themselves.

Acknowledgements. The computer reconstructions were devised and executed by Dr P. B. Moens, York University. His skilful assistance was invaluable and is very much appreciated. This work was supported by the Natural Science and Engineering Research Council of Canada. The excellent technical and secretarial assistance of Karen Rethoret and Dorothy Gunning are gratefully acknowledged.

References

Aist, J. R. (1969). The mitotic apparatus in fungi, *Ceratocystis fagacearum* and *Fusarium oxysporum*. *Journal of Cell Biology*, **40**, 120–35.

Aist, J. R. & Williams, P. H. (1972). Ultrastructure and time course of mitosis in the fungus *Fusarium oxysporum*. *Journal of Cell Biology*, **55**, 368–89.

Allen, R. D. & Taylor,D. L. (1975). The molecular basis of amoeboid movement. In *Molecules and Cell Movements*, ed. S. Inoué & R. E. Stephens, pp. 239–258. New York: Raven Press.

Amos, W. B. (1975). Contraction and calcium binding in the vorticellid ciliates. In *Molecules and Cell Movements*, ed. S. Inoué & R. E. Stephens, pp. 411–36. New York: Raven Press.

Bajer, A. S. (1973). Interaction of microtubules and the mechanism of chromosome movement (zipper hypothesis). 1. General principle. *Cytobios*, **8**, 139–60.

Behnke, O. (1975). Studies on isolated microtubules. Evidence for a clear space component. *Cytobiologie*, **11**, 366–81.

Berezney, R., Basler, J., Hughes, B. B. & Kaplan, S. C. (1979). Isolation and characterization of the nuclear matrix from *Zajdela ascites* hepatoma cells. *Cancer Research*, **39**, 3031–9.

Berezney, R. & Coffey, D. S. (1977). Nuclear matrix. *Journal of Cell Biology*, **73**, 616–37.

Berg, H. C. (1975). Flagellar rotation. In *Proceedings of the First Intersectional Congress of IAMS*, ed. T. Hasegawa, vol. 1, pp. 665–73. Tokyo: Science Council of Japan.

Bergen, L. G. & Borisy, G. G. (1980). Head-to-tail polymerization of microtubules *in vitro*. *Journal of Cell Biology*, **84**, 141–50.

Bland, C. E. & Lunney, C. Z. (1975). Mitotic apparatus of *Pilobulus crystallinus*. *Cytobiologie*, **11**, 382–91.

Borisy, G. G. (1978). Polarity of microtubules of the mitotic spindle. *Journal of Molecular Biology*, **124**, 565–70.

Braselton, J. P., Miller, C. E. & Pechak, D. G. (1975). The ultrastructure of cruciform nuclear division in *Sorosphaera veronicae* (Plasmodiophoromycete). *American Journal of Botany*, **62**, 349–58.

Brokaw, C. J. (1975). Cross-bridge behaviour in a sliding filament model for flagella. In *Molecules and Cell Movement*, ed. S. Inoué & R. E. Stephens, pp. 165–79. NewYork: Raven Press.

Burnett, J. H. (1968). *Fundamentals of Mycology*. London: Edward Arnold.
Comings, D. E. & Okada, T. A. (1976). Nuclear proteins. III. The fibrillar nature of the nuclear matrix. *Experimental Cell Research*, 103, 342–60.
Connolly, J. A., Kalnins, V. I., Cleveland, D. W. & Kirschner, M. W. (1978). Intracellular localization of the high molecular weight microtubule accessory protein by indirect immunofluorescence. *Journal of Cell Biology*, 76, 781–6.
Dentler, W. L., Granett, S. & Rosenbaum, J. L. (1975). Ultrastructural localization of the high molecular weight (MAPs) associated with in-vitro assembled brain microtubules. *Journal of Cell Biology*, 65, 237–41.
Dietz, R. (1972). Die assembly-hypothese der chromosomenbewegung und die veranderungen der spindellänge während der anaphase I in spermatocyten von *Pales ferruginea* (Tipulidae, Diptera). *Chromosoma, Berlin*, 38, 11–76.
Filner, P. H. & Yadav, N. S. (1979). Role of microtubules in intracellular movements. In *Physiology of movement*, ed. W. Haupt & M. E. Feinleib, pp. 95–113. Encyclopedia of Plant Physiology 7. Berlin: Springer-Verlag.
Forer, A. (1969). Chromosome movements during cell division. In *Handbook of Molecular Cytology*, ed. A. Lima-de-faria, pp. 553–601. Amsterdam & London: North-Holland.
Forer, A. (1974). Possible roles of microtubules and actin-like filaments during cell division. In *Cell Cycle Controls*, ed. G. M. Padilla, I. L. Cameron & A. M. Zimmerman, pp. 319–35. London & New York: Academic Press.
Forer, A. (1978). Chromosome movements during cell division: possible involvement of actin filaments. In *Nuclear Division in the Fungi*, ed. I. B. Heath, pp. 21–88. New York & London: Academic Press.
Franke, W. W. & Reau, P. (1973). The mitotic apparatus of a Zygomycete, *Phycomyces blakesleeanus*. *Archiv für Mikrobiologie*, 90, 121–29.
Franke, W. & Scheer, U. (1974). Structures and functions of the nuclear envelope. In *The Cell Nucleus*, ed. H. Busch, Vol. 1, pp. 220–347. New York & London: Academic Press.
Fukui, Y. (1978). Intranuclear actin bundles induced by dimethyl sulfoxide in interphase nucleus of *Dictyostelium*. *Journal of Cell Biology*, 76, 146–57.
Fuller, M. S. (1976). Mitosis in fungi. *International Review of Cytology*, 45, 113–53.
Garber, R. C. & Aist, J. R. (1979). The ultrastructure of mitosis in *Plasmodiophora brassicae* (Plasmodiophorales). *Journal of Cell Science*, 40, 89–110.
Gerace, L. & Blobel, G. (1980). The nuclear envelope lamina is reversibly depolymerized during mitosis. *Cell*, 19, 277–87.
Girbardt, M. (1968). Ultrastructure and dynamics of the moving nucleus. In *Aspects of Cell Motility*, ed. P. L. Miller, pp. 249–59. Symposia of the Society for Experimental Biology 22. Cambridge University Press.
Girbardt, M. (1978). Historical review and introduction. In *Nuclear Division in the Fungi*, ed. I.B. Heath, pp. 1–20. New York & London: Academic Press.
Girbardt, M. & Hädrich, H. (1975). Ultrastruktur des Pilzkernes. *Zeitschrift für allgemeine Mikrobiologie*, 15, 157–73.
Gordon, C. N. (1977). Chromatin behaviour during the mitotic cell cycle of *Saccharomyces cerevisiae*. *Journal of Cell Science*, 24, 81–93.
Haimo, L. T., Telzer, B. R. & Rosenbaum, J. L. (1979). Dynein binds to and crossbridges cytoplasmic microtubules. *Proceedings of the National Academy of Sciences, USA*, 76, 5759–63.
Hanzely, L. & Schjeide, O. A. (1973). Structural and functional aspects of the anastral mitotic spindle in *Allium sativum* root tip cells. *Cytobios*, 7, 147–62.
Hartwell, L. H. (1974). *Saccharomyces cerevisiae* cell cycle. *Bacteriological Reviews*, 38, 164–98.

Hauser, M., Beinbresch, G., Gröschel-Stewart, U. & Jockusch, B. M. (1975). Localization by immunological techniques of myosin in nuclei of lower eukaryotes. *Experimental Cell Research*, **95**, 127–35.

Heath, I. B. (1974*a*). Genome separation mechanisms in prokaryotes, algae, and fungi. In *The Cell Nucleus*, ed. H. Busch, vol. 2, pp. 487–515. New York & London: Academic Press.

Heath, I. B. (1974*b*). Mitosis in the fungus *Thraustotheca clavata*. *Journal of Cell Biology*, **60**, 204–20.

Heath, I. B. (1975*a*). The possible significance of variations in the mitotic systems of the aquatic fungi ('Phycomycetes'). *Biosystems*, **7**, 351–9.

Heath, I. B. (1975*b*). The role of cytoplasmic microtubules in fungi. *Proceedings of the First Intersectional Congress of the International Association of Microbiological Societies*, ed. T. Hasegawa, vol. 2, pp. 92–106. Tokyo: Tokyo University Press.

Heath, I. B. (1976). Ultrastructure of fresh water Phycomycetes. In *Recent Advances in Aquatic Mycology*, ed. E. B. G. Jones, pp. 603–50. London: Paul Elek Press.

Heath, I. B. (1978). Experimental studies of mitosis in the fungi. In *Nuclear Division in the Fungi*, ed. I. B. Heath, pp. 89–176. London & New York: Academic Press.

Heath, I. B. (1980*a*). Variant mitoses in lower eukaryotes: indicators of the evolution of mitosis? *International Review of Cytology*, **64**, 1–80.

Heath, I. B. (1980*b*). The behaviour of kinetochores during mitosis in the fungus *Saprolegnia ferax*. *Journal of Cell Biology*, **84**, 531–46.

Heath, I. B. (1980*c*). Apparent absence of chromatin condensation in mitotic nuclei of *Saprolegnia* as revealed by mithramycin staining. *Experimental Mycology*, **4**, 105–15.

Heath, I. B. (1981). Nucleus associated organelles in fungi. *International Review of Cytology*, **69**, 191–221.

Heath, I. B. & Greenwood, A. D. (1968). Electron microscopic observations of dividing somatic nuclei in *Saprolegnia*. *Journal of General Microbiology*, **53**, 287–289.

Heath, I. B. & Greenwood, A. D. (1970). Centriole replication and nuclear division in *Saprolegnia*. *Journal of General Microbiology*, **62**, 139–48.

Heath, I. B. & Heath, M. C. (1976). Ultrastructure of mitosis in the cowpea rust fungus *Uromyces phaseoli* var. *vignae*. *Journal of Cell Biology*, **70**, 592–607.

Heath, I. B. & Heath, M. C. (1978). Microtubules and organelle movements in the rust fungus *Uromyces phaseoli* var. *vignae*. *Cytobiologie*, **16**, 393–411.

Heath, I. B. & Rethoret, K. A. (1980). Temporal analysis of the nuclear cycle by serial section electron microscopy of the fungus, *Saprolegnia ferax*. *European Journal of Cell Biology*, **21**, 208–13.

Holloway, S. A. & Heath, I. B. (1977). Morphogenesis and the role of microtubules in synchronous populations of *Saprolegnia* zoospores. *Experimental Mycology*, **1**, 9–29.

Huang, H. C. & Patrick, Z. A. (1972). Nuclear distribution and behaviour in *Thielaviopsis basicola*. *Canadian Journal of Botany*, **50**, 2423–29.

Ichida, A. A. & Fuller, M. A. (1968). Ultrastructure of mitosis in the aquatic fungus *Catenaria anguillulae*. *Mycologia*, **60**, 141–55.

Inoué, S. & Ritter Jr., H. (1975). Dynamics of mitotic spindle organization and function. In *Molecules and Cell Movement*, ed. S. Inoué & R. E. Stephens, pp. 3–30. New York: Raven Press.

Inoué, S. & Sato, H. (1967). Cell motility by labile association of molecules. *Journal of General Physiology*, **50** (6, Pt.2, Suppl.), 259–92.

Jockusch, B. M., Ryser, U. & Behnke, O. (1973). Myosin-like protein in *Physarum* nuclei. *Experimental Cell Research*, **76**, 464–6.

Kazama, F. Y. (1974). The ultrastructure of nuclear division in *Thraustochytrium* sp. *Protoplasma*, **82**, 155–75.

Kim, H., Binder L. I. & Rosenbaum, J. L. (1979). The periodic association of MAP$_2$ with brain microtubules *in vitro*. *Journal of Cell Biology*, **80**, 266–76.

Künkel, W. & Hädrich, H. (1977). Ultrastrukturelle Untersuchungen zur antimitotischen aktivität von methylbenzimidazol-2-ylcarbamat (MBC) und seinen Einfluß die replikation des Kern-assoziierten organells ('centriolar plaque', 'MTOC' 'KCE') bei *Aspergillus nidulans*. *Protoplasma*, **92**, 311–23.

McCully, E. K. & Robinow, C. F. (1972). Mitosis in heterobasidiomycetous yeasts. *Journal of Cell Science*, **11**, 1–31.

McCully, E. K. & Robinow, C. F. (1973). Mitosis in *Mucor hiemalis*. *Archiv für Mikrobiologie*, **94**, 133–48.

McDonald, K. L., Edwards, M. K. & McIntosh, J. R. (1979). Cross-sectional structure of the central mitotic spindle of *Diatoma vulgare*. *Journal of Cell Biology*, **83**, 443–61.

McIntosh, J. R., Hepler, P. K. & Van Wie, D. G. (1969). Model for mitosis. *Nature, London*, **224**, 659–63.

McNitt, R. (1973). Mitosis in *Phlyctochytrium irregulare*. *Canadian Journal of Botany*, **51**, 2065–74.

Maniloff, J. (1979). Mycoplasma gliding may be a biased Brownian motion. *Journal of Theoretical Biology*, **81**, 617–20.

Margolis, R. L., Wilson, L. & Keifer, B. I. (1978). Mitotic mechanism based on intrinsic microtubule behaviour. *Nature, London*, **272**, 450–2.

Moens, P. B. (1973). Mechanisms of chromosome synapsis at meiotic prophase. *International Review of Cytology*, **35**, 117–34.

Moens, P. B. (1976). Spindle and kinetochore morphology of *Dictyostelium discoideum*. *Journal of Cell Biology*, **68**, 113–22.

Morris, N. R. (1976). Mitotic mutants of *Aspergillus nidulans*. *Genetical Research*, **26**, 237–54.

Nicklas, R. B. (1971). Mitosis. In *Advances in Cell Biology*, ed. D. M. Prescott, L. Goldstein & E. McConkey, vol. 2, pp. 225–97. New York: Appleton-Century-Crofts.

Oakley, B. R. & Heath, I. B. (1978). The arrangement of microtubules in serially sectioned spindles of the alga *Cryptomonas*. *Journal of Cell Science*, **31**, 53–70.

Oakley, B. R. & Morris, N. R. (1980). Nuclear movement is β-tubulin dependent in *Aspergillus nidulans*. *Cell*, **19**, 255–62.

Olive, L. S. (1953). The structure and behaviour of fungus nuclei. *Botanical Review*, **19**, 439–586.

Oppenheim, D. S., Hauschka, B. T. & McIntosh, J. R. (1973). Anaphase motions in dilute colchicine. *Experimental Cell Research*, **79**, 95–105.

Peterson, J. B. & Ris, H. (1976). Electron-microscopic study of the spindle and chromosome movement in the yeast *Saccharomyces cerevisiae*. *Journal of Cell Science*, **22**, 219–42.

Pickett-Heaps, J. D. & Tippit, D. H. (1978). The diatom spindle in perspective. *Cell*, **14**, 455–67.

Powell, M. J. (1975). Ultrastructural changes in nuclear membranes and organelle associations during mitosis of the aquatic fungus *Entophlyctis* sp. *Canadian Journal of Botany*, **53**, 627–46.

Rebhun, L. I. (1972). Polarized intracellular particle transport: saltatory movements and cytoplasmic streaming. *International Review of Cytology*, **32**, 93–137.

Rickards, G. K. (1975). Prophase chromosome movements in living house cricket spermatocytes and their relationship to prometaphase, anaphase and granule movements. *Chromosoma, Berlin*, **49**, 407–55.

Robinow, C. F. (1963). Observations on cell growth, mitosis, and division in the fungus *Basidiobolus ranarum*. *Journal of Cell Biology*, **17**, 123–52.

Robinow, C. F. & Bakerspigel, A. (1965). Somatic nuclei and forms of mitosis in fungi. In *The Fungi: an Advanced Treatise*, ed. G. C. Ainsworth & A. S. Sussman, pp. 119–42. New York & London: Academic Press.

Roos, U.-P. (1975). Mitosis in the cellular slime mould *Polysphondylium violaceum*. *Journal of Cell Biology*, **64**, 480–91.

Roos, U.-P. (1976). Light and electron microscopy of rat kangaroo cells in mitosis. *Chromosoma, Berlin*, **54**, 363–85.

Ryser, U. Die ultrastruktur der mitosekerne in den plasmodien von *Physarum polycephalum*. *Zeitschrift für Zellforschung und mikroskopische Anatomie*, **110**, 108–30.

Schatten, G. & Thoman, M. (1978). Nuclear surface complex as observed with the high resolution scanning electron microscope. *Journal of Cell Biology*, **77**, 517–35.

Schulz, D. & Jarosch, R. (1980). Rotating microtubules as a basis for anaphase spindle elongation in diatoms. *European Journal of Cell Biology*, **20**, 249–53.

Sherline, P. & Schiavone, K. (1978). High molecular weight MAPs are part of the mitotic spindle. *Journal of Cell Biology*, **77**, R9–R12.

Subirana, J. A. (1968). Role of spindle microtubules in mitosis. *Journal of Theoretical Biology*, **20**, 117–23.

Taylor, E. W. (1965). Brownian and saltatory movements of cytoplasmic granules and the movement of anaphase chromosomes. *Symposium of Biorheology.* Proceedings of the Fourth International Congress on Rheology, part 4, ed. A. L. Copley, pp. 175–91. New York: Wiley Interscience.

Thielke, C. (1974). Intranukleäre spindeln und reduktion des kernvolumens bei der meiose von *Coprinus radiatus* (Bolt) Fr. *Archiv für Mikrobiologie*, **98**, 225–37.

Tippit, D. H. & Pickett-Heaps, J. D. (1977). Mitosis in the pennate diatom *Surirella ovalis*. *Journal of Cell Biology*, **73**, 705–27.

Wise, D. (1978). On the mechanism of prometaphase congression: chromosome velocity as a function of position on the spindle. *Chromosoma, Berlin*, **69**, 231–41.

Wolosewick, J. J. & Porter, K. R. (1979). Microtrabecular lattice of the cytoplasmic ground substance. Artefact or reality? *Journal of Cell Biology*, **82**, 114–39.

Wolpert, L. (1965). Cytoplasmic streaming and amoeboid movement. In *Function and Structure in Micro-organisms*, ed. M. R. Pollock & M. H. Richmond. Symposium of the Society for General Microbiology 15. Cambridge University Press, **15**, 270–293.

Wunderlich, F., Giese, G. & Bucherer, C. (1978). Expansion and apparent fluidity decrease of nuclear membranes induced by low Ca/Mg. *Journal of Cell Biology*, **79**, 479–90.

Wunderlich, F. & Herlan, G. (1977). A reversibly contractile nuclear matrix. *Journal of Cell Biology*, **73**, 271–8.

Wunderlich, F. & Speth, V. (1970). Antimitotic agents and macronuclear division of ciliates. IV. Reassembly of microtubules in macronuclei of *Tetrahymena* adapting to colchicine. *Protoplasma*, **70**, 139–52.

Zickler, D. (1970). Division spindle and centrosomal plaques during mitosis and meiosis in some Ascomycetes. *Chromosoma, Berlin*, **30**, 287–304.

Microtubules and microtubule proteins in fungi

K. GULL

Biological Laboratory, University of Kent,
Canterbury CT2 7NJ, UK

Introduction

Microtubules are a ubiquitous feature of eukaryotic cells and are essential to a range of cellular activities such as spindle formation and chromosome movement in mitosis, maintenance of cell shape, secretory processes and cell movement. A large proportion of our knowledge of the protein chemistry and drug sensitivity of microtubules derives from studies with mammalian brain microtubule proteins, because these proteins are easily purified by successive cycles of temperature-dependent assembly and disassembly (Dustin, 1978; Roberts & Hyams, 1979). The major component of these microtubules ($\sim$85%) is tubulin which exists as a heterodimer of one α subunit (mol. wt. around 55 000), and one β subunit (mol. wt. around 53 000). Tubulin alone does not readily form microtubules. The remaining protein components are termed microtubule associated proteins (MAPs) and these stimulate the assembly of microtubules *in vitro*, and some MAPs are present as projections on the walls of the microtubules (Amos, 1979).

There are a large number of reports of microtubules in fungal cells. Sometimes these microtubules are present as part of highly ordered structures such as the flagellar axoneme and centrioles of fungal zoospores (Reichle, 1969; Beckett, Heath & McLaughlin, 1974). However, in most fungal cells single microtubules are present in the cytoplasm and in the dividing nucleus. The cytoplasmic microtubules of fungal cells are usually orientated parallel to the long axis of the cell and there are numerous observations of an intimate association between microtubules and various cell organelles. The most often cited association is between microtubules and nuclei (Girbardt, 1968; Aist &

Williams, 1972; Heath, I. B. & Heath, M. C., 1976). Various authors have suggested that fungal nuclei may be self-motile, and recent experimental evidence suggests a role for microtubules in nuclear movement in *Aspergillus nidulans* (Oakley & Morris, 1980).

Comparisons with other organisms and experimental systems suggest that microtubules in fungi may be involved in the directed movement of other organelles (Hyams & Stebbings, 1979). There is, as yet, little direct evidence for an involvement of microtubules in the movement of vesicles to the growing hyphal apex although in animal cells there is a well-documented function of microtubules in the export of golgi-derived materials. It has recently been shown that conventional fixations may not preserve all the microtubules in a fungal cell. Howard & Aist (1979) have shown that microtubules are present in the apical zone of the growing hyphal tip and are best seen in cells prepared for transmission electron microscopy using freeze substitution techniques. They also noted the close association of mitochondria with cytoplasmic micro-tubules in freeze-substituted cells.

The microtubule organizing centres of fungal nuclei

In most cells, microtubules appear to be attached to specific organelles or sites from which individual microtubules originate and grow (Tucker, 1979). These sites have been termed microtubule orga-nizing centres (MTOCs) (Pickett-Heaps, 1969) and include structures such as basal bodies of cilia and flagella, centrioles, pericentriolar material and kinetochores. The MTOCs of lower organisms are often little more than discrete areas of electron-dense material. The major structure that may act as an MTOC in fungal cells is the nucleus associated organelle (NAO) or spindle pole body (SPB). The variable structure and terminology of this organelle has been discussed in detail elsewhere (Fuller, 1976). Recent experimental evidence has shown that at least in the yeast *Saccharomyces cerevisiae* these organelles can nucleate the assembly of microtubules *in vitro*. The SPBs of yeast duplicate at the commencement of bud emergence. As the bud enlarges, microtubules form at the intranuclear faces of these SPBs. These microtubules interdigitate to form the short spindle after the two SPBs have moved around the nuclear membrane to their normal polar positions. High-voltage electron microscopy of serial sections through the yeast spindle has shown that there is a one-to-one correlation between the number of linkage groups per nucleus (17 in haploid cells) and the number of discontinuous microtubules in the spindle (Peterson

& Ris, 1976). The number of pole-to-pole continuous microtubules varies slightly but is usually between five and ten.

Recently, Hyams & Borisy (1978) have prepared sphaeroplasts of yeast and then lysed these sphaeroplasts using an adaption of the Kleinschmidt spreading technique. These spread preparations were then picked up onto electron microscope grids and used in a series of experiments to determine if structures in the yeast were capable of nucleating the assembly of microtubules *in vitro*. Grids were transferred to solutions of mammalian brain tubulin under conditions which minimized the spontaneous formation of microtubules. After incubation it was found that the yeast SPBs were consistently associated with long microtubules indicating the potential of these organelles to initiate microtubule assembly *in vitro*. Microtubules were mainly associated with only one face of the SPB – presumably the face which is normally in contact with the nucleoplasm. The morphology of the SPB was unaffected by treatment with DNase I, RNase A or phospholipase A and these enzymes did not affect the potential of the SPBs to initiate microtubule assembly. However, trypsin treatment destroyed the SPBs and removed all capacity for microtubule initiation. One important finding of this study (Hyams & Borisy, 1978) was that individual SPBs appear to possess the capability of nucleating only a certain specific number of microtubules. SPBs from exponentially growing cells initiated 8.5 ± 2 microtubules whilst SPBs from stationary cultures (in the G1 state of the cell cycle) initiated 15 ± 4 microtubules.

The conditions used in these experiments suggest that the tubulin appears to polymerize onto the SPB and not onto persisting short fragments of yeast microtubules. However, it seems very possible that short stubs of microtubules may remain associated with the electron-dense SPBs. In this context Byers, Shriver & Goetsch (1978) have shown that the end of a yeast spindle microtubule proximal to the SPB has a distinct structural modification, in that it has a closed surface. Furthermore, in experiments in which SPBs from lysed sphaeroplasts were incubated with mammalian brain tubulin it was clear that the resulting microtubules polymerized *in vitro* also possessed this closed end proximal to the SPB. Byers *et al.* were also able to show that the number of microtubules arising on each SPB *in vitro* achieved saturation early in the incubation in spite of the continued elongation of the microtubules that were present. This saturation phenomenon again suggests that each SPB possesses a limited number of discrete sites for microtubule assembly. An explanation of this phenomenon might involve the closed end structures persisting on the SPBs after the main microtubule has depolymerized.

Microtubule inhibitors

A number of drugs that possess an anti-mitotic activity have been shown to bind specifically to tubulin, and so inhibit the assembly of microtubules. The best known and characterized of these compounds is the plant alkaloid colchicine (Fig. 1). Colchicine binds selectively to

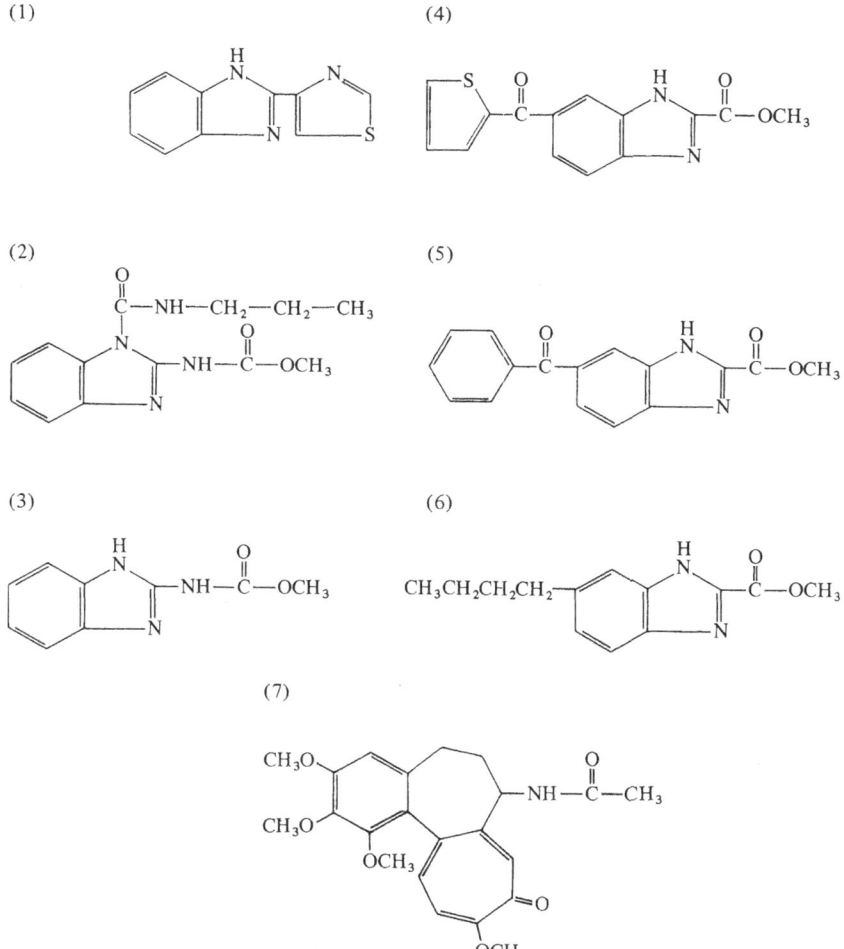

Fig. 1. Chemical structures of some antimicrotubule agents. (1) Thiabendazole; (2) benomyl; (3) benzimidazol-2-yl carbamate; (4) nocodazole; (5) mebendazole; (6) parbendazole; (7) colchicine.

Table 1. *Differential sensitivity of organisms to colchicine*

Tissue/organism	Colchicine conc. to inhibit mitosis (mM)	Reference
Mammalian cells	1×10^{-4}	Taylor, 1965
Pea roots	0.4	
Chlamydomonas reinhardii	5	Adams & Warr, 1972
Tetrahymena pyriformis	12.5	Rosenbaum & Carlson, 1969
		Wunderlich & Speth, 1970
Schizosaccharomyces pombe	100	Lederberg & Stetton, 1970
Saccharomyces cerevisiae	50	Haber *et al.*, 1973

tubulin, there being one binding site per tubulin dimer (mol. wt. 110 000). Radio-labelled colchicine has been used in many mammalian and higher eukaryotic systems as a 'tag' to identify tubulin. However, there is evidence that many lower eukaryotes are considerably less sensitive to colchicine. The concentration of colchicine necessary to inhibit mitosis in a variety of cells and tissues is given in Table 1. In many cases this insensitivity of lower eukaryotes to colchicine has been shown not to be due to reduced uptake, suggesting that it is a direct feature of the microtubule system of these organisms (Haber *et al.*, 1972). Davidse & Flach (1977) were able to find some binding of colchicine to extracts of *Aspergillus nidulans* although the binding reaction had unusual features. The rate of complex formation was rapid and was independent of temperature – both features contrast markedly with the binding of colchicine to mammalian tubulin. Also, the affinity of the *A. nidulans* tubulin to colchicine was very low compared with that of mammalian tubulin. Other workers have been unable to detect colchicine binding in *Schizosaccharomyces pombe* (Burns, 1973) and *Physarum polycephalum* (Jockusch, Brown & Rusch, 1971). Colchicine binding activities have been detected in extracts of *Saprolegnia ferax* (Heath, 1975) and *Allomyces moniliformis* (Olson, 1973). However, Davidse & Flach (1977) have pointed out that in both these cases at least some of the apparent binding may be attributable to an impurity in the [³H]-colchicine preparation. The general view, therefore, is that the fungi are insensitive to colchicine, presumably due to differences in their tubulins.

A group of compounds (Fig. 1) based on benzimidazole derivatives have recently emerged as important tools in the study of the lower eukaryote tubulins. Benomyl, thiabendazole and methyl benzimidazol-

2-yl carbamate (MBC) are important fungicides (Davidse, 1975), whilst mebendazole, cambendazole and albendazole are used as anthelmintics (Borgers *et al.*, 1975). A similar compound, nocodazole, has been shown to have anti-tumour activity (de Brabander *et al.*, 1975). There is increasing evidence that compounds in this range have a similar mechanism of action, involving an inhibition of microtubule functions.

Work from this laboratory has shown that a range of benzimidazole carbamates can inhibit the polymerization of microtubules *in vitro* and that mebendazole competes with colchicine for binding to purified mammalian brain tubulin (Ireland *et al.*, 1979). Essentially, similar results have been obtained with other benzimidazole carbamates (Friedman & Platzer, 1978). In addition, three benzimidazole carbamates have been shown to bind to purified mammalian brain tubulin (Hoebeke, van Nijen & de Brabander, 1976; Laclette, Guerra & Zetina, 1980; Havercroft, Quinlan & Gull, 1980), there being a single binding site per tubulin dimer.

The fungicide benomyl breaks down in aqueous solution to the active component MBC. Hammerschlag & Sisler (1973) have shown that MBC affects the overall rate of DNA, RNA and protein synthesis in *Ustilago maydis* and *Saccharomyces cerevisiae*. However, these effects on macromolecular synthesis are secondary to a primary inhibition of nuclear division and cytokinesis. In *S. cerevisiae*, MBC treatment produces large doublet cells and there appears to be a specific execution point within the cell cycle (Quinlan, Pogson & Gull, 1980). Davidse (1973) was able to show that MBC caused inhibition of mitosis in *Aspergillus nidulans* and that the treated nuclei possessed contracted and abnormal chromatin configurations. Also, Davidse & Flach (1977) showed that *A. nidulans* possesses a protein which will bind MBC and has many properties characteristic of tubulin. The protein had an estimated mol. wt. of 110 000 and was retained on DEAE-Sephadex. This MBC binding activity was competitively inhibited by known inhibitors of microtubule function, such as nocodazole and colchicine. Polyacrylamide gel electrophoresis of partially purified preparations of the MBC binding fractions showed the presence of proteins with similar mobilities to mammalian tubulins. Furthermore, Davidse & Flach (1977) were able to show that a benomyl-resistant mutant had a low binding affinity for MBC whilst a super-sensitive mutant had a high binding affinity when compared with the original strain of *A. nidulans*. These results raised the possibility that these mutants might be mutations in the tubulin gene itself – both mutants (the sensitive 186 and resistant R strains) mapped to a site in one gene located on linkage group VIII (van Tuyl, Davidse & Dekker, 1974). More recently these and other mutants have been extensively investigated in the laboratory of Dr N. R. Morris. This group determined that the benomyl resistant mutant used by Davidse & Flach – *ben*A15 (BEN-13,R) – had an electrophoretically abnormal β tubulin. Over 20 independently isolated mutants which mapped to the *ben*A site were examined and 17 were shown to have abnormalities in β tubulins as assessed by two-dimensional polyacrylamide gel electrophoresis. None of these mutants possessed an abnormal α tubulin (Sheir-Neiss, Lai & Morris, 1978; Morris, 1980). It therefore appears likely that the benomyl binding site of *A. nidulans* tubulin is located, in whole or in part, on the β subunit of tubulin.

Identification of tubulins in fungi

The purification of tubulin from fungal cells has, until very recently, proved to be a rather intractable problem. A number of research groups have, however, attempted to identify fungal tubulins by copolymerizing radiolabelled cell extracts with purified mammalian brain microtubule protein. After two or more cycles of copolymerization the product is run on SDS polyacrylamide gels and any tubulin-like proteins from the fungal cells detected by autoradiography or fluorography. This approach to the identification of tubulins does have some major technical problems. It is well known from studies with animal cells that many cell extracts are often inhibitory to microtubule assembly *in vitro* (Bryan & Nagle, 1976). Some of these problems can be circumvented but not all published reports of copolymerization experiments are satisfactory in this respect. Up to the present, four fungi have been used in copolymerization experiments: *Dictyostelium discoideum*, *Physarum polycephalum*, *Saccharomyces cerevisiae* and *Aspergillus nidulans*.

In *Dictyostelium discoideum*, Capuccinelli, Martinotti & Hames (1978) found that copolymerization experiments purified two radiolabelled proteins which co-migrated with α and β tubulin from mammalian brain. They also identified three bands of lower molecular weight which may be proteolytic artefacts, as these authors were unable to abolish the proteolytic activities in the *D. discoideum* extracts.

In *Saccharomyces cerevisiae*, Water & Kleinsmith (1976) detected two yeast proteins which co-migrated with α and β neurotubulin. However, in similar copolymerization experiments, Baum, Thorner & Honig

(1978) found that the major yeast proteins had mol. wts. of 46 000 and 45 000 and did not co-migrate with the mammalian brain tubulins (mol. wts. 50 000 and 48 000 on their gels). Shriver (1978) has shown that after four cycles of copolymerization the predominant yeast proteins (mol. wts. 48 000 and 50 000) did not co-migrate with either α or β neurotubulin. These proteins yielded peptide maps which were highly dissimilar to those of the neurotubulins. These predominant proteins, therefore, are not yeast tubulins. However, Shriver did report a 56 000 mol. wt. protein which co-migrated with α neurotubulin and thus appeared to be a likely candidate for a yeast tubulin.

We have also performed copolymerization experiments with *Saccharomyces cerevisiae* (Clayton, Pogson & Gull, 1979). In these experiments great care was taken to overcome the inhibition effects of the yeast cell extracts. These were effectively abolished by using RNase and DNase treatments, effective proteolysis inhibitors and a strain of yeast which was deficient in proteases A, B and C. Autoradiograms of SDS gels of twice-cycled copolymers showed two main bands (49 000 and 55 000 mol. wt.). The 55 000 mol. wt. protein migrated as a diffuse band on the same position as brain α tubulin. The other major band in the tubulin area was the 49 000 mol. wt. protein. However, this protein was retained on a phosphocellulose column and was eluted with 0.5 M KCl, indicating that it was not a tubulin. Therefore, this study, using phosphocellulose resolution of the copolymer material, suggests that there is a yeast protein that comigrates with α neurotubulin, but that no yeast protein was purified that comigrated with β neurotubulin. Very recently, Kilmartin (1980) has purified the tubulins of *Saccharomyces uvarum* using column chromatography and polymerization methods, and shown that the yeast tubulin is composed of two subunits that run as a tight doublet on SDS-gels, comigrating with α-neurotubulin. This suggests that the protein, identified in the copolymerization experiments of Clayton *et al.* (1979) and Shriver (1978), which runs with α-neurotubulin, is the *S. cerevisiae* tubulin which, because of the gel loading conditions in these experiments, is not resolved into its two components.

Although not yet purified, the :bulins of *Aspergillus nidulans* have been very well characterized by two-dimensional polyacrylamide gel electrophoresis and peptide mapping of copolymerized material. *A. nidulans* appears to have eight electrophoretically separable tubulins (using two-dimensional resolution): four α-tubulins and four β-tubulins (Morris, 1980).

Roobol, Pogson & Gull (1980*a*) have used copolymerization techniques to identify the tubulins of *Physarum polycephalum*. Again, great care was taken to ensure that cell extracts of *P. polycephalum* myxamoebae were prepared in such a way that they did not inhibit the assembly of brain microtubule protein *in vitro*, even at high extract protein concentrations. Copolymerization of these extracts and brain tubulin were purified to constant stoichiometry and amoebal components identified by autoradiography. The *P. polycephalum* extracts contained a protein which migrated with the neurotubulin subunits (not separated by the gel system used), and the α subunit of *Tetrahymena* ciliary tubulin. There was another protein which comigrated with the β subunit of *Tetrahymena* ciliary tubulins. This initial identification of *P. polycephalum* tubulins has been greatly extended in that an assembly purification of tubulin from this organism has now been done *in vitro*.

Microtubules and the MTOC of *Physarum polycephalum* myxamoebae

The *Physarum polycephalum* myxamoeba possesses one nucleus. The interphase nucleus tapers slightly at a position close to the two centrioles (Fig. 2*a*). Between the centrioles and the nucleus is an amorphous region with an electron-dense centre: the cytoplasmic MTOC. Microtubules radiate in all directions from this MTOC and transverse sections of the centrioles indicate that some of the microtubules which radiate from this region are organized into a distinct band. The centriole–MTOC complex is surrounded by an electron transparent zone which is distinct, in that small structures such as ribosomes are excluded (Fig. 2*b*). Dictyosomes and small vesicles are present, both in this zone of exclusion and on its edge.

We have recently used a gentle method of cell breakage involving lysis in non-ionic detergent to isolate nuclei from *P. polycephalum* myxamoebae. Under phase contrast illumination isolated nuclei were seen to be pear-shaped, the narrow end possessing a 'granule' from which two long fibres were seen to diverge. Negative staining of preparations of whole isolated nuclei showed that the centriole–MTOC–microtubule complex remains closely associated with the nucleus after this isolation technique. Low power electron micrographs of these nuclei show that the cytoplasmic microtubules are extremely long (up to 14 μm) and are often formed into groups (Fig. 3). The microtubules isolated with the *P. polycephalum* nuclei originate from the distinct MTOC and not from the centrioles themselves. This observation is

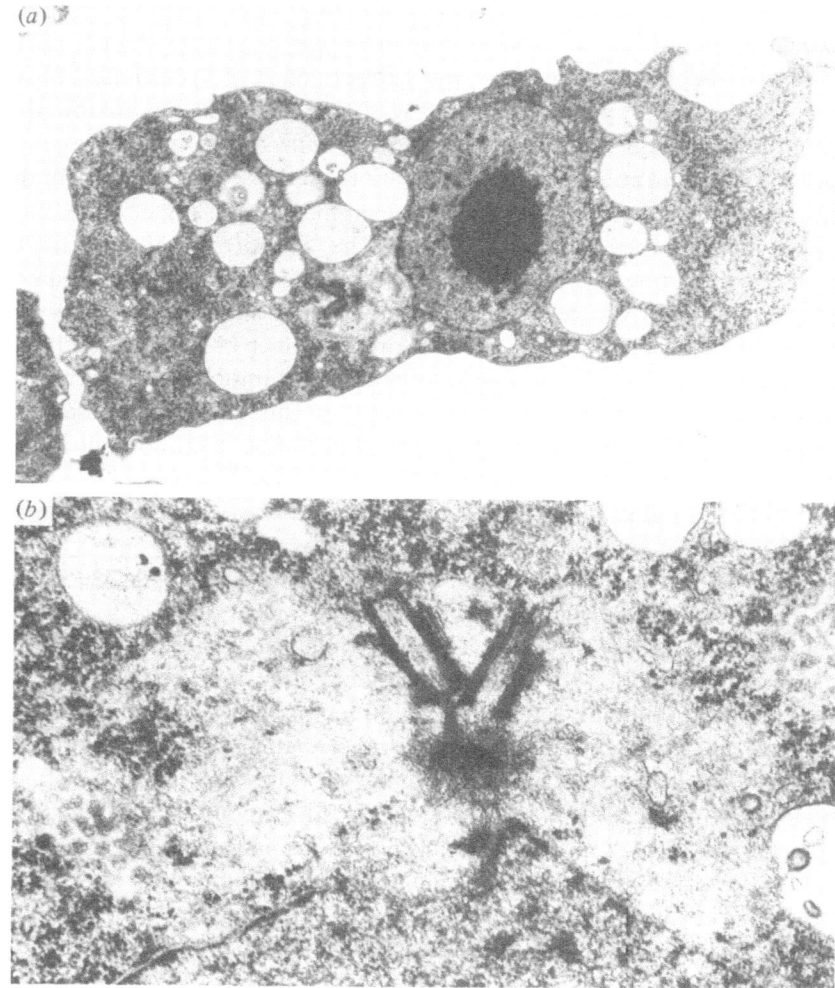

Fig. 2 (*a*) Electron micrograph of a myxamoeba of *P. polycephalum.* ×
8000. (*b*) Electron micrograph showing the centriole–MTOC complex
of a myxamoeba. × 25 000.

consistent with experimental results from animal cells that it is the
pericentriolar material which initiates cytoplasmic microtubule assembly rather than the centrioles themselves (Gould & Borisy, 1977).

Thin-section electron micrographs show that the isolated centriole–
MTOC complex is situated close to the tapered end of the nucleus. The
complex is not linked to the nucleus by any unique membrane or an

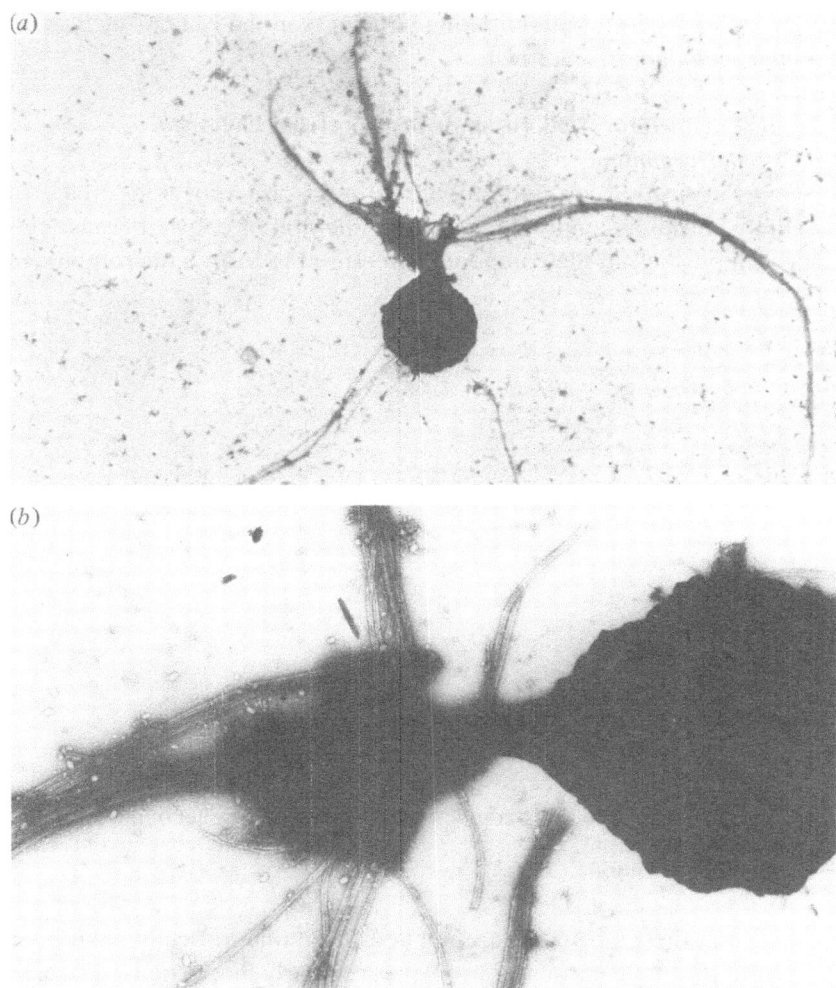

Fig. 3. Whole mounts of isolated nuclei, negatively stained with uranyl acetate. (*a*) Single nucleus isolated from a myxamoeba showing the centriole–MTOC complex and cytoplasmic microtubules. × 6000. (*b*) Higher magnification of the microtubules attached to the nuclear-associated centriole–MTOC complex. × 40 000.

extension of the nuclear envelope. Some microtubules which originate in the MTOC run alongside the nucleus and are in intimate contact with the nuclear envelope. These may link the MTOC to the nucleus, or alternatively other as yet undetected fibrillar proteins may be responsible for the close association of the two organelles. The biochemical nature of the microtubule-initiating structures in the MTOC of *Physarum polycephalum* is unknown.

Purification of microtubule proteins from *Physarum polycephalum*

Purification of microtubule proteins by repeated cycles of assembly and disassembly *in vitro* is the method of choice because this approach purifies all the components required to form a microtubule *in*

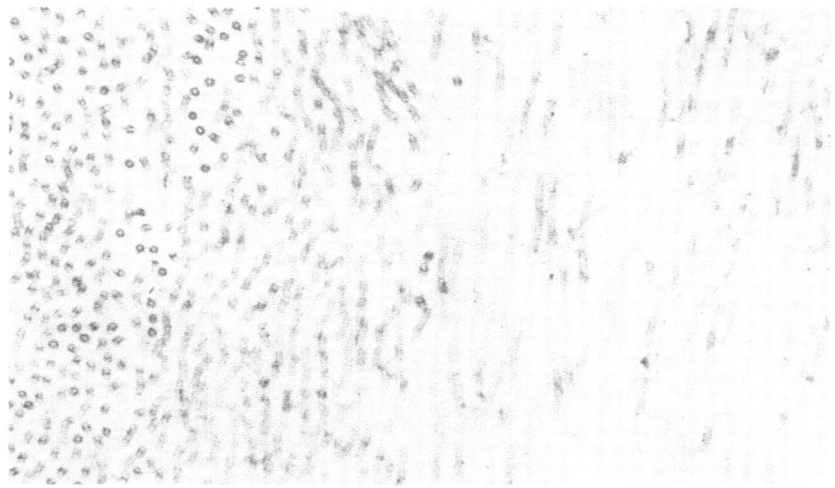

Fig. 4. Electron micrograph of a thin section of a pellet of microtubules produced from *Physarum polycephalum* myxamoebae by *in vitro* polymerization.

vitro. When these protocols are applied to non-neural cells, inhibitory factors in the cell extracts become particularly important. Also, to exceed the critical concentration for microtubule assembly, extremely high protein concentrations have to be achieved because tubulin is not an abundant protein in these cell extracts (often less than 1% of total soluble protein). Recently we have been able to achieve the purification of microtubule proteins from *Physarum polycephalum* using an *in vitro* polymerization technique (Roobol, Pogson & Gull, 1980*b*). This is the first report of a lower eukaryote microtubule protein preparation that is capable of true self-assembly. Preliminary experiments, in which the inhibition of polymerization of mammalian brain microtubule protein by amoebal extracts was studied, served as guide-lines in developing the conditions required for in-vitro assembly of *P. polycephalum* micro-

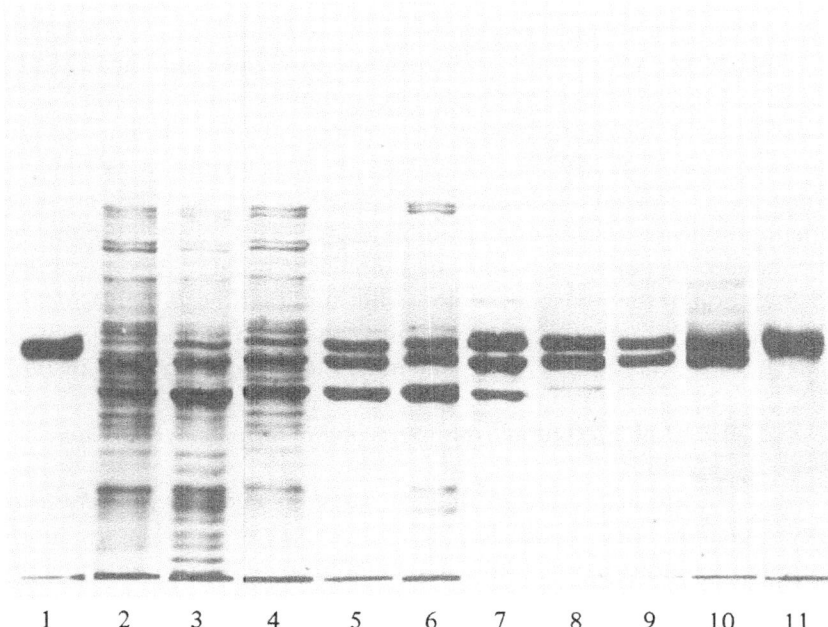

1 2 3 4 5 6 7 8 9 10 11

Fig. 5. SDS–polyacrylamide gel electrophoresis of samples taken during preparation of myxamoebal microtubule protein. Electrophoresis was on a 7.5% acrylamide slab gel which was then stained with Coomassie blue. 1, 2.5 μg brain tubulin; 2, 15 μg myxamoebal extracts; 3, 15 μg HP_1; 4, 15 μg CS_1; 5, 10 μg HP_2; 6, 10 μg CP_2; 7, 10 μg CS_2; 8, 5 μg HP_3; 9, 4 μg CS_3; 10, 4 μg *Tetrahymena* ciliary outer doublet tubulin; 11, 2.5 μg brain tubulin.

tubules. The inhibitory components which were defined in this way were proteolytic activity, GTPase activity and the presence of RNase- and DNase-sensitive materials. Each of these problems was overcome: *P. polycephalum* extracts prepared in the presence of leupeptin (a protease inhibitor), and treated with RNase and DNase and provided with a GTP-regenerating system, were capable of forming endogenous microtubules. Three cycles of temperature-dependent assembly and disassembly produced a very pure preparation of microtubules (Figs. 4 and 5). SDS–polyacrylamide gel analysis of this purification procedure showed that the third cold supernatant (CS_3) contained more than 95% tubulin. The only other protein present had an apparent mol. wt. of 43 000 and was probably a trace of actin which is abundant in *P. polycephalum* extracts.

This microtubule protein purified from *Physarum polycephalum* will assemble in the absence of glycerol to form microtubules of characteristic appearance (Fig. 4). The microtubule associated proteins (MAPs) which are present in microtubule preparations from mammalian brain can be separated from tubulin by phosphocellulose chromatography (Weingarten *et al.*, 1975). This tubulin is then unable to assemble *in vitro*. A *P. polycephalum* tubulin prepared in this way does not form microtubules on warming in assembly buffer. This purified tubulin can be induced to polymerize into microtubules if brain MAPs are added. Salt elution of the phosphocellulose indicated that the *P. polycephalum* MAPs were present in small amounts, the most prominent proteins having mol. wts. of 49 000, 57 000 and 59 000 (Roobol *et al.*, 1980*b*).

Effect of microtubule inhibitors on *Physarum polycephalum* tubulin

Having achieved an in-vitro assembly method for *Physarum polycephalum* microtubules it was of obvious interest to determine the sensitivity of this assembly to known inhibitors of microtubule function (Quinlan *et al.*, 1981). Colchicine at 100 μM did not inhibit the growth of myxamoebae of *P. polycephalum* and was completely ineffective in preventing the in-vitro assembly of *P. polycephalum* microtubule protein. This concentration of colchicine is ten-fold greater than that required to completely inhibit brain microtubule protein assembly under the same conditions. Therefore, the insensitivity of *P. polycephalum* (and presumably other fungi) towards colchicine is a direct consequence of the sensitivity of the organism's tubulin and has nothing to do with uptake phenomena or metabolic inactivation of the drug. Preliminary colchicine binding assays (Roobol *et al.*, 1980*b*) have shown that the binding ratio for *P. polycephalum* tubulin at 100 μM is some 50-fold lower than the binding ratio for brain tubulin determined under the same conditions. Quinlan *et al.* (1981) were also able to show that the benzimidazole carbamates, nocodazole and parbendazole were very effective inhibitors of both growth and of microtubule assembly *in vitro* (Figs. 6–8).

Analysis of *Physarum polycephalum* tubulin by SDS–polyacrylamide gel electrophoresis

The phosphocellulose purified *Physarum polycephalum* tubulin has been compared with mammalian brain tubulin and *Tetrahymena* tubulin by SDS–polyacrylamide gel electrophoresis (SDS-PAGE) and peptide mapping (Roobol, Pogson and Gull, 1980*a*; Clayton *et al.*, 1980). Differences in migration of the proteins on SDS- and urea-SDS-PAGE followed by peptide mapping have demonstrated that *P. polycephalum* tubulin consists of a tubulin that is similar to brain β-tubulin, and a tubulin that is dissimilar to both brain subunits (Fig. 9). In addition, the analogous proteins from *P. polycephalum* and brain demonstrated micro-heterogeneity during urea-SDS-PAGE. This finding that *P. polycephalum* myxamoebal tubulin is slightly different to mammalian tubulin, in that it does not contain a protein identical to brain α-tubulin, may explain the insensitivity of *P. polycephalum* tubulin to colchicine.

These results highlight the fact that differences do exist between higher and lower eukaryote tubulins, confirming the indirect evidence suggested by differences in apparent drug sensitivity and binding. Purification and characterization of tubulins from other fungi should

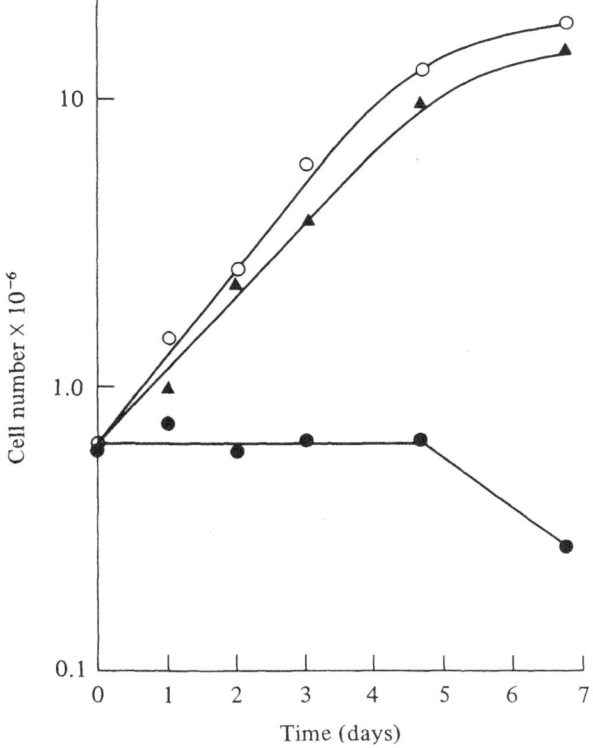

Fig. 6. Influence of nocodazole on growth of *P. polycephalum* myxamoebae. (○), control; (▲), 0.4 μM; (●), 2.0 μM.

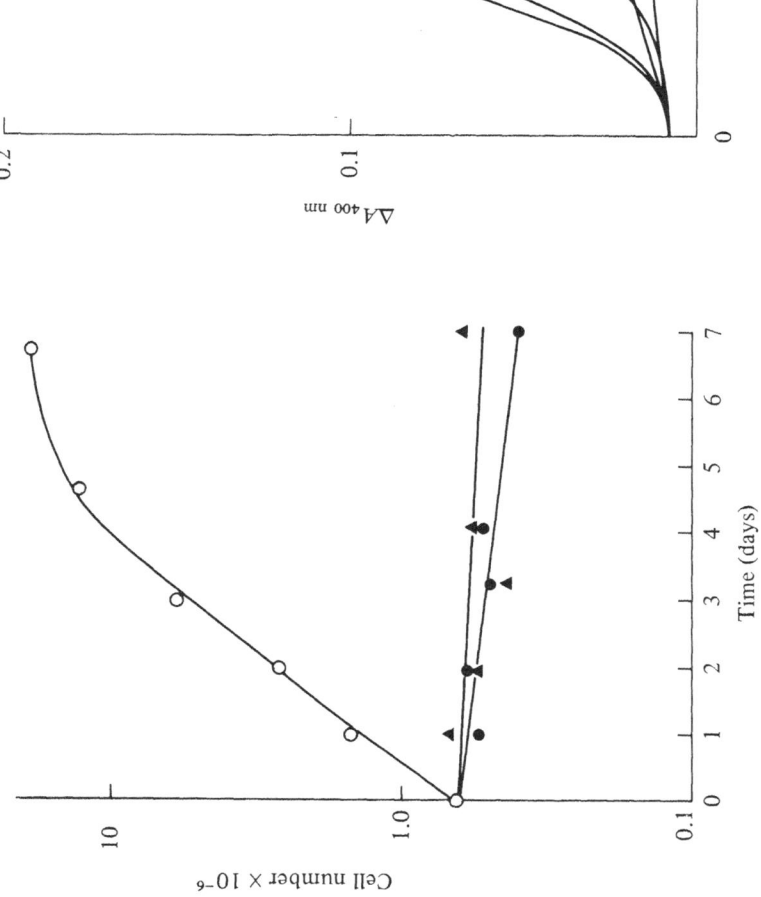

Fig. 8. Influence of nocodazole and parbendazole on the *in-vitro* assembly of myxamoebal microtubule protein. A, control; B, 0.4 μM nocodazole; C, 0.4 μM parbendazole; D, 2.0 μM nocodazole; E, 2.0 μM parbendazole.

Fig. 7. Influence of parbendazole on growth of *P. polycephalum* myxamoebae. (○), control; (▲), 0.4 μM; (●), 2.0 μM.

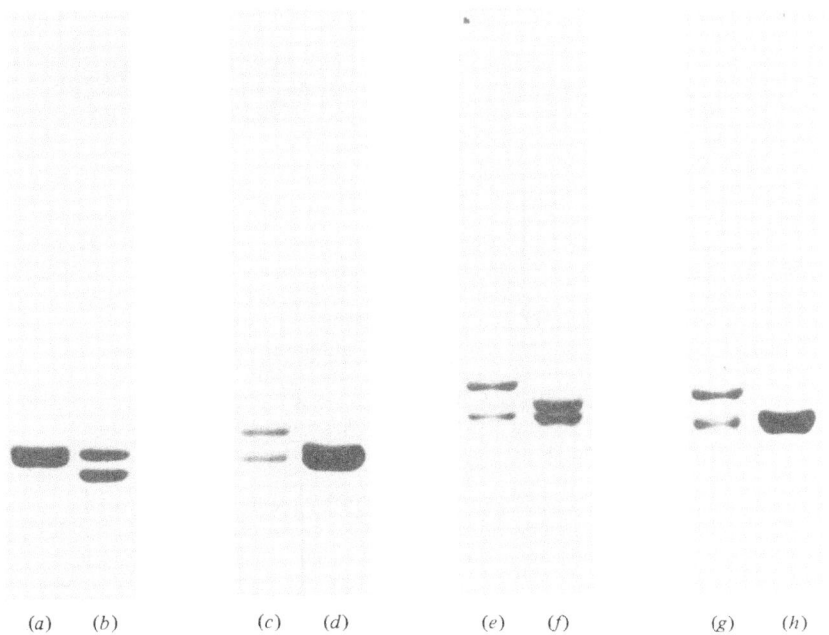

(a)   (b)   (c)   (d)   (e)   (f)   (g)   (h)

Fig. 9. Electrophoresis of sheep brain tubulin (*a, c, e, g*) and *P. polycephalum* tubulin (*b, d, f, h*) on different gel systems. (*a, b*) Fisons SDS-PAGE; (*c, d*) Sigma SDS-PAGE; (*e, f*) Fisons urea-SDS-PAGE; (*g, h*) Sigma urea-SDS-PAGE.

now be possible and such studies will no doubt then exploit the potential of fungi as experimental systems.

Acknowledgements. Work in this laboratory has been supported by grants from The Wellcome Trust, The Medical Research Council and Scientific Investigation Grants from The Royal Society.

References

Adams, M. & Warr, J. R. (1972). Colchicine resistant mutants of *Chlamydomonas reinhardi. Experimental Cell Research*, **71**, 473–75.

Aist, J. R. & Williams, P. H. (1972). Ultrastructure and time course of mitosis in the fungus *Fusarium oxysporum. Journal of Cell Biology*, **55**, 368–89.

Amos, L. A. (1979). Structure of microtubules. In *Microtubules*, ed. K. Roberts & J. S. Hyams, pp. 1–64. New York & London: Academic Press.

Baum, P., Thorner, J. & Honig, L. (1978). Identification of tubulin from the yeast *Saccharomyces cerevisiae. Proceedings of the National Academy of Sciences, USA*, **75**, 4962–66.

Beckett, A., Heath, I. B. & McLaughlin, D. J. (1974). *An Atlas of Fungal Ultrastructure.* London: Longman.

Borgers, M., de Nollin, S., Verheyen, A., de Brabander, M. & Thienpont, D. (1975). Effects of new anthelmintics on the microtubular system of parasites. In *Microtubules and Microtubule Inhibitors*, ed. M. Borgers & M. de Brabander, pp. 497–508. Amsterdam: North-Holland.

Bryan, J. & Nagle, B. W. (1976). In *Molecular Basis of Motility*, ed. L. Heilmeyer, J. C. Ruegg & T. Weiland, pp. 161–74. Berlin: Springer-Verlag.

Burns, R. G. (1973). [^{3}H]-Colchicine binding. Failure to detect any binding to soluble proteins from various lower organisms. *Experimental Cell Research*, **81**, 285–92.

Byers, B., Shriver, K. & Goetsch, L. (1978). The role of spindle pole bodies and modified microtubule ends in the initiation of microtubule assembly in *Saccharomyces cerevisiae. Journal of Cell Science*, **30**, 331–53.

Clayton, L., Pogson, C. I. & Gull, K. (1979). Microtubule proteins in the yeast *Saccharomyces cerevisiae. FEBS Letters*, **106**, 67–70.

Clayton, L., Quinlan, R. A., Roobol, A., Pogson, C. I. & Gull, K. (1980). A comparison of tubulins from mammalian brain and *Physarum polycephalum* using SDS-polyacrylamide gel electrophoresis and peptide mapping. *FEBS Letters*, **115**, 301–5.

Capuccinelli, P., Martinotti, G. & Hames, B. D. (1978). Identification of cytoplasmic tubulin in *Dictyostelium discoideum. FEBS Letters*, **91**, 153–7.

Davidse, L. C. (1973). Antimitotic activity of methyl benzimidazol-2-yl carbamate in *Aspergillus nidulans. Pesticide Biochemistry and Physiology*, **3**, 317–25.

Davidse, L. C. (1975). Antimitotic activity of methyl benzimidazol-2-yl carbamate in fungi and its binding to cellular protein. In *Microtubules and Microtubule Inhibitors*, ed. M. Borgers & M. de Brabander, pp. 483–95. Amsterdam: North-Holland.

Davidse, L. C. & Flach, W. (1977). Differential binding of methyl benzimidazol-2-yl carbamate to fungal tubulin as a mechanism of resistance to this antimitotic agent in mutant strains of *Aspergillus nidulans. Journal of Cell Biology*, **72**, 174–93.

de Brabander, M., van de Beire, R., R., Aerts, F., Guens, G., Borgers, M., Desplenter, L. & de Cree, J. (1975). Oncodozole (R17934); a new anti-cancer drug interfering with microtubules. Effects on neoplastic cells cultures *in vitro* and *in vivo*. In *Microtubules and Microtubule Inhibitors*, ed. M. Borgers & M. de Brabander, pp. 509–21. Amsterdam: North-Holland.

Dustin, P. (1978). *Microtubules*. Berlin: Springer-Verlag.

Friedman, P. A. & Platzer, E. G. (1978). Interaction of anthelmintic benzimidazoles and benzimidazole derivatives with bovine brain tubulin. *Biochimica et Biophysica Acta*, **544**, 605–14.

Fuller, M. S. (1976). Mitosis in fungi. *International Review of Cytology*, **45**, 113–53.

Girbardt, M. (1968). Ultrastructure and dynamics of the moving spindle. In *Aspects of Cell Motility*, ed. P. L. Miller, pp. 247–59. Symposia of the Society for Experimental Biology 22. Cambridge University Press.

Gould, R. R. & Borisy, G. G. (1977). The pericentriolar material in Chinese hamster cells nucleates microtubule formation. *Journal of Cell Biology*, **73**, 601–15.

Haber, J. E., Peloquin, J. G., Halvorson, H. O. & Borisy, G. G. (1972). Colcemid inhibition of cell growth and the characterisation of a colcemid-binding activity in *Saccharomyces cerevisiae. Journal of Cell Biology*, **55**, 355–67.

Hammerschlag, R. S. & Sisler, H. D. (1973). Benomyl and methyl-2-benzimidazole carbamate (MBC): biochemical, cytological and chemical aspects of toxicity to *Ustilago maydis* and *Saccharomyces cerevisiae. Pesticide Biochemistry and Physiology*, **3**, 42–52.

Havercroft, J. C., Quinlan, R. A. & Gull, K. (1980). Binding of parbendazole to tubulin and its influence on microtubules in tissue culture cells as revealed by immunofluorescence microscopy. *Journal of Cell Science*, in press.

Heath, I. B. (1975). Colchicine and colcemid binding components of the fungus *Saprolegnia ferax. Protoplasma*, **85**, 177–92.

Heath, I. B. & Heath, M. C. (1976). Ultrastructure of mitosis on the cowpea rust fungus *Uromyces phaseoli* var. *vignae. Journal of Cell Biology*, **70**, 592–607.

Hoebeke, J., van Nijen, G. & de Brabander, M. (1976). Interaction of oncodazole (R 17934), a new antitumoral drug with rat brain tubulin. *Biochemical and Biophysical Research Communications*, **69**, 319–24.

Howard, R. J. & Aist, J. R. (1979). Hyphal tip preservation of the fungus *Fusarium*: improved preservation by freeze substitution. *Journal of Ultrastructure Research,* **66**, 224–34.

Hyams, J. S. & Borisy, G. G. (1978). Nucleation of microtubules *in vitro* by isolated spindle pole bodies of the yeast *Saccharomyces cerevisiae. Journal of Cell Biology*, **78**, 401–14.

Hyams, J. S. & Stebbings, H. (1979). Microtubule-associated cytoplasmic transport. In *Microtubules*, ed. K. Roberts & J. S. Hyams, pp. 487–530. New York & London: Academic Press.

Ireland, C. M., Gull, K., Gutteridge, W. E. & Pogson, C. I. (1979). The interaction of benzimidazole carbamates with mammalian microtubule protein. *Biochemical Pharmacology*, **28**, 2680–2.

Jockusch, B. M., Brown, D. F. & Rusch, H. P. (1971). Synthesis and some properties of an actin-like nuclear protein in the slime mould *Physarum polycephalum. Journal of Bacteriology*, **108**, 705–14.

Kilmartin, J. V. (1980). Benzimidazole fungicides inhibit the self assembly of yeast tubulin. Abstracts 2nd International Congress on Cell Biology. *European Journal of Cell Biology*, **22**, 298.

Laclette, J. P., Guerra, G. & Zetina, C. (1980). Inhibition of tubulin polymerisation by mebendazole. *Biochemical and Biophysical Research Communications*, **92**, 417–423.

Lederberg, S. & Stetton, G. (1970). Colcemid sensitivity of fission yeast and the isolation of colcemid-resistant mutants. *Science*, **168**, 485–7.

Morris, R. N. (1980). Chromosome structure and the molecular biology of mitosis in eukaryotic microorganisms. In *The Eukaryotic Microbial Cell*, ed. G. W. Gooday, D. Lloyd and A. P. J. Trinci, pp. 41–75. Symposium of the Society for General Microbiology 30. Cambridge University Press.

Oakley, B. R. & Morris, N. R. (1980). Nuclear movement is β-tubulin dependent in *Aspergillus nidulans. Cell*, **19**, 255–62.

Olson, L. W. (1973). A low molecular weight colchicine bonding protein from the aquatic phycomycete *Allomyces neo-moniliformis. Archiv für Mikrobiologie*, **91**, 281–97.

Peterson, J. B. & Ris, H. (1976). Electron microscopic study of the spindle and chromosome movement in the yeast *Saccharomyces cerevisiae. Journal of Cell Science*, **22**, 219–42.

Pickett-Heaps, J. D. (1969). The evolution of the mitotic apparatus; an attempt at comparative ultrastructural cytology in dividing plant cells. *Cytobios*, **3**, 257–80.

Quinlan, R. A., Pogson, C. I. & Gull, K. (1980). The influence of the microtubule inhibitor, methyl benzimidazol-2-yl carbamate (MBC) on nuclear division and the cell cycle in *Saccharomyces cerevisiae. Journal of Cell Science*, **46**, 341–52.

Quinlan, R. A., Roobol, A., Pogson, C. I. & Gull, K. (1981). A correlation between *in vivo* and *in vitro* effects of the microtubule inhibitors colchicine, parbendazole and nocodazole on myxamoebae of *Physarum polycephalum. Journal of General Microbiology*, **122**, 1–6.

Reichle, R. E. (1969). Fine structure of *Phytophthora parasitica* zoospores. *Mycologia*, **61**, 30–50.

Roberts, K. & Hyams, J. S. (1979). *Microtubules*. New York & London: Academic Press.

Roobol, A., Pogson, C. I. & Gull, K. (1980a). Identification and characterisation of microtubule proteins from myxamoebae of *Physarum polycephalum. Biochemical Journal*, **189**, 305–12.

Roobol, A., Pogson, C. I. & Gull, K. (1980b). *In vitro* assembly of microtubule proteins from myxamoebae of *Physarum polycephalum. Experimental Cell Research*, **129**, 455–67.

Rosenbaum, J. L. & Carlson, A. (1969). Cilia regeneration in *Tetrahymena* and inhibition by colchicine. *Journal of Cell Biology*, **40**, 415–25.

Sheir-Neiss, G., Lai, M. H. & Morris, N. R. (1978). Identification of a gene for β-tubulin in *Aspergillus nidulans. Cell*, **15**, 638–47.

Shriver, M. K. J. (1978). Biochemistry and assembly of yeast microtubules. PhD Thesis, University of Washington, Seattle.

Taylor, E. W. (1965). The mechanism of colchicine inhibition of mitosis. I. Kinetics of inhibition and the binding of ^{3}H-colchicine. *Journal of Cell Biology*, **25**, 145–60.

Tucker, J. N. (1979). Spatial organisation of microtubules. In *Microtubules*, ed. K. Roberts, J. S. Hyams, pp. 315–57. London & New York: Academic Press.

van Tuyl, J. M., Davidse, L. C. & Dekker, J. (1974). Lack of cross resistance of benomyl and thiabendazole in some strains of *Aspergillus nidulans. Netherlands Journal of Plant Pathology*, **80**, 165–73.

Water, R. D. and Kleinsmith, L. J. (1976). Identification of α and β tubulin in yeast. *Biochemical and Biophysical Research Communications*, **70**, 704–8.

Weingarten, M. D., Lockwood, A. H., Hwo, S. Y. & Kirschner, M. W. (1975). A protein factor essential for microtubule assembly. *Proceedings of the National Academy of Sciences, USA*, **72**, 1858–62.

Wunderlich, F. & Speth, V. (1970). Anti-mitotic agents and macronuclear division of ciliates. IV. Reassembly of microtubules in macronuclei of *Tetrahymena* adapting to colchicine. *Protoplasma*, **70**, 139–49.

Single-stranded DNA-binding proteins: their ubiquity and functions

G. R. BANKS

*National Institute for Medical Research, The Ridgeway,
Mill Hill, London NW7 1AA, UK*

Introduction

The study of interactions of proteins with deoxyribonucleic acid(s) (DNA) constitutes a central area of research in the biological sciences. These interactions control and organize the ordered expression of the genome during cellular growth, development, DNA replication, recombination and repair. Although we now understand in a general way how some of these processes are controlled, the detailed molecular rules that govern the interactions of such a wide range of DNA-binding proteins with DNA and lead to such diverse consequences are not well understood. This chapter will describe a class of these proteins that bind preferentially to single-stranded DNA and thus destabilize double-stranded DNA. They are known as single-stranded DNA-binding (SSB), helix-destabilizing or DNA-melting proteins (Alberts & Sternglanz, 1977; Champoux, 1978; Coleman & Oakley, 1980; Kornberg, 1980). The biochemistry of the proteins isolated from both prokaryotic and eukaryotic cells, in particular the basidiomycete fungus *Ustilago maydis* protein, will be described and compared, followed by evidence for their involvement in DNA replication, recombination and repair deduced primarily from bacterial and phage systems.

The SSB proteins of prokaryotic organisms

Bacteriophage T4

The bacteriophage T4 gene *32* product was the first SSB protein to be isolated, through the development of DNA–cellulose affinity chromatography (Alberts & Frey, 1970). The asymmetric protein consists of a single polypeptide chain of mol. wt. 35 000, but in the absence of polynucleotides it can aggregate indefinitely at high protein concentrations, and it may be that the monomer–monomer interactions responsible are also involved in its co-operative binding to single-stranded DNA (Alberts & Frey, 1970; Carroll, Neet & Goldthwait, 1972, 1975). At high ionic strengths, dimerization occurs that is independent of protein concentration, but the dimers are unable to bind to polynucleotides and are, therefore, not intermediates in the binding reaction (Carroll *et al.* 1972, 1975; Kelly & von Hippel, 1976). The protein binds weakly and non-co-operatively to native DNA with a binding constant of $<10^4$ M^{-1} at 0.1 M sodium ion concentration (Jensen, Kelly & von Hippel, 1976). In contrast, it binds tightly and co-operatively to single-stranded DNA and polynucleotides with a co-operative binding constant of $>10^9$ M^{-1} under similar ionic conditions. The binding constant for the binding to single-stranded DNA at a site adjacent to a pre-existing bound protein molecule is 10^3 greater than binding at an isolated site (the co-operativity parameter), and is largely independent of ionic strength (Alberts & Frey, 1970; Jensen *et al.*, 1976; Kelly, Jensen & von Hippel, 1976; Peterman & Wu, 1978). Both monomer–monomer protein interactions and distortion of the polynucleotide chain induced by the protein, resulting in increased affinity for contiguous binding, may be responsible for the co-operativity. Although the binding constants are similar for polynucleotides of differing base sequences, there is evidence that some sequence specificity may exist (Bobst & Pan, 1975; Kelly *et al.*, 1976). A monomer of the protein covers 5–10 nucleotides at saturation, although only two adjacent nucleotides appear to interact directly with each monomer for both single- and double-stranded DNA (Alberts & Frey, 1970; Kelly *et al.*, 1976; Jensen *et al.*, 1976; Record, Lohman & de Haseth, 1976).

On the basis of sedimentation studies of single-stranded bacteriophage fd DNA saturated with the SSB protein, it has been concluded that the protein holds the normally folded DNA in a rigid and extended conformation (Alberts & Frey, 1970). Electron microscopy of these protein–DNA complexes confirms this, as well as the co-operativity of binding. A 50% increase in contour length of the DNA was recorded when saturating amounts of the protein were bound, and this produced an internucleotide distance of about 5.3 Å as opposed to about 7 Å for completely extended single-stranded DNA (Delius, Mantell & Alberts, 1972). Ultraviolet spectroscopy has revealed that the protein can increase the hyperchromicity of both single-stranded polydeoxyribonucleotides and polyribonucleotides above that produced by heat, leading to almost complete disruption of adjacent base–base stacking (Jensen *et al.*, 1976).

Because the SSB protein binds preferentially to single-stranded DNA, it should destabilize double-stranded DNA and lower its equilibrium melting temperature (T_m). This has indeed been observed for poly(d[A–T]·d[A–T]), the T_m of which is lowered more than 40 °C by the protein. All attempts to make the protein denature natural native DNAs have failed, although it can locally destabilize AT-rich regions (Morrow & Berg, 1973; Brack, Bickle & Yuan, 1975). These observations suggest that the function of the bacteriophage T4 SSB protein *in vivo* is not to unwind a DNA helix, but a 27 000 dalton proteolytic product is able to denature natural DNAs and also destabilizes poly(d[A−T]·d[A−T]) to a greater extent than the intact protein (Moise & Hosoda, 1976; Greve *et al.*, 1978; Williams & Konigsberg, 1978). There is no evidence for proteolysis of the protein within the T4-infected host cell, but if the missing fragment of the proteolytic product inhibits denaturation in the intact protein, this inhibition might be controlled by interactions with other proteins of, for instance, the T4 DNA replication complex (Moise & Hosoda, 1976; Williams & Konigsberg, 1978).

Complementary single strands of natural DNAs do not renature appreciably under physiological conditions because intra-strand hydrogen-bonded helices contained within the single strands present a kinetic block to renaturation. Increasing the temperature can remove these helices and increase the renaturation rate dramatically. This can also be achieved under physiological conditions by the T4 SSB protein, again by denaturation of the intra-strand helices, exposing their bases for inter-strand helix formation (Alberts & Frey, 1970).

Biochemical studies of the effects of the T4 SSB protein on the action of enzymes involved in DNA metabolism have provided clues about its function *in vivo*. A role in replication is suggested because it stimulates some five-fold the rate of DNA synthesis specifically by the T4 DNA polymerase, with which it can complex, using a variety of single-stranded DNA templates (Huberman, Kornberg & Alberts, 1971). It also enables synthesis to be initiated at a single-strand break in duplex DNA, probably by facilitating displacement of the non-template strand under ionic conditions that weaken base-pair interactions (Nossal, 1974). Indeed, the SSB protein is absolutely required in concert with the purified proteins of the T4 genes *41*, *43* (DNA polymerase), *44*, *45*, *61* and *62* to replicate extensively both single- and double-stranded DNA molecules. With single-stranded circular DNA, initiation *de novo* is followed by elongation to generate rolling circles (Morris, Sinha & Alberts, 1975; Liu *et al.*, 1978). The SSB protein also inhibits the 3′→5′

exonuclease activity associated with the T4 DNA polymerase (Huang & Lehman, 1972).

Escherichia coli

The *Escherichia coli* SSB protein has a monomeric mol. wt. of 22 000, but under a range of ionic conditions it exists as a tetramer. It binds tightly and co-operatively to single-stranded DNA and poly-nucleotides with a co-operative binding constant of 3.8–7.6×10^{10} M^{-1} under physiological conditions, which is greater than that for non-co-operative binding (to an isolated site) by a factor of about 10^5. There is good evidence that the tetramer binds to DNA with a stoichiometry of one tetramer per 32 nucleotides. In contrast to the T4 SSB protein, it decreases the contour length of single-stranded DNA by about 35%, resulting in an internucleotide distance of 1.8 Å, suggesting that DNA is folded by the protein. It also denatures T4 duplex DNA at low ionic strengths in the absence of magnesium ions, a process that is initiated at AT-rich regions (Sigal *et al.*, 1972; Molineux, Friedman & Gefter, 1974; Winer, Bertsch & Kornberg, 1975; Ruyechan & Wetmur, 1975). It also renatures complementary single strands of DNA, but the mechanism appears to be more complex than removal of intra-strand helices because the reaction rate drops sharply above pH 6.0 in the presence of magnesium ions. In the presence of 2 mM spermidine or spermine, however, the rate increases by 5×10^3, or more at physiological pHs over that in the absence of the protein (Christiansen & Baldwin, 1977).

The *Escherichia coli* SSB protein exhibits an interesting spectrum of effects on purified proteins involved in DNA metabolism. It inhibits the rate of DNA synthesis by the *E. coli* DNA polymerases I and III and also the bacteriophage T4-induced polymerase, but stimulates DNA polymerase II and the T7-induced DNA polymerase (Sigal *et al.*, 1972; Molineux *et al.*, 1974). Similar effects also obtain for the $3' \rightarrow 5'$ exonuclease activities associated with these polymerases. Detailed studies have established that the $3' \rightarrow 5'$ exonuclease of *E. coli* DNA polymerase II is itself unable to degrade a DNA–SSB protein complex, but in the presence of an excess of a protein (over that required to saturate the DNA), it forms a polymerase II–SSB protein complex which degrades the DNA–SSB protein in a processive manner, in contrast to the distributive degradation observed in the absence of the protein (Molineux & Gefter, 1974, 1975). It also inhibits DNA degradation by pancreatic DNase I, snake venom phosphodiesterase, and specific single-stranded DNA endonucleases from *Neurospora crassa* and *Aspergillus* S1, probably by steric hindrance (Molineux & Gefter, 1975). The polymerization reaction of *E. coli* polymerase II also becomes processive in the presence of the SSB protein and removes intra-strand helices in the DNA template, which are kinetic barriers to elongation (Sherman & Gefter, 1976). Because no requirement for DNA polymerase II in DNA replication *in vivo* has yet been demonstrated the relevance of these findings in relation to the function of the *E. coli* SSB protein *in vivo* is unknown, although they do point to possible functions. The protein is required, however, for the replication in all phases of bacteriophages M13, G4 and ϕX174 DNAs by purified proteins (Weiner *et al.*, 1975; Kornberg, 1980), a reaction requiring host DNA polymerase III.

The protein increases several fold the fidelity of DNA polymerases. With a poly(d[A–T]·d[A–T]) template in the presence of 5 mM $MnCl_2$, the misincorporation rate of dGTP by *Escherichia coli* DNA polymerase III decreased from 1/32 000 in the absence of the SSB protein to 1/132 500 when enough protein was present to saturate 40% of the polynucleotide. The fact that fidelity was also increased with DNA polymerases lacking a $3' \rightarrow 5'$ (proof-reading) exonuclease suggests that the protein maintains the template nucleotides rigidly in a conformation optimal for accurate base-pairing with the incoming deoxyribonucleoside triphosphate at the active site of the polymerase (Kunkel, Meyer & Loeb, 1979).

Bacteriophage T7

A 25 000–31 000 dalton SSB protein is synthesized between 5 and 18 minutes after infection of *Escherichia coli* by bacteriophage T7 and accounts for more than 0.1% of the soluble protein in the infected cell. Although it can exist under physiological conditions as a monomer, it aggregates at high protein concentrations in the presence of magnesium ions. It binds tightly and co-operatively to single-stranded DNA and reduces the T_m of poly(d[A–T]·d[A–T]) by 40 °C. It stimulates the rate of DNA synthesis by the T7 DNA polymerase with extensively single-stranded DNA templates, but does not allow it to utilize intact or nicked double-stranded templates. Although this stimulation is specific for the T7 DNA polymerase, no polymerase–SSB protein complex has been detected (Scherzinger, Litfin & Jost, 1973; Reuben & Gefter, 1974). The protein also stimulates RNA primer synthesis by the T7 gene 4 protein on the displaced lagging strand of a duplex T7 DNA template when the T7 DNA polymerase, deoxy- and ribonucleoside triphosphates are present. Again, the *E. coli* SSB protein substitutes just as efficiently (Romano & Richardson, 1979).

Filamentous bacteriophages

The *Escherichia coli* filamentous bacteriophages (fd, f1 and M13) code their own SSB protein, known as the gene 5 protein. A mol. wt. of 9690 has been established from the amino acid sequence, but it exists primarily as a dimer in solution, which is apparently dissociated by sucrose. It binds tightly and co-operatively to single-stranded DNA and polynucleotides with a stoichiometry of one monomer per four nucleotides, holding the phosphodiester backbone in a rigid and extended conformation. Electron microscopy of its complexes with circular single-stranded fd DNA reveals branching rod-like structures which are believed to consist of two single strands of DNA held together by the dimeric protein. It denatures poly(d[A−T]·d[A−T]) and natural DNA duplexes, the T_m of T4 DNA being 35.5 °C under conditions at which it is 73 °C in the absence of protein. It does not, however, renature complementary strands of DNA, but conditions under which the *E. coli* SSB protein renatures DNA have not been investigated. It inhibits the *E. coli* DNA polymerases I, II and the T7 DNA polymerase with extensively single-stranded DNA templates, but increases the amount of synthesis (but not the rate) two- to three-fold by *E. coli* DNA polymerase II when replicative intermediates extracted from fd-infected *E. coli* cells are substituted as templates. It also inhibits DNA degradation by *E. coli* exonuclease I partially and exonuclease III completely (Alberts, Frey & Delius, 1972; Oey & Knippers, 1972; Pretorius, Klein & Day, 1975; Dunker & Anderson, 1975; Cavalieri, Neet & Goldthwait, 1976). The gene 5 SSB protein has been extensively characterized physico-chemically (Coleman & Oakley, 1980) and its structure has been determined to 23 nm resolution by X-ray diffraction of its crystals. The polypeptide folds into three distinct regions, two two-strand antiparallel β-ribbons and one three-strand antiparallel sheet. These have tentatively been assigned as a site for maintenance of the dimer in solution, a site for co-operative binding on DNA, and the major DNA-binding interface, respectively. Crystals of protein–oligonucleotide complexes reveal that 12 monomers constitute an asymmetric unit. It is suggested that this aggregate forms a squat cylinder, around the outside of which the DNA chains are wound to generate a 'lock washer' unit which can be extended to the limit of the DNA (McPherson *et al.*, 1978, 1979).

The SSB proteins of eukaryotic organisms
Mammalian cells

Proteins that bind to single-stranded DNA have been detected in cells of many eukaryotic organisms, but only in some cases does their range of biochemical properties suggest functions analogous to the prokaryotic SSB proteins. Of the proteins isolated from mammalian sources, those from calf thymus tissue have been characterized most thoroughly. Three major groups exist, one of which, UP1, has many properties similar to the prokaryotic SSB proteins. It is present in about 8×10^5 copies per cell, possesses a monomer mol. wt. of 19 000–24 000, but it is composed of four or five sub-species with isoelectric points clustered around neutrality. This heterogeneity may be genetic in origin or a result of limited proteolysis. Although UP1 binds preferentially to single-stranded DNA, it also has substantial affinity for double-stranded DNA and RNA. Its stoichiometry with the former is about one protein monomer per seven nucleotides, but binding is non-co-operative. It removes intra-strand helices from single-stranded DNA and promotes the denaturation of both poly(d[A−T]·d[A−T]) and the low GC-containing *Clostridium perfringens* DNA, reducing their T_ms by 45 and 17 °C respectively, in 35 mM NaCl. It also reduces the T_m of poly(r[A−U]·r[A−U]), but the different isoelectric fractions differ in their effectiveness. Electron microscopy of UP1–fd DNA complexes reveals that the DNA is rigidly held, with no indication of co-operative binding. The contour length of the complexes increases about 16% over that of the naked DNA. The UP1 stimulates DNA synthesis by the calf thymus DNA α-polymerase but not the β-polymerase using gapped native DNA and poly(dC)·oligo(dG) templates, whereas the T4 SSB protein has no effect (Herrick & Alberts, 1976*a*, *b*; Herrick, Delius & Alberts, 1976).

Two different SSB proteins have been characterized from rat tissues. One is abundant (about 10^6 copies per cell) in regenerating rat liver and has a monomeric mol. wt. of 25 000, but exists as a tetramer under low ionic strength. It depresses the T_m of poly(d[A−T]·d[A−T]) by 40 °C in 20 mM KCl and also those of poly(rA)·poly(dT) and poly(rA)·poly(rU), suggesting that it binds preferentially to both single-stranded DNA and RNA. The rates of DNA synthesis by both rat liver DNA α- and β-polymerases are stimulated by this protein, using either poly(d[A−T]·d[A−T]) or poly(dC)·oligo(dG) templates. The degree of stimulation depends on the polymerase, primer : template-nucleotide and SSB-protein : DNA ratios. At high ratios of the latter, inhibition of DNA synthesis is found, perhaps because the polymerase is stereo-chemically excluded from the template (Duguet *et al.*, 1977; Duguet & Recondo, 1978). The other rat SSB protein, named the R-protein (for reassociation), has been isolated from a lipoprotein complex of rat spermatocyte nuclei and is absent from rat liver tissue. It is a 30 000–35 000 dalton protein which partially denatures T7 DNA in the absence of magnesium ions, but in their presence renatures complementary strands of T7 DNA. An interesting property from the point of view of its possible functional regulation is its ability to be phosphorylated. The modified protein then no longer binds to single-stranded DNA or renatures it. De-phosphorylation re-establishes both activities. Similar R-proteins have been isolated from bovine and human testicular tissues (Hotta & Stern, 1971*a*; Mather & Hotta, 1977).

An SSB protein from mouse ascites·cells has a monomeric mol. wt. of 30 000–35 000 and exists as such even at low ionic strengths in solution. It binds preferentially to single-stranded DNA and RNA but nitrocellu-lose filter binding assays also reveal binding to double-stranded DNA. This is much more sensitive to inhibition by increasing ionic strength than single-stranded DNA binding. It stimulates the rate of DNA synthesis four-fold by mouse DNA α-polymerase with a denatured DNA template, but only marginally so by mouse DNA β-polymerase, *Escherichia coli* DNA polymerase I and T7 DNA polymerase, by all four enzymes with a nicked double-stranded DNA template. This protein can also be phosphorylated, and as a result its binding specificity to single-stranded DNA is increased below 200 mM NaCl and the stimulation of mouse DNA α-polymerase is drastically reduced (Otto, Baynes & Knippers, 1977).

Lymphocytes from patients with chronic lymphocytic leukaemia contain a protein which binds preferentially to single-stranded and ultraviolet-irradiated double-stranded DNAs. Its native mol. wt. is 25 000, but it appears to consist of two polypeptides of mol wts. 13 000 and 11 000. It reduces the T_m of poly(d[A−T]·d[A−T]) by 20 °C in 15 mM NaCl and also that of sheared ultraviolet-irradiated calf thymus DNA by 17 °C. Interestingly, it increases the rate of nicking of ultraviolet-irradiated DNA by the ultraviolet-specific endonuclease from *Micrococcus luteus*, which might suggest a function in the repair of ultraviolet-damaged DNA (Huang, Riddle & Koons, 1975).

The only plant source of an SSB protein described so far is *Lilium speciosum*. An R-protein has been isolated, again specifically from a lipoprotein fraction in meiotic nuclei in prophase stages. In the presence of magnesium or calcium ions this 35 000 dalton protein binds preferentially to single-stranded DNA and, like the R-proteins described above, efficiently renatures complementary single strands of DNA (Hotta & Stern, 1971*b*, *c*).

Lower eukaryotic organisms

Tetrahymena pyriformis contains three proteins which preferentially bind to single-stranded DNA. One of these, a 47 000 dalton protein, binds co-operatively and if oligo(dT)$_{100}$ is preincubated with it for a comparatively long time (120 min), it becomes partially resistant to degradation by the $3' \rightarrow 5'$ exonuclease of *E. coli* DNA polymerase I. However, no further properties typical of the prokaryotic SSB proteins have yet been described (Donnelly, Westergaard & Klenow, 1975).

An SSB protein has been isolated from the yeast *Saccharomyces cerevisiae* (W. L. Crosby & D. Livingston, personal communication). It is a 38 000 dalton protein which exists as a monomer in solution, but precipitates out of solution at low ionic strength. It binds specifically to single-stranded DNA, does not stimulate significantly yeast DNA polymerases I or II, but inhibits the degradation of denatured T7 DNA by the *Aspergillus* S1 endonuclease. Also, 37 000 dalton SSB protein, which might be identical to the protein just described, has been isolated from yeast. It forms insoluble aggregates at low ionic strength or high protein concentrations (S. G. Labonne & L. B. Dumas, personal communication). It binds specifically to single-stranded DNA with a stoichiometry of 1 monomer per 10–20 nucleotides, decreases the T_m of poly(d[A−T]·d[A−T]) by 30 °C in 100 mM NaCl and renatures complementary DNA strands. In the presence of an excess of the protein, DNA degradation by pancreatic DNase I is inhibited. DNA synthesis on extensively single-stranded DNA templates by both yeast DNA polymerase I and bacteriophage T4 DNA polymerase, but not *E. coli* polymerase I or calf thymus α-polymerase, is stimulated.

An SSB protein from *Ustilago maydis* has been extensively studied in this laboratory (Banks & Spanos, 1975). It is purified from log. phase cells by removal of endogenous DNA followed by single-stranded DNA-cellulose and then carboxymethylcellulose chromatography. The homogeneous protein has a mol. wt. of 20 000, determined by SDS–urea polyacrylamide gel electrophoresis, and a sedimentation coefficient of

2.6 S, suggesting that under the sedimentation conditions it exists as the monomer. When assayed by its ability to bind nucleic acids to nitrocellulose filters, it binds specifically to single-stranded DNA (Fig. 1). The sigmoidal binding curve indicates co-operative binding or a requirement for more than one protein molecule to bind a DNA strand to the filter. The stoichiometry of binding calculated from the curve is 1 protein monomer per 7–10 nucleotide bases.

Binding to RNA homopolymers can also be detected by this method, but it is 20–50-fold less efficient than binding to single-stranded DNA, a two-fold preference for poly(rA) or poly(rU) over poly(rI) or poly(rC) being observed.

The melting curve of poly(d[A−T]·d[A−T]) in the presence of the *Ustilago maydis* SSB protein and magnesium ions is shown in Fig. 2. The T_m is 12.5 °C, which is an equilibrium value because the transitions generated by increasing or decreasing the temperature are superimposable. Magnesium ions increase the renaturation rate. Thus the protein reduces the T_m of this copolymer by 50 °C, but it is not possible to determine if it reduces the T_m of duplex T7 DNA because heat

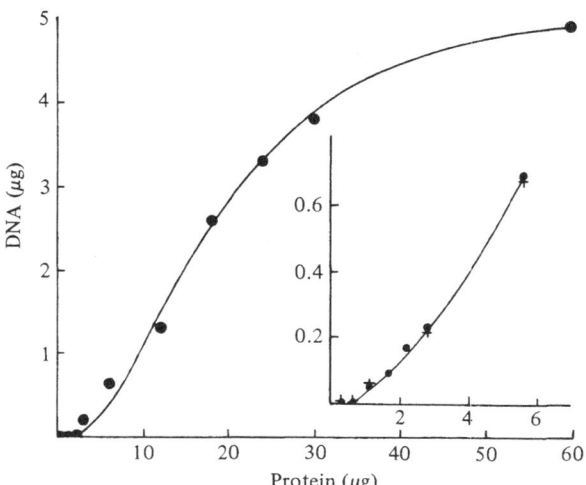

Fig. 1. DNA-binding curve for the *U. maydis* SSB protein. Heat denatured T7 [³H]DNA (5 µg) was incubated with amounts of the protein indicated at 20 °C for 10 min. The solution (1 ml) was filtered through a nitrocellulose filter which was washed and dried; the DNA retained was determined by its radioactivity (●—●). Omission of the filter washing did not eliminate the sigmoidal nature of the curve (+ — +). In parallel experiments, no native T7 [³H]DNA was retained.

denatures the protein before the helix–coil transition temperature is reached. On the other hand, complementary strands of T7 DNA are renatured efficiently at 37 °C by the protein in the presence of magnesium ions (Fig. 3).

The protein increases the rate, but not the final extent of DNA synthesis by the *Ustilago maydis* replicative DNA polymerase with extensively single-stranded T7 DNA templates (Fig. 4). It has no effect on *Escherichia coli* polymerase I, *Micrococcus luteus*, T4 or T7 DNA polymerases, and the T4 and T7 SSB proteins have no effect on the *U. maydis* DNA polymerase.

Stimulation of the *Ustilago maydis* DNA polymerase also occurs with a heat-denatured template, but not with a nicked native one. Kinetic experiments have revealed that the protein reduces the apparent K_m and

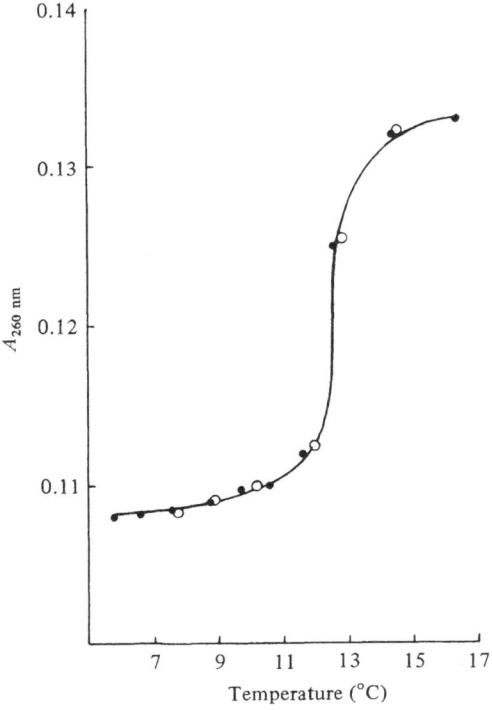

Fig. 2. Melting temperature of poly(d[A − T]·d[A − T]) in the presence of the *U. maydis* SSB protein. The copolymer (6.3 μg) and protein (60 μg) in a volume of 1 ml were incubated for 15 min at the temperature indicated and then the absorbance at 260 nm of the solution was determined (●—●). After the transition, the temperature cycle was reversed (○—○).

increases the V_{max} of the polymerase for deoxyribonucleoside triphosphates with the denatured, but not with the nicked native DNA templates (Table 1) (Yarranton, Moore & Spanos, 1976). The apparent K_m and V_{max} of the polymerase for the denatured DNA template are also affected in a similar way by the SSB protein.

These results suggest that the DNA polymerase interacts with an SSB protein complexed to single-stranded DNA rather than with free SSB protein because the K_m and V_{max} changes are only observed with single-stranded DNA templates. The resulting complex may then have a

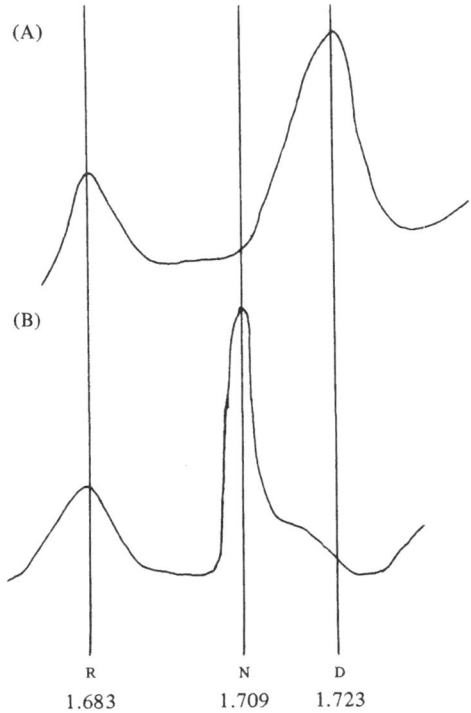

Fig. 3. Renaturation of complementary single strands of DNA catalysed by the *U. maydis* SSB protein. Alkali-denatured T7 DNA (7.5 μg in 1 ml) was incubated (A) alone or (B) with the protein (46 μg) at 37 °C for 30 min. Detergent was added to each to disrupt DNA–protein complexes and the DNA analysed by analytical CsCl gradient sedimentation. The gradients were photographed at equilibrium and the photographs scanned by a densitometer to determine the buoyant densities of the DNAs. The scans are shown along with R = yeast mitochondrial DNA reference, $\varrho = 1.683$ g cm^{-3}; N, native T7 DNA reference, $\varrho = 1.709$ g cm^{-3} and D, denatured T7 DNA reference, $\varrho = 1.723$ g cm^{-3}.

Table 1. *Effects of the* U. maydis *helix destabilizing protein on the kinetic constants of the homologous DNA polymerase for (A) deoxyribonucleoside triphosphates and (B) heat-denatured DNA template*

Template	HDP	Apparent K_m (μM)	V_{max} (pmol incorp. per 20 min)
(A)			
Denatured DNA	−	1.1	26.7
	+	0.3	41.3
Activated DNA	−	1.2	62.5
	+	1.2	62.5
(B)			
Denatured DNA	−	125.0	40.0
	+	31.3	50.0

greater affinity for the triphosphate substrates compared to a binary polymerase–DNA complex. The increased affinity of the polymerase for the template DNA when coated with SSB protein may explain the observed six-fold increase in processivity of the polymerase over that

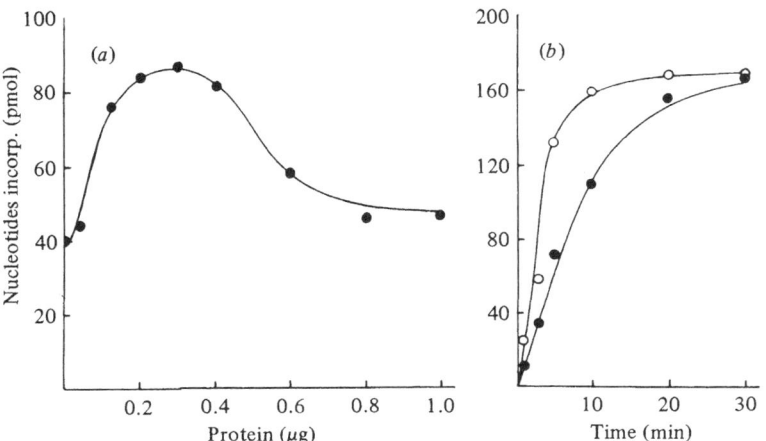

Fig. 4. SSB protein stimulation of DNA synthesis by the replicative *U. maydis* DNA polymerase. (*a*) DNA polymerase was incubated at 37 °C for 5 min in a polymerase assay mixture (0.15 ml, Banks & Spanos, 1975) containing T7 DNA template from *E. coli* degraded by exonuclease III (0.75 μg) in the presence of the amounts of SSB protein indicated. (*b*) Assays as for (*a*) except that SSB protein was absent (●—●) or present (0.2 μg) (○—○) and incubations were for the times indicated.

measured in the absence of the protein, although it might also accomplish this by melting out intra-strand helices which impede the polymerase (Yarranton, 1976). Nucleotide turnover (nucleotides incorporated into DNA but subsequently removed by the 3′→5′ exonuclease activity of the polymerase as monophosphates) during synthesis on a denatured DNA template was halved when the SSB protein was present, which could indicate an increase in polymerase fidelity, because no inhibition of the exonuclease activity was observed in the absence of DNA synthesis (Yarranton, 1976; G. R. Banks, unpublished observations). In contrast, the protein does inhibit the degradation of native T7 DNA by *Escherichia coli* exonuclease III and of denatured DNA by the *Ustilago maydis* DNase I, a single-stranded DNA-specific endonuclease (Fig. 5).

Electron microscopy of complexes of the *Ustilago maydis* SSB protein with circular single-stranded ϕX174 DNA reveals that their contour lengths are about 40% shorter than those of the naked DNA (Fig. 6). The protein prevents the spermine-induced collapse of DNA, indicating that the DNA is held rigidly within the complexes. When subsaturating levels of the protein are present, completely naked and completely complexed DNA molecules coexist on the same grid (Fig. 6). Thus a

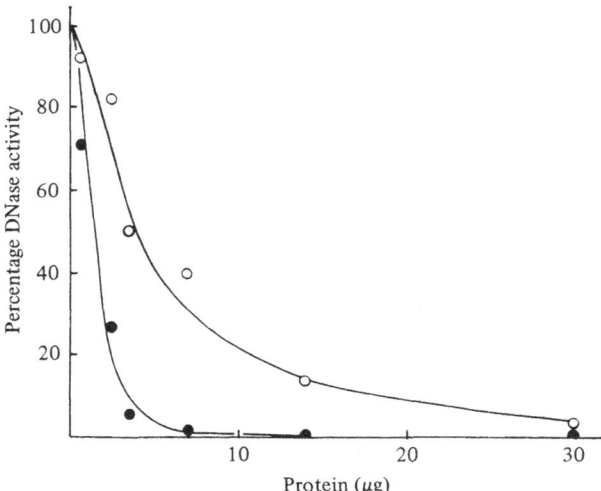

Fig. 5. Inhibition of deoxyribonucleases by the *U. maydis* SSB protein. The reaction mixtures (0.15 μl) contained native T7 [³H]DNA (0.18 μg) and *E. coli* exonuclease III (0.19 units) (●—●) or heat-denatured T7 [³H]DNA (0.2 μg) and *U. maydis* DNase I (0.7 units) (○—○) along with the amounts of the SSB protein indicated.

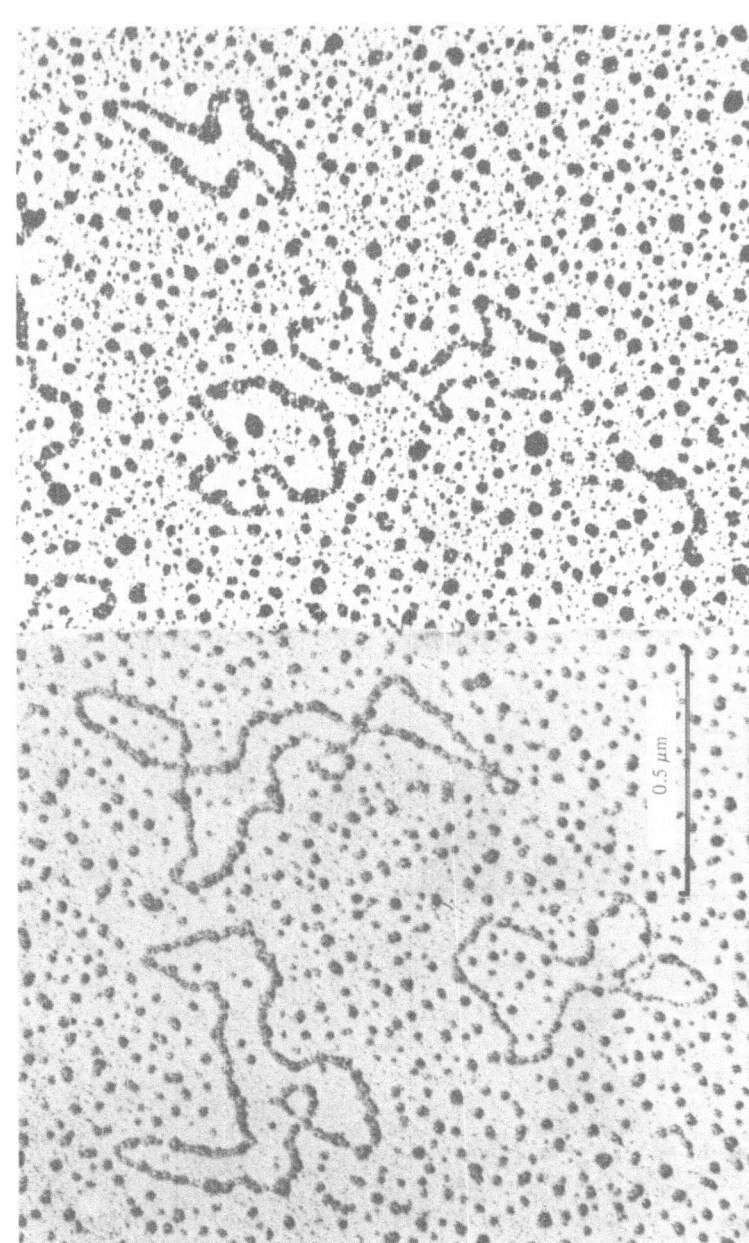

0.5 μm

Fig. 6. Electron micrographs of single-stranded circular φX174 DNA incubated with subsaturating amounts of the T4 (*left*) or *U. maydis* (*right*) SSB proteins. The complexes were prepared for electron microscopy as described by Delius *et al.* (1972). In each case, the top two molecules are fully complexed with proteins and the bottom one is naked DNA.

histogram of the number of molecules versus their contour lengths is biphasic, the two phases being centred on the contour lengths of naked and complexed DNA molecules (Fig. 7). This result is confirmation of highly co-operative binding of the protein to single-stranded DNA. When the protein was incubated at 50 °C with native λ DNA and then treated with glutaraldehyde, there was no hint of any DNA denaturation, although under identical conditions denaturation was induced by the *Escherichia coli* SSB protein, as reported previously (Sigal *et al.*, 1972).

Functions of SSB proteins

Genetical evidence for the functions of the SSB proteins is available only in bacteriophage T4 and its host *Escherichia coli*. Mutants

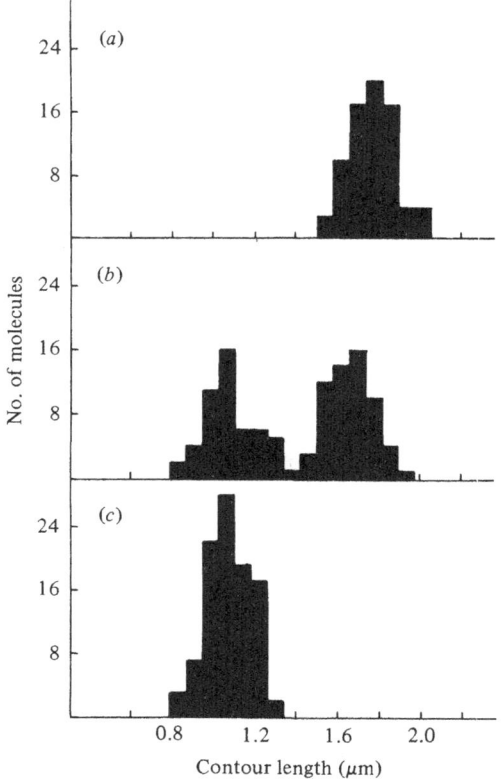

Fig. 7. Histograms of the frequency distribution of contour lengths of ϕX174 DNA molecules determined by electron microscopy. (*a*) no *U. maydis* SSB protein present, (*b*) subsaturating and (*c*) saturating amounts of the protein present.

of gene *32* of T4 were isolated before the protein product was characterized, and recently an *E. coli* mutant, SSB-1 (*lex*C), was shown to possess a thermolabile SSB protein. Nevertheless, comparative biochemical studies of the analogous proteins from other organisms suggest they might have functions in common with the gene *32* and *lex*C proteins. Our detailed biochemical knowledge of the T4 and *E. coli* proteins has provided an adequate explanation of the mechanisms by which they fulfil the in-vivo roles that had previously been deduced from genetical analyses.

Studies with T4 gene *32* mutants have established that the T4 SSB protein is required in stoichiometric amounts throughout the infectious cycle for DNA replication (Epstein *et al.*, 1963; Kozinski & Felgenhauer, 1967; Riva, Cascino & Geiduschek, 1970, Alberts, 1970; Sinha & Snustad, 1971). The protein from a temperature-sensitive mutant, tsL171 is deficient in binding to single-stranded DNA and that of another, tsP7 stimulates the T4 DNA polymerase less efficiently at 37 °C and at high ionic strength than the wild-type protein (Alberts, 1970; Huberman *et al.*, 1971). The absolute requirement for the protein in concert with six other purified proteins to replicate natural DNA molecules also leads to the conclusion that it plays a key role in DNA replication (Liu *et al.*, 1978). Because mutations in gene *32* increase virtually all base-pair substitution mutation rates, it undoubtedly also plays a role in maintaining high fidelity of DNA replication, as indeed the *Escherichia coli* SSB protein can *in vitro* (Bernstein *et al.*, 1972; Koch, McGraw & Drake, 1976; Kunkel *et al.*, 1979). The fidelity of replication by the T4 seven-protein complex is, in fact, greater than that by the DNA polymerase alone by two orders of magnitude, but the contribution of the SSB protein itself is not yet known (Liu *et al.*, 1978). The protein inhibits the $3' \rightarrow 5'$ exonuclease (proof-reading) activity of the T4 DNA polymerase and so presumably other proteins of the complex may mask this inhibition of proof-reading. The role, if any, of the proteolysis of the protein described above is unknown, but the functions of the protein could be modulated either by proteolysis or by specific interaction of other proteins with the cleavable oligopeptide in the intact protein. The details of its involvement in DNA replication remain to be elucidated, but it could maintain the DNA at the replication fork in a single-stranded conformation, favourable for maximum fidelity. (Each infected cell contains enough SSB protein for each T4 replication fork to be associated with about 170 molecules; Alberts & Frey, 1970). It could also protect this DNA from nucleolytic degradation and provide interaction sites for other proteins (see below).

T4 gene *32* mutants are also defective in DNA recombination and in the formation of intermediate joint and branched DNA molecules (Tomizawa, Anraku & Iwama, 1966; Berger, Warren & Fry, 1969; Broker & Lehman, 1971; Broker, 1973). Because the purified SSB protein is unable to denature T4 DNA, it is unlikely to perform this function *in vivo* in order to promote strand crossing-over during recombination, although, as noted above, a specific fragment of the protein can denature such DNA. Its ability to renature complementary DNA strands may, therefore, be important in recombination. The SSB protein is also required for the repair of ultraviolet-irradiated DNA and it may be significant that it does not function in excision repair, but in post-replicational (recombination?) repair (Wu & Yeh, 1973; Maynard-Smith & Symonds, 1973).

The surprising complexity of the different SSB protein interactions with DNA, replication, recombination and membrane proteins has been clearly demonstrated (Mosig *et al.*, 1978). Mutations in gene *32* that prevent DNA binding might be expected to affect most or all steps in replication and recombination, whereas those that prevent specific protein–protein interactions might affect only certain steps in one or both processes. Mosig and collaborators determined if certain steps are differentially blocked under restrictive conditions for several gene *32* mutants, and so constructed a functional gene *32* map (Fig. 8). This

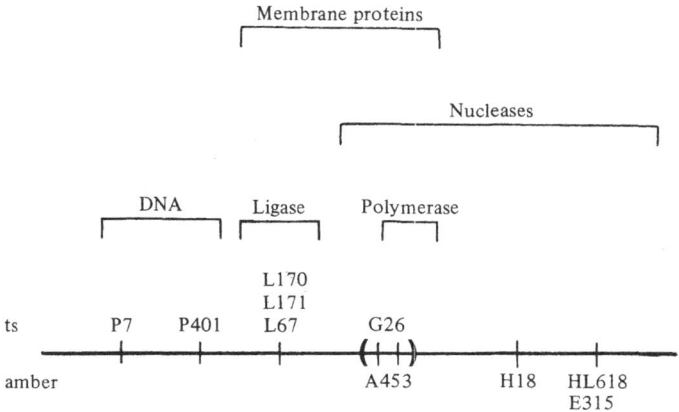

Fig. 8. Map of bacteriophage T4 gene *32* mutations and the interactions of the SSB protein which are affected by these mutations. (After Mosig *et al.*, 1978.)

shows that interaction sites with DNA and proteins can indeed be localized on the genetic map. The importance of the SSB protein protection of DNA from degradation by host recombination nucleases is illustrated by the gene *32* amber and tsG26 mutant strains, which on infection of the host under non-permissive conditions replicate as much parental DNA as the wild-type bacteriophage, but the DNA is subsequently completely degraded (Breschkin & Mosig, 1977; Mosig, 1978).

It is now established that the T4 SSB protein regulates its own expression. There is good evidence that the amount of protein in an infected cell is directly related to the amount of single-stranded DNA present. When all the latter is saturated by the protein, free protein binds specifically to its own messenger RNA, which represses further translation, resulting in regulation at the translational level (Krisch, Bolle & Epstein, 1974; Gold, O'Farrell & Russel, 1976; Russel *et al.*, 1976).

A temperature-sensitive *E. coli* mutant, SSB-1, possessing a thermolabile SSB protein, has been identified recently (Meyer, Glassberg & Kornberg, 1979; Glassberg, Meyer & Kornberg, 1979). On shifting a culture to the restrictive temperature, it immediately ceases chromosomal DNA replication (and thus the SSB protein is required for DNA chain elongation) and it restricts the growth of λ and single-stranded DNA bacteriophages. Protein extracts from the mutant cells convert single-stranded bacteriophage G4 DNA to duplex form only when wild-type SSB protein is included. High levels of the purified mutant protein, along with the other required purified proteins, are also temperature-sensitive for this conversion. These results complement those in which an absolute requirement for the *Escherichia coil* SSB protein in all phases of the replication of single-strand bacteriophage DNAs with purified proteins has been demonstrated (Kornberg, 1980). In the in-vitro conversion of viral to duplex DNA, but not in the subsequent replication of the latter, one function of the SSB protein is to bind to the DNA except at an intra-strand duplex loop, which is then the only site available for the RNA initiation of DNA synthesis. The *ssb*-1 strain is also extremely sensitive to ultraviolet irradiation (even at the permissive temperature for replication) compared to the wild-type strain, so that this SSB protein is undoubtedly also involved in DNA repair processes (Glassberg *et al.*, 1979). On the other hand, DNA recombination by the mutant is depressed only five-fold, but if the protein possesses functional domains like the T4 SSB protein, the examination of other mutants will be required before a direct involvement in recombination is established. The fact that the *ssb*-1 locus appears to be identical to the *lexC* locus also points to an involvement in the inducible *Escherichia coli rec*A-dependent error-prone repair system (Glassberg *et al.*, 1979).

Further evidence for the involvement of SSB proteins in recombination has been gathered recently. It has been established that the bacteriophage T7 SSB protein plays an important role in a pathway for the generation of recombinant T7 DNA by cell-free extracts of infected host cells. Both its renaturing activity and its ability to stimulate the T7 gene 6 exonuclease may be required (P. Sadowski, personal communication).

The major role of the gene 5 protein of filamentous bacteriophages, deduced by the study of gene 5 temperature-sensitive mutants, is in effecting the switch from double-stranded to single-stranded DNA synthesis (Salstrom & Pratt, 1971; Mazur & Model, 1973). After injection of the viral single-stranded DNA circle into the host, it is converted to its duplex form (accomplished by host cell proteins). The parental duplex molecule serves as a source of more duplexes by a rolling circle replication mechanism, which generates intermediate single strands on which the complementary strand is immediately synthesized. When some 200 duplexes have accumulated, enough gene 5 protein has been synthesized (about 10^5 molecules per cell) to bind co-operatively to the intermediate single strands and prevent synthesis of their complements. These complexed strands migrate to the host cell membrane through which the DNA is extruded, packaged by coat proteins with the gene 5 protein remaining in the host cytoplasm. Thus after infection of *Escherichia coli* at the restrictive temperature, gene 5 temperature-sensitive mutants accumulate duplex DNA molecules and fail to produce viral ones.

The SSB or R-protein from meiotic cells has been implicated in chromosome pairing or synapsis during meiosis (Hotta & Stern, 1971*b*, *c*; Stern & Hotta, 1974; Stern & Hotta, 1977). The evidence is circumstantial although persuasive, resting on the biochemical properties of the protein, time of appearance and correlations with other phenomena characteristic of the zygotene stage of the meiotic cell cycle, when chromosome pairing occurs in *Lilium*. Some 0.3–0.4% of nuclear DNA of microsporocytes of *Lilium* remains unreplicated after premeiotic S phase until zygotene (Hecht & Stern, 1971). Such regions are interspersed throughout the genome and although the parental DNA strands are intact, the newly replicated ones are gapped and probably hold sister chromatids together. They are then replicated at zygotene, although nicks persist. This DNA is transiently associated with a nuclear lipoprotein complex, which only appears at zygotene. Inhibition of zygotene DNA synthesis or lipoprotein synthesis arrests ongoing chromosome synapsis and progression of the cell through zygotene. During zygotene the R-protein is also associated with this lipoprotein complex (Hotta & Stern, 1971*b*). There is evidence that the synthesis of the *Lilium* R-protein precedes this association and in the presence of colchicine which rather specifically prevents synapsis, the only observed biochemical change is that most R-protein remains unassociated with the lipoprotein complex (Hotta & Shepard, 1973). Thus Stern and co-workers conclude that the efficient DNA renaturing activity of the R-protein is involved in chromosome synapsis by the alignment of complementary sequences (Stern & Hotta, 1974, 1977). The fact that an R-protein is also associated with a lipoprotein complex in meiotic, but not somatic cells of several mammals adds weight to these suggestions (Hotta & Stern, 1971*a*).

We have no direct evidence about the functions of the fungal *Ustilago maydis* SSB protein. The fact that it is found in stoichiometric amounts in mitotic cells, and its many striking biochemical similarities with the prokaryotic SSB proteins, argues against the role that has been proposed for R-proteins, but argues for a role in DNA replication, recombination and repair. The isolation of mutants in the structural gene of the protein and a study of their effects on these processes would provide such evidence. Because these proteins are ubiquitous, it is probable that they play a key role in these processes as determined for the T4 and *E. coli* SSB proteins. Despite their apparent simplicity and relatively small size, it is evident that their interactions with DNA and other proteins are diverse, complex and subtle. Further studies on them will not only increase our understanding of the mechanisms of DNA replication, recombination and repair, but also of the way in which proteins interact with and regulate the expression of DNA, and modify the activities of other proteins.

Acknowledgements. I am indebted to Ad Spanos, who isolated and characterized the *Ustilago maydis* SSB protein; Geoffrey Yarranton, who studied its interaction with DNA polymerase; Michael Kairis, who prepared the electron micrographs; Robin Holliday for constant encouragement and stimulation; Drs W. L. Crosby and S. G. Labonne for details of the yeast SSB protein and Dr P. Sadowski for communicating the involvement of the T7 SSB protein in recombination, before publication.

References

Alberts, B. M. (1970). Function of gene *32*-protein, a new protein essential for the genetic recombination and replication of T4 bacteriophage DNA. *Federation Proceedings*, **29**, 1154–63.

Alberts, B. M. & Frey, L. (1970). T4 bacteriophage gene *32*: a structural protein in the replication and recombination of DNA. *Nature, London*, **227**, 1313–18.

Alberts, B., Frey, L. & Delius, H. (1972). Isolation and characterisation of gene *5* protein in filamentous bacterial viruses. *Journal of Molecular Biology*, **68**, 139–52.

Alberts, B. & Sternglanz, R. (1977). Recent excitement in the DNA replication problem. *Nature, London*, **269**, 655–61.

Banks, G. R. & Spanos, A. (1975). The isolation and properties of a DNA unwinding protein from *Ustilago maydis*. *Journal of Molecular Biology*, **93**, 63–77.

Berger, H., Warren, A. J. & Fry, K. E. (1969). Variations in genetic recombination due to amber mutations in T4D bacteriophage. *Journal of Virology*, **3**, 171–5.

Bernstein, C., Bernstein, H., Mufti, S. & Strom, B. (1972). Stimulation of mutation in phage T4 by lesions in gene *32* and by thymidine imbalance. *Mutation Research*, **16**, 113–19.

Bobst, A. M. & Pan, Y.-C. E. (1975). A spin probe approach for measuring the nucleic acid affinity of gene *32* protein. *Biochemical and Biophysical Research Communications*, **67**, 562–70.

Brack, C., Bickle, T. A. & Yuan, R. (1975). The relation of single-stranded regions in bacteriophage PM2 supercoiled DNA to the early melting sequences. *Journal of Molecular Biology*, **96**, 693–702.

Breschkin, A. M. & Mosig, G. (1977). Multiple interactions of a DNA-binding protein *in vivo*. *Journal of Molecular Biology*, **112**, 279–94.

Broker, T. R. (1973). An electron microscopic analysis of pathways for bacteriophage T4 DNA recombination. *Journal of Molecular Biology*, **81**, 1–16.

Broker, T. R. & Lehman, I. R. (1971). Branched DNA molecules: intermediates in T4 recombination. *Journal of Molecular Biology*, **60**, 131–49.

Carroll, R. B., Neet, K. F. & Goldthwait, D. A. (1972). Self-associations of gene-*32* protein of bacteriophage T4. *Proceedings of the National Academy of Sciences, USA*, **69**, 2741–4.

Carroll, R. B., Neet, K. F. & Goldthwait, D. A. (1975). Studies of the self-association of bacteriophage T4 gene *32* protein by equilibrium sedimentation. *Journal of Molecular Biology*, **91**, 275–91.

Cavalieri, S. J., Neet, K. E. & Goldthwait, D. A. (1976). Gene *5* protein of bacteriophage fd: a dimer which interacts cooperatively with DNA. *Journal of Molecular Biology*, **102**, 697–711.

Champoux, J. J. (1978). Proteins that affect DNA conformation. *Annual Review of Biochemistry*, **47**, 449–79.

Christiansen, C. & Baldwin, R. L. (1977). Catalysis of DNA reassociation by the *Escherichia coli* DNA binding protein. *Journal of Molecular Biology*, **115**, 441–54.

Coleman, J. E. & Oakley, J. L. (1980). Physical chemical studies of the structure and function of DNA binding (helix-destabilising) proteins. *CRC Critical Reviews in Biochemistry*, **7**, 247–89.

Delius, H., Mantell, N. J. & Alberts, B. (1972). Characterization by electron microscopy of the complex formed between T4 bacteriophage gene *32*-protein and DNA. *Journal of Molecular Biology*, **67**, 341–50.

Donnelly, T. E., Westergaard, O. & Klenow, H. (1975). Isolation and characterization of a DNA-binding non-histone protein from *Tetrahymena pyriformis*. *Biochimica et Biophysica Acta*, **402**, 150–60.

Duguet, M. & de Recondo, A.-M. (1978). A deoxyribonucleic acid unwinding protein isolated from regenerating rat liver. *Journal of Biological Chemistry*, **253**, 1660–6.

Duguet, M., Soussi, T., Rossignol, J.-M., Mechali, M. & de Recondo, A.-M. (1977). Stimulation of rat liver α- and β-type DNA polymerases by an homologous DNA-unwinding protein. *FEBS Letters*, **79**, 160–4.

Dunker, A. K. & Anderson, E. A. (1975). The binding of the fd gene-*5* protein to single-stranded nucleic acid. *Biochimica et Biophysica Acta*, **402**, 31–4.

Epstein, R. H., Bolle, A., Steinberg, C. M., Kellenberger, E., Boy de la Tour, E., Chevalley, R., Edgar, R. S., Susman, M., Denhardt, G. H. & Lielausis, A. (1963). Physiological studies of conditional lethal mutants of bacteriophage T4D. *Cold Spring Harbor Symposia on Quantitative Biology*, **28**, 375–92.

Glassberg, J., Meyer, R. R. & Kornberg, A. (1979). Mutant single-strand binding protein of *Escherichia coli*: genetic and physiological characterization. *Journal of Bacteriology*, **140**, 14–19.

Gold, L., O'Farrell, P. Z. & Russell, M. (1976). Regulation of gene *32* expression during bacteriophage T4 infection of *Escherichia coli*. *Journal of Biological Chemistry*, **251**, 7251–62.

Greve, J., Moestre, M. F., Moise, H. & Hosoda, J. (1978). Circular dichroism studies of the interaction of a limited hydrolysate of T4 gene *32* protein with T4 DNA and poly[d(A − T)]·poly[d(A − T)]. *Biochemistry*, **17**, 893–8.

Hecht, N. B. & Stern, H. (1971). A late replicating DNA protein complex from cells in meiotic prophase. *Experimental Cell Research*, **69**, 1–10.

Herrick, G. & Alberts, B. (1976*a*). Purification and physical characterization of nucleic acid helix-unwinding proteins from calf thymus. *Journal of Biological Chemistry*, **251**, 2124–32.

Herrick, G. & Alberts, A. (1976*b*). Nucleic acid helix-coil transitions mediated by helix-unwinding proteins from calf thymus. *Journal of Biological Chemistry*, **251**, 2133–41.

Herrick, G., Delius, H. & Alberts, B. (1976). Single-stranded DNA structure and DNA polymerase activity in the presence of nucleic acid helix-unwinding proteins from calf thymus. *Journal of Biological Chemistry*, **251**, 2142–6.

Hotta, Y. & Shepard, J. (1973). Biochemical aspects of colchicine action on meiotic cells. *Molecular and General Genetics*, **122**, 243–60.

Hotta, Y. & Stern, H. (1971*a*). Meiotic protein in spermatocytes of mammals. *Nature New Biology*, **234**, 83–6.

Hotta, Y. & Stern, H. (1971*b*). A DNA-binding protein in meiotic cells in *Lilium*. *Developmental Biology*, **26**, 87–99.

Hotta, Y. & Stern, H. (1971*c*). Analysis of DNA synthesis during meiotic prophase in *Lilium*. *Journal of Molecular Biology*, **55**, 337–55.

Huang, W. M. & Lehman, I. R. (1972). On the exonuclease activity of phage T4 deoxyribonucleic acid polymerase. *Journal of Biological Chemistry*, **247**, 3139–46.

Huang, A. T.-F., Riddle, M. M. & Koons, L. S. (1975). Some properties of a DNA-unwinding protein unique to lymphocytes from chronic lymphocytic leukemia. *Cancer Research*, **35**, 981–6.

Huberman, J. A., Kornberg, A. & Alberts, B. M. (1971). Stimulation of T4 bacteriophage DNA polymerase by the protein product of T4 gene *32*. *Journal of Molecular Biology*, **62**, 39–52.

Jensen, D. E., Kelly, R. C. & von Hippel, P. H. (1976). DNA 'melting' proteins. II. Effects of bacteriophage T4 gene *32*-protein binding on the conformation and stability of nucleic acid structures. *Journal of Biological Chemistry*, **251**, 7215–28.

Kelly, R. C., Jensen, D. E. & von Hippel, P. H. (1976). DNA 'melting' proteins. IV. Fluorescence measurements of binding parameters for bacteriophage T4 gene *32*-protein to mono-, oligo-, and polynucleotides. *Journal of Biological Chemistry*, **251**, 7240–50.

Kelly, R. C. & von Hippel, P. H. (1976). DNA 'melting' proteins, III. Fluorescence 'mapping' of the nucleic acid binding site of bacteriophage T4 gene *32*-protein. *Journal of Biological Chemistry*, **251**, 7229–39.

Koch, R. E., McGraw, M. K. & Drake, J. W. (1976). Mutator mutations in bacteriophage T4 gene *32* (DNA unwinding protein). *Journal of Virology*, **19**, 490–94.

Kornberg, A. (1980). *DNA replication*. San Francisco: Freeman.

Kozinski, A. W. & Felgenhauer, Z. Z. (1967). Molecular recombination in T4 bacteriophage DNA. II. Single-strand breaks and exposure of uncomplemented areas as a prerequisite for recombination. *Journal of Virology*, **1**, 1193–202.

Kunkel, T. A., Meyer, R. & Loeb, L. A. (1979). Single-strand binding protein enhances fidelity of DNA synthesis *in vitro*. *Proceedings of the National Academy of Sciences, USA*, **76**, 6331–5.

Krisch, H. M., Bolle, A. & Epstein, R. H. (1974). Regulation of the synthesis of bacteriophage T4 gene *32* protein. *Journal of Molecular Biology*, **88**, 89–104.

Liu, C. C., Burke, R. L., Hibner, U., Barry, J. & Alberts, B. (1978). Probing DNA replication mechanisms with the T4 bacteriophage in-vitro system. *Cold Spring Harbor Symposia on Quantitative Biology*, **43**, 469–87.

McPherson, A., Jurnak, F., Wang, A., Kolpak, F., Molineux, I. & Rich, A. (1978). Gene-*V* product of bacteriophage fd: structure of a DNA-unwinding protein and its complexes with DNA. *Cold Spring Harbor Symposia on Quantitative Biology*, **43**, 21–33.

McPherson, A., Jurnak, F. A., Wang, A. H. J., Molineux, I., & Rich, A. (1979). Structure at 2.3 Å resolution of the gene *5* product of bacteriophage fd: a DNA unwinding protein. *Journal of Molecular Biology*, **134**, 379–400.

Mather, J. & Hotta, Y. (1977). A phosphorylatable DNA-binding protein associated with a lipoprotein fraction from rat spermatocyte nuclei. *Experimental Cell Research*, **109**, 181–9.

Maynard-Smith, S. & Symonds, N. (1973). Involvement of bacteriophage T4 genes in radiation repair. *Journal of Molecular Biology*, **74**, 33–44.

Mazur, B. J. & Model, P. (1973). Regulation of coliphage F1 single stranded DNA synthesis by a DNA-binding protein. *Journal of Molecular Biology*, **78**, 285–300.

Meyer, R. R., Glassberg, J. & Kornberg, A. (1979). An *Escherichia coli* mutant defective in single-strand binding protein is defective in DNA replication. *Proceedings of the National Academy of Sciences, USA*, **76**, 1702–5.

Moise, H. & Hosoda, J. (1976). T4 gene *32* protein modal for control of activity at replication fork. *Nature, London*, **259**, 455–8.

Molineux, I., Friedman, S. & Gefter, M. L. (1974). Purification and properties of the *Escherichia coli* deoxyribonucleic acid-unwinding protein. *Journal of Biological Chemistry*, **249**, 6090–8.

Molineux, I. J. & Gefter, M. L. (1974). Properties of the *Escherichia coli* DNA binding (unwinding) protein: Interaction with DNA polymerase and DNA. *Proceedings of the National Academy of Sciences, USA*, **71**, 3858–62.

Molineux, I. J. & Gefter, M. L. (1975). Properties of the *Escherichia coli* DNA-binding (unwinding) protein interaction with nucleolytic enzymes and DNA. *Journal of Molecular Biology*, **98**, 811–25.

Morris, C. F., Sinha, N. K. & Alberts, B. M. (1975). Reconstruction of bacteriophage T4 DNA replication apparatus from purified components: rolling circle replication following de-novo chain initiation on a single-stranded circular DNA template. *Proceedings of the National Academy of Sciences, USA*, **72**, 4800–04.

Morrow, J. F. & Berg, P. (1973). Location of the T4 gene *32* protein binding site on simian virus 40 DNA. *Journal of Virology*, **12**, 1631–2.

Mosig, G., Luder, A., Garcia, G., Dannenberg, R. & Bock, S. (1978). In-vivo interactions of genes and proteins in DNA replication and recombination of phage T4. *Cold Spring Harbor Symposia on Quantitative Biology*, **43**, 501–15.

Nossal, N. (1974). DNA synthesis on a double-stranded DNA template by the T4 bacteriophage DNA polymerase and the T4 gene *32* DNA unwinding protein. *Journal of Biological Chemistry*, **249**, 5668–76.

Oey, J. L. & Knippers, R. (1972). Properties of the isolated gene *5* protein of bacteriophage fd. *Journal of Molecular Biology*, **68**, 125–38.

Otto, B., Baynes, M. & Knippers, R. (1977). A single-strand-specific DNA-binding protein from mouse cells that stimulates DNA polymerases. *European Journal of Biochemistry*, **73**, 17–24.

Peterman, B. F. & Wu, C.-W.(1978). Ionic strength perturbation kinetics of gene *32* protein dissociation from its complex with single-stranded DNA. *Biochemistry*, **17**, 3889–92.

Pretorius, H. T., Klein, M. & Day, L. A. (1975). Gene V protein of fd bacteriophage. Dimer formation and the role of tyrosyl groups in DNA binding. *Journal of Biological Chemistry*, **250**, 9262–9.

Record, M. T., Lohman, T. M. & de Haseth, P. (1976). Ion effects on ligand–nucleic acid interactions. *Journal of Molecular Biology*, **107**, 145–58.

Reuben, R. C. & Gefter, M. L. (1974). A deoxyribonucleic acid-binding protein induced by bacteriophage T7. *Journal of Biological Chemistry*, **249**, 3843–50.

Riva, S., Cascino, A. & Geiduschek, E. P. (1970). Coupling of late transcription to viral replication in bacteriophage T4 development. *Journal of Molecular Biology*, **54**, 85–102.

Romano, L. J. & Richardson, C. C. (1979). Requirements for synthesis of ribonucleic acid primers during lagging strand synthesis by the DNA polymerase and gene *4* protein of bacteriophage T7. *Journal of Biological Chemistry*, **254**, 10476–82.

Russel, M., Gold, L., Morrissett, H. & O'Farrell, P. Z. (1976). Translational, autogenous regulation of gene *32* expression during bacteriophage T4 infection. *Journal of Biological Chemistry*, **251**, 7263–70.

Ruyechan, W. T. & Wetmur, J. G. (1975). Studies on the cooperative binding of *Escherichia coli* DNA unwinding protein to single-stranded DNA. *Biochemistry*, **14**, 5529–4.

Salstrom, J. S. & Pratt, D. (1971). Role of coliphage M13 genes in single-stranded DNA production. *Journal of Molecular Biology*, **61**, 489.

Scherzinger, E., Litfin, F. & Jost, E. (1973). Stimulation of T7 DNA polymerase by a new phage-coded protein. *Molecular and General Genetics*, **123**, 247–67.

Sherman, L. A. & Gefter, M. L. (1976). Studies on the mechanism of enzymatic DNA elongation by *Escherichia coli* DNA polymerase II. *Journal of Molecular Biology*, **103**, 61–76.

Sigal, N., Kornberg, T., Gefter, M. L. & Alberts, B. (1972). A DNA-unwinding protein isolated from *Escherichia coli*: its interaction with DNA and with DNA polymerases. *Proceedings of the National Academy of Sciences, USA*, **69**, 3537–41.

Sinha, N. K. & Snustad, D. P. (1971). DNA synthesis in bacteriophage T4-infected *Escherichia coli*: evidence supporting a stoichiometric role for gene *32*-product. *Journal of Molecular Biology*, **62**, 267–71.

Stern, H. & Hotta, Y. (1974). Biochemical controls of meiosis. *Annual Reviews of Genetics*, **7**, 37–66.

Stern, H. & Hotta, Y. (1977). Biochemistry of meiosis. *Philosophical Transactions of the Royal Society of London*, **277**, 277–94.

Tomizawa, J., Anraku, N. & Iwana, Y. (1966). Molecular mechanisms of genetic recombination in bacteriophage. VI. A mutant defective in the joining of DNA molecules. *Journal of Molecular Biology*, **21**, 247–53.

Weiner, J. H., Bertsch, L. L. & Kornberg, A. (1975). The deoxyribonucleic acid unwinding protein of *Escherichia coli*. *Journal of Biological Chemistry*, **250**, 1972–80.

Williams, K. R. & Konigsberg, W. (1978). Structural changes in the T4 gene 32 protein induced by DNA and polynucleotides. *Journal of Biological Chemistry*, **253**, 2463–70.

Wu, J.-R. & Yeh, Y.-C. (1973). Requirement of a functional gene *32* product of bacteriophage T4 in UV repair. *Journal of Virology*, **12**, 758–65.

Yarranton, G. T. (1976). The enzymology of DNA replication in *Ustilago maydis*. PhD thesis. CNAA, London.

Yarranton, G. T., Moore, P. D. & Spanos, A. (1976). The influence of DNA binding protein on the substrate affinities of DNA polymerase from *Ustilago maydis*: one polymerase implicated in both DNA replication and repair. *Molecular and General Genetics*, **145**, 215–18.

Structure and function of chromatin

E. M. BRADBURY AND H. R. MATTHEWS

Department of Biological Chemistry, School of Medicine, University of California, Davis, California 95616, USA

Introduction

Probably one of the most important series of unsolved problems in biology concerns the organization and expression of the eukaryotic genome. There are several interrelated structure–function problems: (i) the sequence organization of eukaryotic DNA; (ii) the structure of inactive chromatin; (iii) the structural transition which occurs when genes are activated; (iv) the determination, maintenance and control of active genes in differentiated tissues; (v) the disassembly and reassembly of chromatin during DNA replication; (vi) the control of chromosome structure through the cell cycle, and (vii) the structure of metaphase chromosomes. Recently much excitement has been generated in this area of research through the realization that solutions to several of these problems are now in sight. This fortunate situation is a consequence of major advances in biochemical strategies developed over the past five years.

Components of the eukaryotic genome

There are four major components of the eukaryotic genome: (i) DNA; (ii) a group of five basic proteins, histones, which are now regarded as major structural proteins of the chromosome; (iii) a large number of other so far poorly characterized, chromosomal proteins called non-histone proteins (NHP); and (iv) RNA. Some general rules have emerged from studies of the compositions of the nuclei of different cells; firstly the ratio of the amount of histones to that of DNA is relatively constant for all tissues and lies in the range 1.1 to 1.3. Secondly although the ratios of NHP to DNA are very variable, from 0.2 to 0.8, there appears to be a rough positive correlation between the amounts of NHP and the metabolic activities of cells, and a similar correlation appears to hold for the variable low amounts of RNA associated with chromatin. Although the functions of most of the NHP are unknown at present, included in this group of proteins are the enzymes required for essential functions of the nucleus and, it is thought, the proteins involved in the control of gene expression.

A major enigma in understanding the molecular biology of eukaryotes is posed by the extraordinarily large amounts of DNA found in their genomes, amounts far in excess of the coding requirements of the organisms. In addition to this genetic complexity, these enormous lengths of DNA have to be physically controlled in the nucleus through the cell cycle, and we shall discuss these problems in this chapter. To understand the current concepts concerning chromosome structure and its control it is necessary to have an appreciation of the recent advances in our knowledge of chromatin structure, of the properties of chromosomal proteins, particularly histones, and of the biochemical events thought to be involved in modulating the interactions of these proteins with DNA. In our studies we have utilized several biological systems, choosing those systems particularly suited to the problem under investigation. Thus for studies of the conformations and interactions of histones, calf thymus or chicken erythrocyte histones have been used because they can be isolated in large quantities and easily purified. For structural studies again calf thymus or chicken erythrocyte chromatins have been used because these are from tissues that are relatively inactive genetically and the proportions of NHP are substantially less than, for example, in liver. In our studies of the control of chromosome structure through the cell cycle we have used the naturally synchronous growth cycle of the multinuclear plasmodia of the true slime mould *Physarum polycephalum*. *P. polycephalum* plasmodia can be grown easily to 15 cm diameter and this size of plasmodium contains in the order of 10^9 nuclei which provides about 1 mg DNA and 2 mg of chromosomal protein. The nuclear division cycle in plasmodia is about 9–11 h; there is no G1 phase, S phase occupies 4–5 h and G2 phase 5–6 h. Within a single plasmodium the nuclei enter metaphase within 3–5 min in the 9–11 h cycle. The nuclear division cycle is therefore extremely precise and is achieved naturally because all the nuclei within a plasmodium are in the same cytoplasm and come under the same biochemical controls.

Physarum polycephalum is being used increasingly as a model eukaryote. Two of the histones, H3 and H4, are very similar to the same histones from higher eukaryotes. This can be seen in Fig. 1 where *P. polycephalum* histones (P) are compared with calf thymus histones (C) by three different types of polyacrylamide gel electrophoresis. The bands labelled have been identified by electrophoresis of purified histones. In all three gel systems *P. polycephalum* H4 comigrates with calf thymus H4; H3 from both organisms migrates in a very similar, though not always identical, manner. Also histones H2A and H2B from both organisms are not very different in their migratory behaviours. Histone H1 shows the greatest differences, and although the amino acid composition of *P. polycephalum* H1 is not very different from calf thymus H1, there are differences in size in the C-terminal region.

Properties of histones

The sequence properties of histones have been recently reviewed in detail by Von Holt and colleagues (Von Holt *et al.*, 1979).

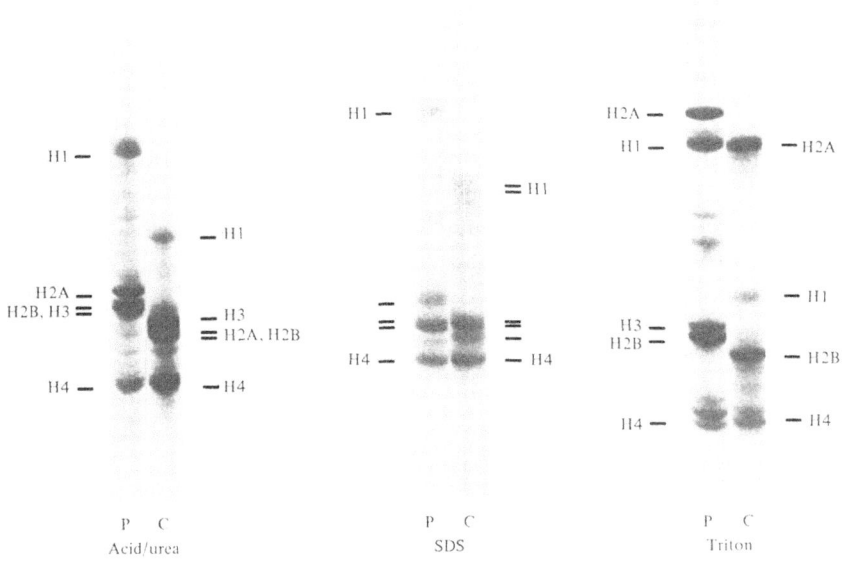

Fig. 1. Comparison of histones from *P. polycephalum* plasmodia (P) and calf thymus (C) by polyacrylamide gel electrophoresis. Gel electrophoresis in the presence of (i) acetic acid urea (acid/urea), (ii) sodium dodecyl sulphate (SDS), and (iii) Triton X–100. The labelled bands were identified by electrophoresis of purified histone fractions. (From Chahal, Matthews & Bradbury, 1980.)

Histones can be classified into three groups: the very lysine-rich histones H1, the moderately lysine-rich histones H2A and H2B, and the arginine-rich histones H3 and H4. Each of these groups is conserved to a different extent. Histones H3 and H4 have the most highly conserved sequences of all proteins found so far in nature with an accepted point mutation rate (PAM) of 0.1 mutations per 100 residues per 10^8 years which implies that each and every residue along the polypeptide chain is essential for the functions of both molecules. Histones H2A and H2B are more variable (PAM = 1.0) while the very lysine-rich histones H1 show the most sequence variability of the histones (PAM = 4.0). In certain cells, e.g. avian erythrocytes, the H1 molecules are largely replaced by another type of very lysine-rich histone, H5. So far there is no strong evidence to suggest that histones H2A, H2B, H3 and H4 recognize specific base sequences in the DNA. The relatively constant ratio of these histones to DNA in many organisms strongly suggests that they are associated with all types of DNA found in the eukaryotic genome, though this has yet to be demonstrated for minor DNA components.

A second major property of histones is that all their sequences are highly asymmetrical with respect to the distribution of net positive charge and apolar residues. The four histones H2A, H2B, H3 and H4 have highly basic N-terminal regions which extend to 25% of the molecules and are rich in lysine residues. In contrast the central and C-terminal regions of these histones contain a high proportion of apolar residues, arginines and acidic residues. The presence of acidic residues in these apolar regions of histones greatly reduces their basicity. Histone sequence and nuclear magnetic resonance (NMR) studies of histone interactions each led to proposals that the apolar central and C-terminal regions of histones H2A, H2B, H3 and H4 contained the sites of histone–histone interactions, while the basic N-terminal regions were probably required for DNA interactions (Bradbury & Rattle, 1972). Support for these proposals has recently been provided by the comparisons of the peptides and sequences of histones H2A and H2B from several sources with the homologous calf thymus histones. For pea H2A and H2B the differences have been located in the basic N-terminal regions whereas the C-terminal regions are highly conserved (Hayashi *et al.*, 1977). The sequence of H2A from sea urchin gonads shows that 11 substitutions and 4 deletions are located in the N-terminal and C-terminal ends while the central apolar region is highly conserved (Von Holt *et al.*, 1979). A similarly conserved region occurs in histones H2B$_3$

from sea urchin sperm (Strickland *et al.*, 1977*a*, *b*) and H2B from the mollusc *Patella granatina* (Van Holde *et al.*, 1979), strongly suggesting that the central apolar regions of these histones are conserved to preserve interaction essential to chromatin structure.

Histone variability

Because the variability in the sequences of histones H2A and H2B is confined to the basic N-terminal regions, the question arises as to the function of this variability and the more extensive variability found for the H1 molecule. Histone H1 consists of a family of H1 subfractions (Rall & Cole, 1971) which are tissue and organism specific and there is a pattern of synthesis of the H1 subfractions during the early stages of development in sea urchins (Cohen *et al.*, 1973). Similarly, subsets of histones H2A and H2B are synthesized at different stages of sea urchin development (Newrock *et al.*, 1978). It has been suggested that variabilities H1 and H2A and H2B are involved in controlling changes in chromatin structure during development, though the level of chromatin structure at which the effects are exerted is not known.

Histone complexes

A major advance in our understanding of histones came with the discovery of specific histone complexes both in the histones dissociated from chromatin at high salt concentrations and in the complexes formed between individual histones in aqueous solution. The first complex to be identified was the dimer (H2A, H2B) (Kelly, 1973), and chemical crosslinking of gently prepared histones H3 and H4 showed the existence of a tetrameric complex $(H3, H4)_2$ (Kornberg & Thomas, 1974). Systematic studies from Isenberg's laboratory of the interactions of pairs of histones showed that the strongest complexes formed were for the tetramer $(H3, H4)_2$ and for the dimer (H2A, H2B) with other weaker complexes also being formed (D'Anna & Isenberg, 1974).

Interactions and conformations of histones

Because the sequences of histones are highly asymmetric their interactions could be studied by high resolution NMR spectroscopy. The resonances from specific residues located in different regions of the histones, e.g. signals from the aromatic residues in the apolar regions, were used to report on the local mobility and interactions of these residues. Studies of the self and cross interactions of histones (Moss *et al.*, 1976*a*,*b*) and of the interactions of large histone peptides (Bohm *et al.*, 1977) have shown that the histone complexes are held together by interactions of structured central and C-terminal apolar regions leaving the N-terminal regions free and mobile. A similar behaviour has also been found for the histone complexes released from chromatin at 2M NaCl (Cary, Moss & Bradbury, 1979). Models for (H3, H4)$_2$ and (H2A, H2B) are illustrated in Fig. 2 and, although the interacting regions are depicted as globular structures, they could also be asymmetric and more extended. Histones are subjected to post-synthetic chemical modifications which have been correlated with very important functions of chromatin. For histones H2A, H2B, H3 and H4 a major reversible modification is acetylation of lysine residues, converting them to neutral acetyl lysines. These sites of acetylation have been identified (DeLange *et al.*, 1969; Candido & Dixon, 1971, 1972*a,b*; Hooper *et al.*, 1973; Santiére *et al.*, 1974; Dixon *et al.*, 1975) and are all located in the basic N-terminal regions of histones which are not directly involved in interactions between core histones (Fig. 1). Acetylation of 4 lysines in the N-terminal regions of H2B, H3 and H4 considerably reduces the basicity of these regions.

Histone H1

The functions of histone H1 are thought to be different from those of the other histones and there is much evidence to suggest that

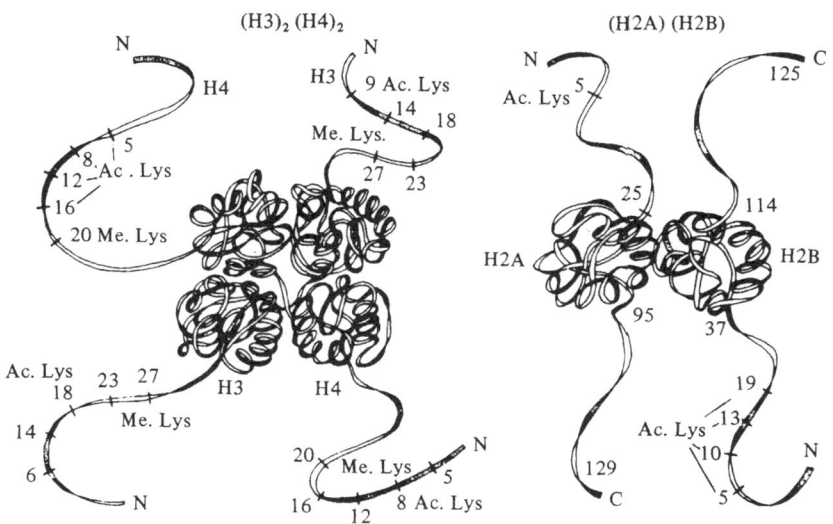

Fig. 2. Proposed models for the structures of isolated specific histone complexes (H3)$_2$(H4)$_2$ and (H2A)(H2B). (From Bradbury, 1978.)

the H1 molecule is required for higher order chromatin structures (Bradbury, Carpenter & Rattle, 1973). The sequence of histone H1 is highly asymmetric (MacLeod, Wong & Dixon, 1977) and contains three well-defined regions: (i) residues 1–40 which has a very basic region between residues 19 and 35 and no apolar residues; (ii) residues 41–121 which contains nearly all (86%) of the apolar residues in the molecule, and (iii) residues 122–216 which is very basic and has a most unusual composition containing more than 90% of lysines, prolines and alanines. Detailed NMR studies of H1 and H1 peptides (Bradbury *et al.*, 1975*b,c,d*; Chapman, Hartman & Bradbury, 1976; Hartman *et al.*, 1977; Chapman *et al.*, 1978) clearly show that the structure of H1 induced by ionic strength is located entirely in the central apolar region from residues 41 to 121, leaving the basic N- and C-terminal regions free and mobile. A model for the structured form of H1 in solution is shown in Fig. 3. It has been shown that the basic C-terminal half of H1 is strongly binding to DNA (Bradbury *et al.*, 1975*d*). Histone H1 is subject to another type of modification: phosphorylation of neutral serine and threonine residues, which converts them to negatively charged residues. These sites are indicated on the model in Fig. 3 and will be discussed later. It is to be noted that the growth-associated sites of phosphorylation are located in the basic N- and C-terminal regions of histone H1.

Structures of chromatin

The use of nucleases to cleave DNA has revolutionized our understanding of chromatin structure. The major breakthrough came from the work of Hewish & Burgoyne (1973) who showed that the addition of Ca^{2+} to rat liver nuclei activated an endogenous endonuclease which digested the chromatin. The DNA digestion products formed a series of DNA lengths, the longer members of the series being approximately integral multiples of the shortest length, shown later to be about 200 base pairs (bp) of DNA (Burgoyne, Hewish & Mobbs, 1974). On

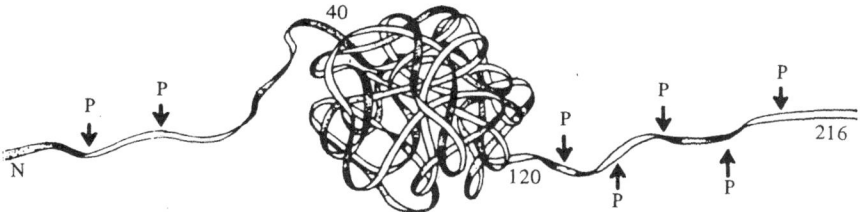

Fig. 3. Proposed model for the structure of isolated histone H1. (From Bradbury, 1978.)

the basis of this result it was proposed that chromatin consisted of a simple repeating structure, with DNA accessible to nuclease action at regularly spaced intervals. Similar results (Fig. 4) were obtained for chromatin in the nuclei of *Physarum polycephalum* (Johnson *et al.*, 1976), showing that *P. polycephalum* chromatin is organized in the same basic structure as higher eukaryotes. Regular subunits were also visualized on electron micrographs in the form of a string of beads (Olins, A. L. & Olins, D. E., 1974; Woodcock, 1973). From the histone tetramer (H3, H4)$_2$ identified by crosslinking histones and the known histone: DNA ratio, Kornberg (1974) proposed that the subunit contained 200 bp of DNA + 2(H2A + H2B + H3 + H4) + H1. This full chromatin subunit is now called the nucleosome. The full nucleosome DNA content varies from 154 bp in a lower eukaryote, *Aspergillus* (Morris, 1976) to 241 bp in sea urchin sperm (Compton *et al.*, 1976). For most somatic tissues, however, the DNA repeat is 195 ± 5 bp. Care has to be exercised in accepting that even within one tissue the nucleosome

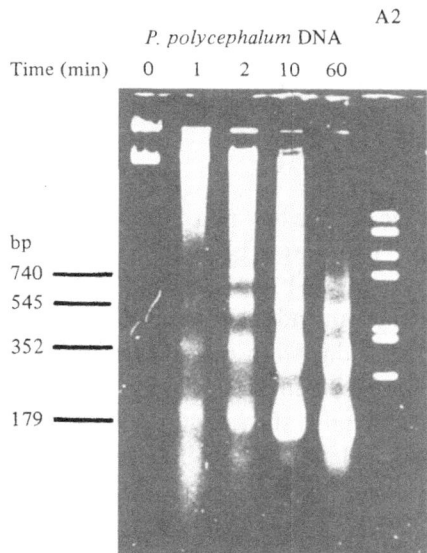

Fig. 4. Double-stranded DNA products from staphylococcal nuclease digestion of chromatin in *P. polycephalum* nuclei at different times of digestion. The DNA band at the bottom of each track comes from the mononucleosome. The higher bands are the DNA pieces from the dinucleosome, trinucleosome, etc. The numbers on the left refer to the DNA lengths of the bands in the track corresponding to 1 min of nuclease digestion. (From Johnson *et al.*, 1976.)

DNA content is constant. Results of time course studies of staphylococcal nuclease digestion of nuclei from *Physarum polycephalum* (Johnson *et al.*, 1976) showed that with increasing times of digestion the nucleosome DNA content varied from 191 to 173 bp as can be seen in Fig. 4. This variability was attributed to a spread of nucleosome repeats existing within the nucleus with the larger repeats located in open regions more accessible to nuclease attack. Similar results were found for yeast (Lohr *et al.*, 1977).

A major finding was that with increasing time of staphylococcal nuclease digestion, a well-defined subnucleosome particle could be obtained from chromatins of different origins (see van Holde & Isenberg, 1975). This was called the 'core particle' and is a highly conserved structural unit containing a nominal '140 bp' of DNA and 2(H2A, H2B, H3 and H4). The H1 histone was lost when the digestion proceeded from 160 to 140 bp. This was taken to indicate that at least part of the H1 molecule is attached to the linker DNA between two core particles (Noll & Kornberg, 1977). Recently it has been shown (Lutter, 1979) that the DNA content of the core particle is 146 ± 2 bp DNA. The core particle isolated from several tissues is so regular in structure that crystals can be grown. For this reason it has been subjected to detailed physical studies.

The use of nucleases in studying chromatin structure has led to several major advances: (i) demonstration of the existence of chromatin subunits, nucleosomes, which contain an extremely regular core particle; (ii) the use of other nucleases as biochemical probes of chromatin structure, and (iii) development of biochemical strategies for the isolation of large amounts of well-defined pieces of chromatin for physical studies.

Shape of the nucleosome and core particle

Neutron scatter and diffraction techniques have particularly powerful application to studies of two-component biological systems such as chromatin or ribosomes. This comes from the difference in the neutron scattering length of ^1H (-0.378×10^{-12} cm) compared to ^2H ($+ 0.65 \times 10^{-12}$ cm) and the neutron scattering lengths of other elements found in biological macromolecules C ($+ 0.66 \times 10^{-12}$ cm), N ($+ 0.94 \times 10^{-12}$ cm), O ($+ 0.58 \times 10^{-12}$ cm) and P ($+ 0.51 \times 10^{-12}$ cm). Because of the anomalous negative neutron scattering length of ^1H, the average neutron scattering length of a biological macromolecule depends on the proportion of hydrogen it contains. Proteins contain a much higher proportion of hydrogen ^1H than DNA and have a neutron scattering length density of $+ 0.18 \times 10^{-12}$ cm compared to $+ 0.37 \times 10^{-12}$ cm for DNA. This very large difference can be exploited by studying solutions of core particles or nucleosomes in mixtures of normal water, ^1H_2O, and heavy water, ^2H_2O; the neutron scattering length density of ^1H_2O is $- 0.66 \times 10^{-12}$ cm, while for ^2H_2O it is $+ 0.63 \times 20^{-12}$ cm. Thus any scattering length density between these two extremes can be obtained by taking suitable mixtures of ^1H_2O and ^2H_2O. Histones have the same average scatter as the water mixture 37% ^2H_2O/63% ^1H_2O and the neutron scatter from chromatin particles in this water mixture will be dominated by the DNA component; in 65% ^2H_2O/35% ^1H_2O the DNA component will be matched and the histones will dominate the neutron scatter. This approach was used to analyse the neutron diffraction of fibres of chromatin which led to the proposal of a model for the nucleosome with DNA coiled around a histone core (Baldwin *et al.*, 1975). From the neutron scatter curves of nucleosomes and core particles in ^1H_2O/^2H_2O solutions the radius of gyration of the DNA component was found to be 4.7 nm while that of the histone component was about 3.3 nm (Bradbury *et al.*, 1975*a*; Pardon *et al.*, 1975). These results clearly showed that in the chromatin particles the DNA was external to a histone core. This histone core has been equated to a complex of the apolar regions of the histones H2A, H2B, H3 and H4 found in NMR studies of the histone complexes (Baldwin *et al.*, 1975).

Information on the shape of the nucleosome was obtained from a full analysis of all the neutron scatter curves obtained for a range of ^1H_2O/^2H_2O mixtures (Hjelm *et al.*, 1977; Suau *et al.*, 1977). Stuhrmann (1975) has developed an analysis whereby it is possible to obtain a scatter function, *Ic*, for the pure shape of the particle and a scatter function, *Is*, resulting from density fluctuations within the particle, i.e. from its internal structure. For chicken erythrocyte nucleosomes the best fit to the shape function was found to be an oblate spheroid of axial ratio 0.5 (Hjelm *et al.*, 1977). More detailed models were obtained for the regular core particle in solution by the analysis outlined above (Suau *et al.*, 1977), and also by fitting models to the core particle scatter curve in ^2H_2O (Richards *et al.*, 1977); both approaches gave models with essentially the same shape, a disc, of dimensions $11 \times 11 \times (5.5–6.0)$ nm. The radius of gyration of the DNA component of 4.7 nm demands that the DNA is wrapped around the edge of the disc and at this radius 140 bp DNA would require 1.7 ± 0.2 turns of DNA. The pitch of the DNA coil that gave the best fit to the shape and internal structure–functions was 3.0 nm (Suau *et al.*, 1977). The first suggestion of a flat disc for the nucleosome came from electron microscopy studies of chromatin by Langmore & Wooley (1977) who proposed a disc 13.5 × 13.5 × 5.7 nm with one and a half to two turns of DNA wrapped around the periphery.

X-ray diffraction and electron microscopy of crystals of core particle

Crystals of core particles have been studied by electron microscopy and low resolution X-ray diffraction (Finch *et al.*, 1977). The model proposed from this study consisted of a disc 11 × 11 × 5.5 nm with one and three-quarter turns of DNA of pitch 2.7 nm on the periphery. Thus the structure of the core particle in solution determined by neutron scatter techniques and the low resolution X-ray and electron microscopy study of small crystals of core particles converge on very similar models, showing that there are no major differences in shape between the solution and crystal structures.

The current low-resolution core particle model from neutron and X-ray studies is given in Fig. 5; the detailed molecular structure of the core particle will eventually come from X-ray and neutron diffraction of single crystals. The nuclease digestion of chromatin in the nuclei of *Physarum polycephalum* (Fig. 1) and the properties of histones H3 and

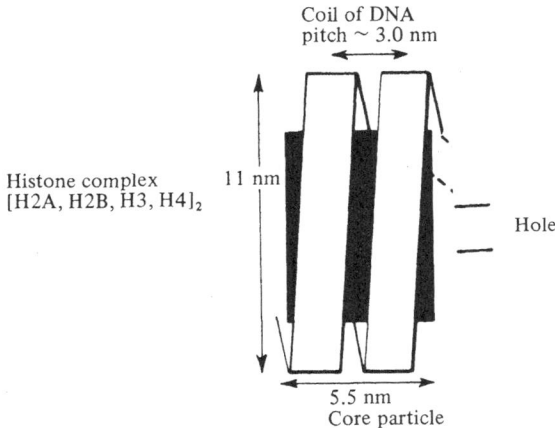

Fig. 5. Model for the structure of the core particle from neutron scatter studies. (G. W. Braddock, J. P. Baldwin, E. M. Bradbury, unpublished.)

H4 both strongly suggest that the *P. polycephalum* core particles will be very similar to the model given in Fig. 5. A direct test of this is being made.

The structural roles of histones

Initial studies have been made of the roles of the different histones in chromatin structure. It was shown some years ago that the specific removal of histone H1 did not affect the structural features of chromatin as revealed by X-ray diffraction. More recently it has been shown by X-ray fibre diffraction, and nuclease digestion (Boseley *et al.*, 1976; Moss *et al.*, 1977) that most of the structural features of chromatin and of the core particle can be generated by complexing DNA with only histones H3 and H4 but not with histones H2A and H2B, although the latter histones are probably required to produce a complete nucleosome. Histones H3 and H4 have the ability to organize a nominal '140 bp' of DNA into a compact, core particle-like structure which probably contains an octamer $(H3, H4)_4$ (Camerini-Otero, Sollner-Webb & Felsenfeld, 1976; Sollner-Webb, Camerini-Otero & Felsenfeld, 1976; Moss *et al.*, 1977). Histones H3 and H4 are clearly major structural proteins of the core particle, and these observations raise the distinct possibility that histones H2A and H2B have additional functions outside the core particle.

Interactions of histones in the core particle

If the octamer of histones $(H2A, H2B, H3, H4)_2$ was fully bound to DNA in the core particle, then this complex of mol. wt. 196 000 would be expected to give a much broadened NMR spectrum. Sharp resonances, however, are observed (Cary, Moss & Bradbury, 1979) which account for 17–20% of the total NMR signal from the dissociated random coil histones. In the core particle, therefore, some regions of the histones are not tightly bound at low ionic strength. An estimate can be made of the amino acid composition of these mobile regions from the sharp NMR spectrum and this corresponds mainly to the composition of the basic N-terminal regions of histones H2A and H2B. Increase in ionic strength to 0.6 M NaCl appears to release the N-terminal regions of H3 and H4, although at this ionic strength the core particle is far from the fully unfolded state and this could be relevant to the role of histone acetylation. These studies and trypsin digestion studies (Whitlock & Simpson, 1977) of histones in the core particle show that the N-terminal regions are free or weakly bound, which strongly suggests that the apolar central and C-terminal regions of H2A, H2B, H3 and H4 are major sites of interaction with the DNA in the core particle.

These observations are consistent with the properties of histones described earlier. Histones H3 and H4, which are rigidly conserved, are essential for the core particle structure. These histones together with the conserved apolar central regions of histones H2A and H2B complex with constant features of the DNA molecule, probably the phosphate-ester chain, to give the core particle structure which is itself highly conserved. The variable N-terminal regions of H2A and H2B appear to have functions outside the core particle which could involve the linker DNA and have a role in higher order chromatin structures.

Higher order chromatin structures

Many electron microscopy studies of chromatin structure have described filaments with diameters of about 10 and 30 nm as well as higher order structures. The 10 nm filament, the nucleofilament, consists of a linear array of nucleosomes and is observed in solutions of low ionic strength. With increase in ionic strength, either monovalent or divalent cations, the nucleofilament undergoes a transition to the 30 nm filament. The widespread occurrence of a 30 nm filament has been reported in interphase and metaphase nuclei.

Arrangement of nucleosomes in the nucleofilament

The finding that the chromatin core particle is a disc 11 × 11 × 5.5 nm has raised questions as to how the discs are arranged with respect to each other in the nucleofilament. Neutron scatter studies (Suau, Bradbury & Baldwin, 1979) of the nucleofilament in solution at ionic strength below 20 mM NaCl give a mass per unit length equivalent to 1 nucleosome per 10–11 nm. This corresponds to a DNA packing ratio of about 7:1. From the neutron scatter curves of the nucleofilament in mixtures of light and heavy water the transverse radius of gyration of the DNA component was found to be 3.4 nm, while that of the protein was 2.1 nm. These values are appreciably lower than would be expected if the core particles were arranged face-to-face along the nucleofilament. The core particle arrangement that agrees best with the neutron scatter parameters has the discs close to edge-to-edge as shown in Fig. 6(a).

30 nm chromatin filament

Various proposals have been made concerning the structures of the 30 nm filament. From neutron diffraction of chromatin fibres (Carpenter *et al.*, 1976) and from an electron microscope study of chromatin (Finch & Klug, 1976), it has been proposed that the 30 nm filament consists of a supercoil or solenoid of the nucleofilament of pitch 10–11 nm and containing 6–7 nucleosomes per turn. Both studies attributed the well-known semi-meridional arc at 10–11 nm in the X-ray and neutron diffraction patterns of fibre of chromatin to the pitch of the coiled nucleofilament. The same 10–11 nm arc is found in the neutron fibre diffraction patterns of H1-depleted chromatin (Baldwin *et al.*, 1975; Carpenter *et al.*, 1976), which strongly suggests that the 30 nm filament can be formed at high concentrations by interactions of histones H2A, H2B, H3 and H4. It is probable that H1 stabilizes the 30 nm filament and is further involved in even higher order structures.

The ionic strength transition induced from the nucleofilament to the 30 nm filament has been studied in solution by neutron scatter (Suau *et al.*, 1979). It was found that above 20 mM NaCl the nucleofilament made an abrupt transition to a higher order structure with a radius of gyration of about 9.2 nm and a packing of 2 nucleosomes per 10 nm. With increasing ionic strength the radius of gyration slightly increased to

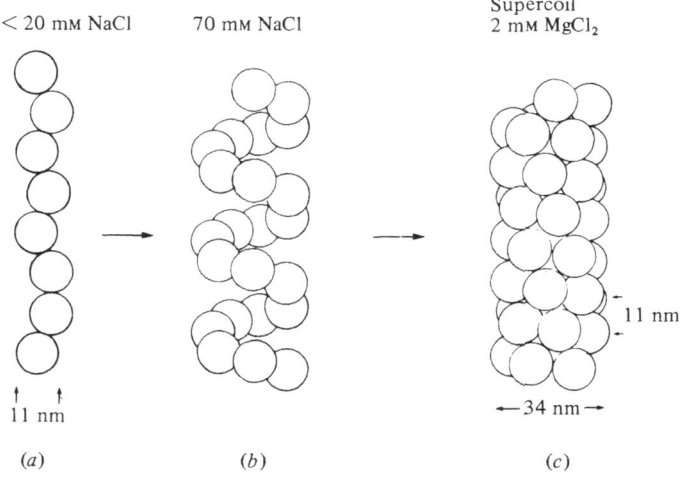

(a)   *(b)*   *(c)*

Fig. 6 (*a*) Model for the extended structure of chromatin at low ionic strength with the nucleosome discs of Fig. 5 arranged edge-to-edge; (*b*) salt-induced transition of chromatin; (*c*) the compact 34 nm diameter supercoil of nucleosomes. The orientations of the nucleosome discs are not known in stages (*b*) and (*c*).

9.5 nm while the packing ratio increased to 6 nucleosomes per 10 nm. This behaviour suggests a family of coils of variable pitch responding to ionic conditions, which in the most compact form corresponds to the supercoil or solenoid structure (Fig. 6*b,c*). The diameter of the hydrated state of the most compact structure was found to be 34 nm which, as might be expected, is larger than the values of 25–30 nm for the dehydrated state obtained by electron microscopy. Questions concerning the 34 nm supercoil which remain unanswered are: (i) the histones and interactions which generate the supercoil, (ii) the arrangement of disc-shaped core particles, and (iii) whether or not there is a hole along the axis of the supercoil.

Control of chromosome structure

Although the structure of chromatin is strongly dependent on ionic strength, as described above, other mechanisms are also thought to play a major role in the control of chromosome structure through the cell cycle; in particular, the post-synthetic reversible chemical modifications of histones. As mentioned earlier there are two major types of modifications: (i) acetylations of lysines in the basic N-terminal regions of the core histones H2A, H2B, H3 and H4, and (ii) phosphorylations of serines and threonines in the basic N- and C-terminal regions of histone H1 and a lower level of phosphorylation of histones H2A and H3 (Gurley, Walters & Tobey, 1975). Both types of modification reduce the overall positive charge of the well-defined basic domains of histones. Acetylation converts lysines from positively charged residues to neutral acetyl lysines and for each acetylation the net positive charge is reduced by one unit. Phosphorylation converts neutral serines and threonines to negatively charged serine and threonine phosphates, and reduces the net positive charge of the basic regions by the introduction of negative charges. These modifications are thought to modulate the interactions of histones with DNA in chromosomes in response to the functional requirements of cells.

Biological functions associated with histone acetylation

Histone acetylation has been correlated with aspects of DNA processing such as transcription, replication and the replacement of histones by protamines during spermiogenesis.

Transcription

Allfrey and coworkers (1964) first suggested that gene activation may require histone acetylation. Since then there have been many reports of correlations between histone acetylation and the onset of transcriptional activity (Allfrey, 1977).

Chromatin replication

Histone acetylation turns over rapidly during S phase of the cell cycle (Jackson *et al.*, 1975). This has been correlated with the disassembly of chromatin prior to DNA replication (Sung & Dixon, 1970; Louie, Candido & Dixon, 1973) with, it is now thought, newly synthesized histones associated with one of the daughter DNA molecules and 'old' histones associated with the other DNA molecule (Leffak, Grainger & Weintraub, 1977). Although histone acetylation has been correlated with DNA replication, it should be pointed out that S-phase acetylation is very difficult to analyse because of the different types of acetylation occurring simultaneously; non-reversible acetylation of the N-terminal serines of histones H1, H2A and H4, and rapid acetylation and deacetylation of old and new histones (Jackson *et al.*, 1976; Reeves & Jones, 1976).

Spermiogenesis

Dixon and coworkers (Candido & Dixon, 1971, 1972*a,b*) have studied histone acetylation during spermiogenesis in trout and showed that four core histones are acetylated in the absence of histone synthesis. Similar results were obtained during spermiogenesis in rat (Grimes, Chae & Irvin, 1975; Marushige, Y. & Marushige, K., 1975). These studies support the idea that histone acetylation is required in the disassembly of chromatin to allow histones to be replaced by protamines.

Histone acetylation through the cell cycle

For the reasons outlined earlier, *Physarum polycephalum* is a very useful organism for precise studies of events that occur during the cell cycle. Recently we (Chahal, Matthews & Bradbury, 1980*a*) have studied in detail the acetylation of histone H4 through the nuclear division cycle of individual plasmodia of *P. polycephalum*. Histone H4 was isolated from plasmodia at different times of the cycle and analysed by electrophoresis on long polyacrylamide gels in acetic acid–urea (Panyim & Chalkley, 1969). The H4 was first purified by chromatography on Biogel P10. Fig. 7 shows scans of stained gels containing H4 from defined stages of the cell cycle. In all cases, five bands were present but the proportion of each band varied. The proportions at a specific stage were reproducible. These and other gel scans have been analysed in terms of five Gaussian peaks and the area of each peak calculated. The areas were thus fully corrected for the overlap between bands. The area corresponding to each species of H4 is plotted in Fig. 7.8 as a function of stage of the cell cycle.

It is clear that at all stages the most abundant species is mono-acetylated H4 (Ac$_1$H4). During G2 phase, the proportion of Ac$_1$H4 rises to about 62% and then falls to about 55%. The fall occurs during the middle part of G2 phase, from 5 h to about 9 h after mitosis in the 11 h cycle. During the period 6 h to about 9 h after mitosis, there are complementary increases in proportion of Ac$_2$H4 (16% to 23%), Ac$_3$H4 (2.5% to 4.3%), and Ac$_4$H4 (1% to 3%). The data imply an overall

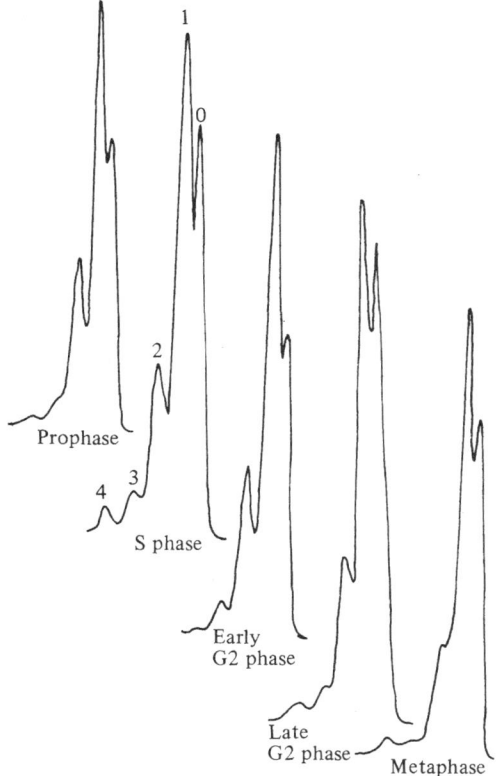

Fig. 7. Polyacrylamide gel electrophoresis of *P. polycephalum* histone H4 at different stages of the nuclear division cycle. Each densitometer trace shows 5 bands corresponding to 4, 3, 2, 1 and 0 acetyl lysines per molecule, as indicated on the S-phase profile. (From Chahal, Matthews & Bradbury, 1980.)

increase in acetylation during mid G2 phase. Late G2 phase is characterized by a sharp drop in Ac_4H4 (3% to 1%) followed in early prophase by a drop in Ac_3H4 (4% to 2%). Ac_2H4 drops sharply through late G2 phase and prophase (23% to 11%). In early prophase the overall proportion of highly acetylated H4 (2 to 4 acetates per molecule) is at its minimum value (13.5%). During late mitosis and early S phase the increase in proportion of highly acetylated H4 is balanced by drops in both non- and mono-acetylated H4 so that in mid S phase the proportion of highly acetylated H4 reaches its maximum (33.6%). In late S phase the proportion of highly acetylated H4 falls to a plateau level (about 26%) before the minimum (19.4%) in early G2 phase. The most dramatic changes in an individual H4 species occur in Ac_4H4 and these are shown on an expanded scale in Fig. 8. There is a clear minimum

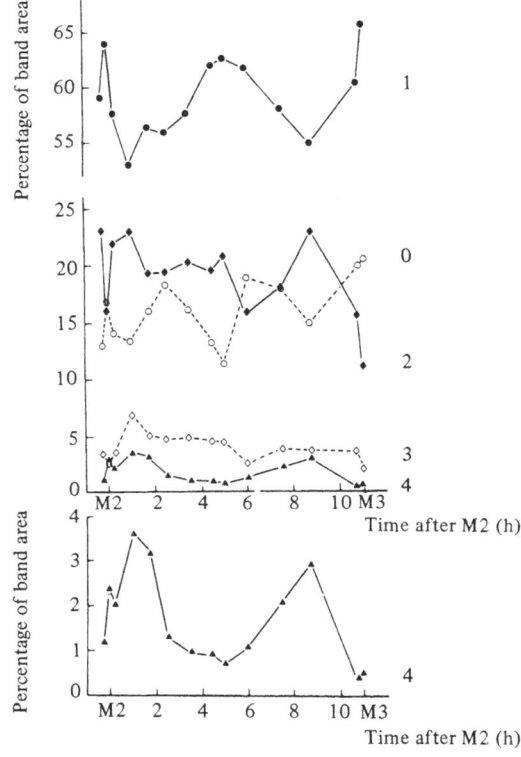

Fig. 8. Changes in the acetylated species [0, 1, 2, 3 and 4] of histone H4 through the nuclear division cycle of *P. polycephalum*. The lower panel shows the variation in the amounts of Ac_4H4, on an expanded scale.

(0.4%) in early prophase, a sharp maximum (3.6%) in mid S phase and a second maximum (up to 3%) in mid G2 phase.

Comparison with acetylation of histone H4 in other systems

Histone H4 is extensively modified in the cytoplasm before entry into the nucleus in duck erythroid cells (Ruiz-Carrillo, Waugh & Allfrey, 1975) and in hepatoma cell culture (Jackson *et al.*, 1976). The N-terminal acetyl serine of H4 is phosphorylated and one lysine residue is acetylated in duck erythroid cells. The phosphate and acetate groups are removed soon after the histone enters the nucleus to leave unmodified H4. (The N-terminal acetyl serine is not reversible.) Further modifications occur thereafter, within the nucleus, where up to 4 acetate groups can be added or removed and appear to turn-over rapidly. Combinations of acetate and phosphate were observed, but since the amounts were very small it was not clear whether phosphorylation occurred in the nucleus as well as the cytoplasm (Ruiz-Carrillo *et al.*, 1975). In hepatoma tissue culture (HTC) cells a similar pattern was observed except that very little phosphorylation was observed and newly synthesized H4 was rapidly converted to a di-acetyl form, presumably in the cytoplasm, and then returned to the unmodified form in the nucleus where further turn-over occurred (Jackson *et al.*, 1975, 1976). In *Physarum polycephalum* the lack of ^{32}P incorporation (Bradbury *et al.*, 1973) argues against the presence of phosphorylated H4 in the nucleus but H4 could still be phosphorylated in the cytoplasm if the modification were rapidly reversed when H4 enters the nucleus. However, Fig. 8 shows a peak of Ac_2H4 in S phase nuclei which is consistent with the notion that newly synthesized H4 enters the nucleus in the di-acetyl form and is then converted to the unmodified form, which shows an increase in late S phase (M2 + 1 h to M2 + 2.5 h) following the increase in Ac_2H4 (M2 to M2 + 1 h). This suggestion is analogous to the situation in HTC cells and preliminary pulse label experiments confirm this interpretation in *P. polycephalum* (S. Chahal, E. M. Bradbury & H. R. Matthews, unpublished).

The overall level of acetylation is determined, most of the time, by the rates of acetylation and deacetylation occurring in the nucleus. The overall levels are substantially different in different systems. For example, dividing avian erythroblasts have most H4 unmodified and very little with more than one acetate (Ruiz-Carrillo *et al.*, 1975); exponentially growing HeLa cells have about equal amounts of unmodified and Ac_1H4; and *Physarum polycephalum* has, on average, about

58% of Ac_1H4 and approximately equal amounts of unmodified and Ac_2H4. In all these systems the amounts of Ac_3H4 and Ac_4H4 are small (about 4% in *P. polycephalum*) and very small (about 2% in *P. polycephalum*), respectively. Increase in acetylation is correlated, in various cell types, with transcription of greater regions of the genome (Allfrey *et al.*, 1964; Allfrey, 1977). The above differences in acetate levels on H4 may be correlated with differences in the amount of the genome being expressed. However, *P. polycephalum* nuclei have a substantial DNA content, 1.2 pg per plasmodial nucleus (Mohberg *et al.*, 1973), and several different cell types, so it is most unlikely that the amount of Ac_1H4 is a direct measure of the proportion of the genome that is active. That would require about 58% of the genome to be active during growth of the plasmodium. It is more likely that transcription is associated with highly acetylated H4, maybe even Ac_4H4, that is turning over rapidly. Then the increases in Ac_1H4 and Ac_2H4 between erythroblasts and *P. polycephalum* reflect the overall increase in net acetylation that might accompany activation of a higher proportion of the genome.

The use of deoxyribonuclease-1 (DNase-1) has also provided evidence linking H4 acetylation with transcription. Firstly, DNase-1 digests chromatin in such a way that transcriptionally active chromatin is destroyed at early digestion times, before the bulk of the chromatin (Weintraub & Groudine, 1976; Garel & Axel, 1976). DNase-1 digestion releases histones and non-histone proteins (Levy, Wong & Dixon, 1977; Vidali, Boffa & Allfrey, 1977) and at early digestion times the histone H4 released is more acetylated than the bulk H4 as if acetylated H4 were associated with transcriptionally active chromatin (Nelson, Perry & Chalkley, 1978). Transcription in *Physarum polycephalum* during the cell cycle is probably biphasic (Mittermayer, Braun & Rusch, 1964; Grant, 1972; Davies & Walker, 1977) with a very low level in mitosis, a high level in S phase, a low in early G2 phase and the second high in mid to late G2 phase. Several lines of evidence suggest that the S phase peak is predominantly mRNA synthesis and the G2 phase peak is predominantly rRNA synthesis although the distinction, if correct, is not absolute. The biphasic pattern correlates closely with the pattern of Ac_4H4 given in Fig. 8. This supports the correlation between acetylated histones and transcription discussed above and suggests the following, more stringent, test of the correlation. About 2% of *P. polycephalum* DNA is genes for rRNA and associated spacer sequences (rDNA) (Molgaard, Matthews & Bradbury, 1976; Vogt & Braun, 1976). If the

G2 phase transcription is predominantly rRNA and associated with Ac_4H4, then the G2 phase increase in Ac_4H4 from 1% to 3% implies that a large proportion of the H4 associated with rDNA would become tetra-acetylated in G2 phase. This prediction has been tested directly and found to hold (S. Chahal, H. R. Matthews & E. M. Bradbury, unpublished observations). It is quite clear from these studies involving *P. polycephalum* that there is a strict correlation between histone acetylation and the requirements of transcription.

Properties of 'active' chromatin

Weintraub & Groudine (1976) made the important observation that the DNA coding for globin genes in chromatin from avian erythrocytes was preferentially digested by DNase-1 but this was not found for cells such as fibroblasts in which the globin genes are inactive. A similar DNase-1 sensitivity was found for the ovalbumin genes in oviduct (Garel & Axel, 1976). These findings have been used to identify proteins associated with active chromatin (Levy *et al.*, 1977; Vidali *et al.*, 1977, 1978), and it has been found that limited DNase-1 digestion of duck erythrocyte chromatin results in the selective release of the more acetylated histones and also a specific group of non-histone proteins, the high mobility group (HMG) of proteins identified and characterized by Johns and collaborators (Goodwin & Johns, 1973; Goodwin, Nicolas & Johns, 1975; Goodwin, Walker & Johns, 1978; Walker, Goodwin & Johns, 1979). Histone acetylation appears, therefore, to play an important role in the structural transition of inactive chromatin to active chromatin, and the HMG proteins, particularly HMG 14 and 17, are also implicated in the structure of active chromatin.

Role of histone acetylation

Although it was originally thought that histone acetylation caused the DNA around the nucleosome to unfold to a more linear form to facilitate transcription, this simple view does not accord with the properties and interactions of histones described earlier. The sites of reversible acetylation are located entirely in the basic N-terminal regions of the histones H2A, H2B, H3 and H4, and it has been shown that these regions can be released from the chromatin core particle without unfolding the particle. It seems unlikely, therefore, that if the N-terminal regions of the core histones are released by acetylation this would lead to the unfolding of the nucleosome. Although not proven it is now thought that acetylation of the N-terminal regions of the core histones results in the destabilization of the 34 nm supercoil of nucleosomes and induces its transition to the extended 10 nm linear array of nucleosomes (Vidali *et al.*, 1978; Chahal *et al.*, 1980).

Phosphorylation of histone H1 and control of chromosome condensation

The properties of isolated histone H1 have been discussed earlier and its structure is shown in Fig. 3. In relation to the three structural domains of H1, the known sites of phosphorylation as found by Langan (1978) are given in Fig. 9. There are three types of phosphorylation: (i) a cyclic AMP-dependent phosphorylation found at serine 37 (Site A); (ii) an in-vitro-only phosphorylation of serine 106 (Site B), and (iii) growth-associated phosphorylations of serine 180 and threonines 16, 136 and 153 located entirely in the basic N- and C-terminal regions of H1. There are additional sites of growth-associated phosphorylations that have not yet been identified. These growth-associated phosphorylations are reversible and have been studied through the cell cycle in single plasmodia of *Physarum polycephalum* and in synchronized cultured cell lines.

The correlation of H1 phosphorylation with cell growth was initially demonstrated by Chalkley's group (Balhorn *et al.*, 1972*a*) who measured the H1 phosphate content in a number of cell types covering a wide range of growth rates and observed a linear correlation between growth rate and phosphate content. Phosphorylation of H1 has been observed to occur during S phase (Cross & Ord, 1970; Balhorn *et al.*, 1972*b*) and this was shown in *Physarum polycephalum* (Bradbury, Inglis & Matthews, 1974*a*) to be in part the phosphorylation of newly synthesized H1 molecules, bringing their phosphate content up to the level of old H1 molecules. This S phase phosphorylation may be

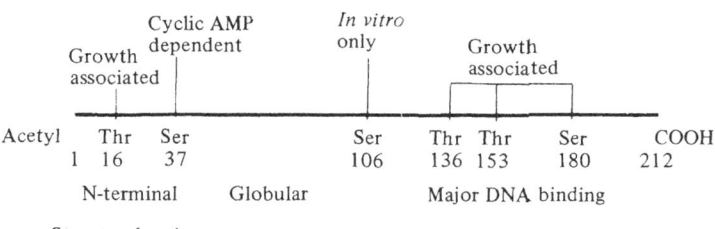

Fig. 9. The major sites of phosphorylation of histone H1 in relation to the 3 structural domains of H1 as shown in Fig. 3.

important in chromatin replication. S phase phosphorylation is, however, much lower than the phosphorylation observed during G2 phase.

Much of the early difficulties in identifying the timing of the major growth-associated phosphorylation of H1 arose from the difficulty of synchronizing mammalian cell cultures, particularly in G2 phase. This led us to the naturally occurring synchronous growth cycle of *Physarum*

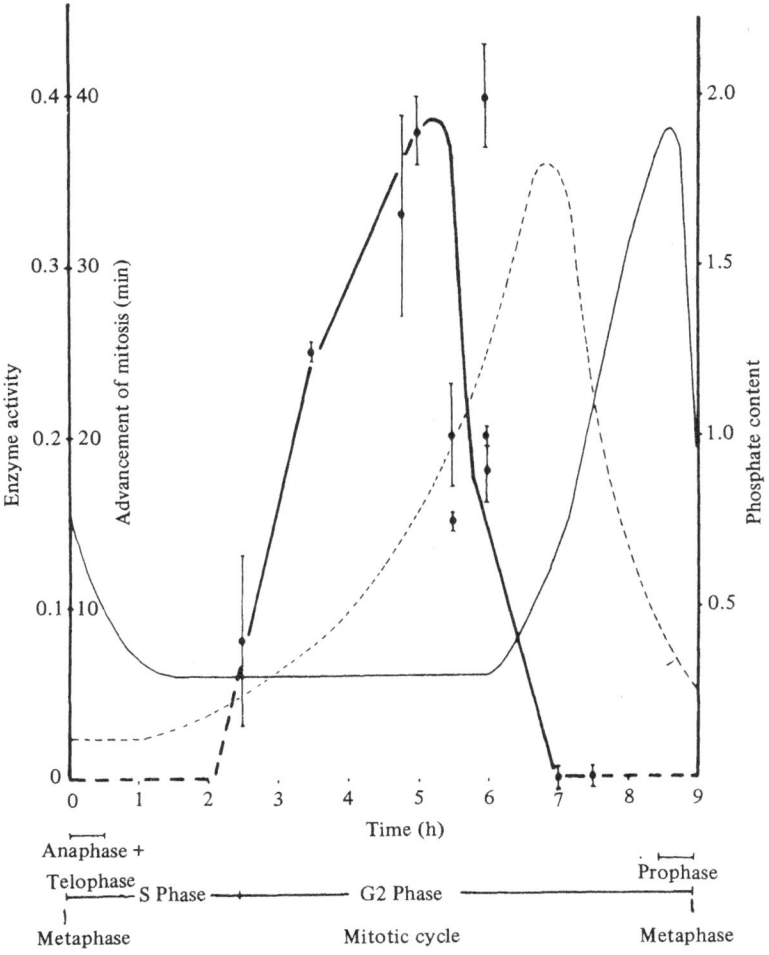

Fig. 10. Nuclear division cycle variations in *P. polycephalum* of (——) the phosphate content of histone H1; (– – –) the levels of activity of *P. polycephalum* histone kinase and (——) the advancement of mitosis induced by the addition of heterologous rat ascites kinase activity. (From Bradbury *et al.*, 1974*b*.)

polycephalum. Measurement of H1 phosphate content through the *P. polycephalum* cell cycle showed that the major increase in phosphate content occurred in late G2 and prophase (Fig. 10). This increase was correlated with the onset of chromosome condensation (Bradbury *et al.*, 1973).

Further evidence for the central role of H1 phosphorylation in mitosis comes from studies of the histone phosphokinase activities. Growth-associated histone kinase is cyclic AMP-independent (Lake, 1973; Ajiro, Borun & Cohen, 1975; Hardie, Matthews & Bradbury, 1976; Langan, 1978). Studies on the function of the growth associated kinase *in vivo* have been carried out in *Physarum polycephalum* (Bradbury *et al.*, 1974a). Kinase activity was measured using sonicated nuclei, excess calf thymus H1 substrate and [γ-^{32}P]ATP. In such crude extracts, phosphatase will also be active, so the following two reactions occur simultaneously:

$$\text{H1} + \text{ATP} \xrightarrow{\text{kinase}} \text{H1}\cdot\text{P} + \text{ADP}, \tag{1}$$

$$\text{H1}\cdot\text{P} \xrightarrow{\text{phosphatase}} \text{H1} + \text{P}_i. \tag{2}$$

The net transfer of ^{32}P from ATP to H1 will depend on the activities of the kinase and phosphatase. However, at short incubation times the H1 and ATP concentrations are high and the H1·P concentration is low so that reaction (1) dominates and it is mainly kinase activity that is measured. This overall nuclear phosphorylating activity showed a very marked change during the cell cycle (Fig. 10), increasing 15-fold during G2 phase to reach a peak in late G2 phase preceding the peak of H1 phosphate content. The activity then dropped throughout prophase to reach the base level in S phase. The kinetics of the increase were shown later to be due to three enzymes (Hardie, Matthews & Bradbury, 1976), but their functions and interrelationships remain to be elucidated. The behaviour of the nuclear histone kinase activity strongly implies that H1 phosphorylation is part of a programmed series of events leading to nuclear division as follows:

    (i) increase of histone kinase activity,

    (ii) increase of H1 phosphate content,

    (iii) initiation of chromosome condensation,

    (iv) full chromosome condensation and metaphase.

Evidence in support of this series of events comes from several sources. Firstly, we have shown (Fig. 10) that mitosis in *Physarum polycephalum* can be advanced by up to 1 h if heterologous histone kinase activity is added to *P. polycephalum* at the time in the cell cycle when the endogenous kinase activity is increasing (Bradbury *et al.*, 1974*b*). Because pure histone kinase from *P. polycephalum* is not available in sufficient quantities partly purified extracts of Ehrlich ascites cells were used. The advancement of mitosis was proportional to the amount of extract added, and extract inactivated by freezing and thawing or by dialysis against low salt was without effect, as were a number of other control substances (Inglis *et al.*, 1976). In spite of these controls the interpretation of the data is limited by the facts that impure kinase was used and that there is no direct demonstration that the added kinase entered the nucleus. It has, however, been very recently shown that if histone H3, specifically labelled at its cysteine with the fluorescent probe, pyrene, is added to the external growth medium of *P. polycephalum*, it is incorporated into its chromatin (Prior *et al.*, 1980). This demonstrates that proteins added externally to *P. polycephalum* plasmodia can be taken up by the organism and utilized. Strong evidence for the importance of histone H1 phosphorylation in chromosome condensation has been provided by the finding of a mutant mammalian cell line that is deficient in growth-associated histone H1 phosphorylation and is consequently blocked in late G2 phase (Matsumoto *et al.*, 1980).

Histone modifications and control of chromosome structure

Two types of reversible post-synthetic modifications of histones have been correlated with major events in chromosome function: (i) acetylation of the core histones H2A, H2B, H3 and H4 with the requirements of DNA processing, and (ii) phosphorylation of histone H1 with chromosome condensation. From our present understanding of the properties and interactions of histones and chromatin structure as discussed earlier, we have proposed (Chahal *et al.*, 1980*a*) a scheme of how these modifications may be involved in controlling chromosome structure. This is shown in Fig. 11. In this scheme the 34 nm diameter supercoil of nucleosomes is regarded as the 'resting' structure of inactive chromatin. Structural transitions below this supercoil to 'active' chromatin involve histone acetylation, possible loss of histone H1 and an involvement of HMG proteins. HMG proteins may bind to sites in chromatin made available by the dissociation of the basic N-terminal regions of the core histones on acetylation and the loss of H1. Chromosome condensation involves higher order organization of the 34 nm supercoil and this is controlled by H1 phosphorylation. How H1

Chromosome condensation   Inactive chromatin   DNA Processing

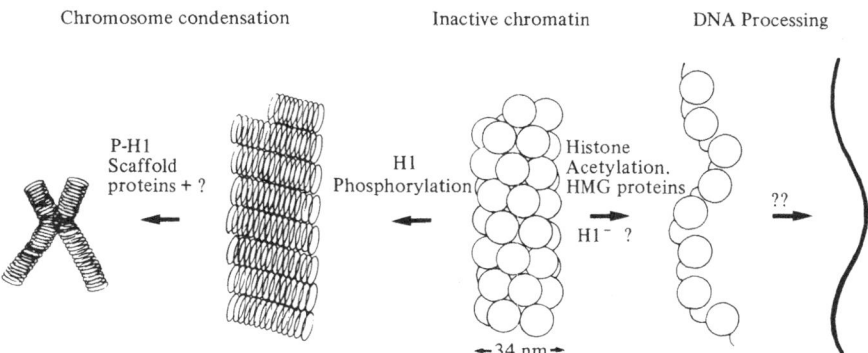

Fig. 11. Overall scheme relating the structure of chromatin to histone modifications and chromatin function. 'DNA processing' includes both replication and transcription. Although both processes involve opening out the chromatin structure, as shown by increased accessibility to nuclease, the structural changes are probably different. For example, turnover of Ac_2H4 seems to be associated with replication, while a high content of Ac_4H4 is associated with transcription. (From Chahal, Matthews & Bradbury, 1980.)

phosphorylation acts is not understood at present though strategies are now available to investigate this question. Also the sensitivity of chromatin structure to the presence of cations, particularly divalent cations, implicates them in a structural role, though whether this is a specific role or they simply provide the physico-chemical milieu of chromosomes remains to be shown.

References

Ajiro, K., Borun, T. & Cohen, L. (1975). Phosphorylation sites of histone 1 (F1) in relation to the cell cycle. *Federation Proceedings,* **34,** 581.

Allfrey, V. G. (1977). Post-synthetic modifications of histone structure. In *Chromatin and Chromosome Structure,* ed. H. J. Li & R. A. Eckhardt, p. 167. New York & London: Academic Press.

Allfrey, V. G., Faulkener, R. M. & Mirsky, A. E. (1964). Acetylation and methylation of histones and their possible role in the regulation of RNA synthesis. *Proceedings of the National Academy of Sciences, USA,* **51,** 786–94.

Baldwin, J. P., Boseley, P. G., Bradbury, E. M. & Ibel, K. (1975). The subunit structure of eukaryotic chromosome. *Nature, London,* **253,** 245–9.

Balhorn, R., Balhorn, M., Morris, H. P. & Chalkley, R. (1972). Comparative high-resolution electrophoresis of tumor histones: variation in phosphorylation as a function of cell replication rate. *Cancer Research,* **32,** 1775–84.

Balhorn, R., Bardwell, J., Sellers, L., Granner, D. & Chalkley, R. (1972). Histone phosphorylation and DNA synthesis are linked in synchronous cultures of HTC cells. *Biochemical and Biophysical Research Communications,* **46,** 1326–33.

Bohm, L., Hayashi, H., Cary, P. D., Moss, T., Crane-Robinson, C. & Bradbury, E. M. (1977). Sites of histone/histone interactions in the H3:H4 complex. *European Journal of Biochemistry*, **77**, 487–93.

Boseley, P. G., Bradbury, E. M., Butler-Brown, G., Carpenter, B. G. & Stephens, R. M. (1976). Physical studies of chromatin. The recombination of histones with DNA. *European Journal of Biochemistry*, **62**, 21–31.

Bradbury, E. M. (1978). La Chromatine. *La Recherche*, **9**, 644–53.

Bradbury, E. M., Baldwin, L. P., Carpenter, G. G., Hjelm, R. P., Hancock, R. & Ibel, K. (1975a). Neutron scattering studies of chromatin. In *Brookhaven Symposia of Biology 27*, vol. 4, ed. B. P. Schoenbora, pp. 97–117. Upton, New York: Brookhaven National Laboratory.

Bradbury, E. M., Carpenter, B. G. & Rattle, H. W. E. (1973). Magnetic resonance studies of deoxyribonucleoprotein chromatin H1 condensation. *Nature, London*, **241**, 123–5.

Bradbury, E. M., Cary, P. D., Chapman, G. E., Crane-Robinson, C., Danby, S. E. & Rattle, H. W. E. (1975b). Studies on the role and mode of operation of the very-lysine-rich histone H1 (F1) in eukaryote chromatin. The conformation of histone H1. *European Journal of Biochemistry*, **52**, 605–13.

Bradbury, E. M., Chapman, G. E., Danby, S. E., Hartman, P. G. & Riches, P. L. (1975c). Studies on the role and mode of operation of the very-lysine-rich histone H1 (F1) in eukaryote chromatin. The properties of the N-terminal and C-terminal halves of histone H1. *European Journal of Biochemistry*, **57**, 521–8.

Bradbury, E. M., Danby, S. E., Rattle, H. W. E. & Giancotti, V. (1975d). Studies on the role and mode of operation of the very-lysine-rich histone H1 (F1) in eukaryote chromatin. Histone H1 in chromatin and in H1 DNA complexes. *European Journal of Biochemistry*, **57**, 97–105.

Bradbury, E. M., Inglis, R. J. & Matthews, H. R. (1974a). Control of cell division by very lysine-rich histone phosphorylation. *Nature, London*, **247**, 257–61.

Bradbury, E. M., Inglis, R. J., Matthews, H. R. & T. A. (1974b). Molecular basis of control of mitotic cell division in eukaryotes. *Nature, London*, **249**, 553–6.

Bradbury, E. M., Inglis, R. J., Matthews, H. R. & Sarner, N. (1973). Phosphorylation of very-lysine-rich histone in *Physarum polycephalum*: correlation with chromosome condensation. *European Journal of Biochemistry*, **33**, 131–9.

Bradbury, E. M. & Rattle, H. W. E. (1972). Simple computer-aided approach for the analysis of the NMR spectra of histones. *European Journal of Biochemistry*, **27**, 270–81.

Burgoyne, L. A., Hewish, D. R. & Mobbs, J. (1974). Mammalian chromatin substructure studies with the calcium–magnesium endonuclease and two-dimensional polyacrylamide gel electrophoresis. *Biochemical Journal*, **143**, 67–72.

Camerini-Otero, R. D., Sollner-Webb, B. & Felsenfeld, G. (1976). The organisation of histones and DNA in chromatin: evidence for an arginine-rich histone kernel. *Cell*, **8**, 333–47.

Candido, E. P. M. & Dixon, G. H. (1971). Sites of *in vivo* acetylation in trout testis histone IV. *Journal of Biological Chemistry*, **246**, 3182–90.

Candido, E. P. M. & Dixon, G. H. (1972). Acetylation of trout testis histones *in vivo*: site of the modification in histone 11b. *Journal of Biological Chemistry*, **247**, 3863–70.

Candido, E. P. M. & Dixon, G. H. (1972). Amino-terminal sequences and sites of in-vivo acetylation of trout testis histones III and IIb2. *Proceedings of the National Academy of Sciences, USA*, **69**, 2015–20.

Carpenter, B. G., Baldwin, J. P., Bradbury, E. M. & Ibel, K. (1976). Organization of subunits in chromatin. *Nucleic Acids Research*, **3**, 1739–46.

Cary, P. D., Moss, T. & Bradbury, E. M. (1979). High resolution proton magnetic resonance studies of chromatin core particles. *European Journal of Biochemistry*, **89**, 475–82.

Chahal, S., Matthews, H. R. & Bradbury, E. M. (1980). Acetylation of histone H4 and role in chromatin structure. *Nature, London*, **287**, 76–9.

Chapman, G. E., Hartman, P. G. & Bradbury, E. M. (1976). Studies on the role and mode of operation of the very-lysine-rich histone H1 in eukaryote chromatin. The isolation of the globular and non-globular regions of the histone H1 molecule. *European Journal of Biochemistry*, **61**, 69–75.

Chapman, G. E., Hartman, P. G., Cary, P. D., Bradbury, E. M. & Lee, D. R. (1978). A nuclear-magnetic-resonance study of the globular structure of the H1 histone. *European Journal of Biochemistry*, **86**, 35–44.

Cohen, L. H., Mahowald, A. P., Chalkley, R. & Zweidler, A. (1973). Histones of early embryos. *Federation Proceedings*, **32**, 588.

Compton, J. L., Bellard, M. & Chambon, P. (1976). Biochemical evidence of variability in the DNA repeat length in the chromatin of higher eukaryotes. *Proceedings of the National Academy of Sciences, USA*, **73**, 4382–6.

Cross, M. E. & Ord, M. G. (1970). Changes in the phosphorylation and thiol content of histones in phytohaemagglutinin-stimulated lymphocytes. *Biochemical Journal*, **118**, 191–4.

D'Anna, J. A. & Isenberg, I. (1974). A histone cross-complexing system. *Biochemistry*, **13**, 4992–7.

Davies, K. E. & Walker, I. O. (1977) In-vitro transcription of RNA in nuclei, nucleoli and chromatin from *Physarum polycephalum*. *Journal of Cell Science*, **26**, 267–79.

DeLange, R. J., Famborough, D. M., Smith, E. L. & Bonner, J. (1969). Calf and pea histone IV. *Journal of Biological Chemistry*, **244**, 319–34.

Dixon, G. H., Candido, E. P. M., Honda, B. M., Louie, A. J., McLeod, A. R. & Sung, M. T. (1975). The structure and function of chromatin. CIBA Foundation Symposium **28**, pp. 220–240, Elsevier, Amsterdam.

Finch, J. T., Lutter, L. C., Rhodes, D., Brown, R. S., Rushton, B., Levitt, M. & Klug, A. (1977). Structure of nucleosome core particles of chromatin. *Nature, London*, **269**, 29–36.

Finch, J. T. & Klug, A. (1976). Solenoidal model for superstructure in chromatin. *Proceedings of the National Academy of Sciences, USA*, **73**, 1897–1901.

Garel, A. & Axel, R. R. (1976). Selective digestion of transcriptionally active ovalbumin genes from oviduct nuclei. *Proceedings of the National Academy of Sciences, USA*, **73**, 3966–70.

Goodwin, G. H. & Johns, E. W. (1973). Isolation and characterization of two calf thymus non histone proteins with high contents of acidic and basic amino acids. *European Journal of Biochemistry*, **40**, 215–19.

Goodwin, G. M., Nicolas, R. H. & Johns, E. W. (1975). An improved large scale fractionation of high mobility group non-histone chromatin proteins. *Biochimica et Biophysica Acta*, **405**, 280–91.

Goodwin, G. H., Walker, J. M. & Johns, E. W. (1978). High mobility group proteins. *The Cell Nucleus,* ed. H. Busch, vol. 6, p. 101. New York & London: Academic Press.

Grant, W. D. (1972). The effect of alpha-amanitin and $(NH_4)_2SO_4$ on RNA synthesis in nuclei & nucleoli isolated from *Physarum polycephalum* at different times during the cell cycle. *European Journal of Biochemistry*, **2**, 94–8.

Grimes, S. R., Jr., Chae, C. B. & Irvin, J. L. (1975). Acetylation of histones of rat testis. *Archives of Biochemistry and Biophysics*, **168**, 425–35.

Gurley, L. R., Walters, R. A. & Tobey, R. A. (1975). Sequential phosphorylation of histone sub-fractions in the Chinese hamster cell cycle. *Journal of Biological Chemistry*, **250**, 3936–3944.

Hardie, D. G., Matthews, H. R. & Bradbury, E. M. (1976). Cell-cycle dependence of two nuclear histone kinase enzyme activities. *European Journal of Biochemistry*, **66**, 37–42.

Hartman, P. G., Chapman, G. E., Moss, T. & Bradbury, E. M. (1977). Studies on the role and mode of operation of the very-lysine-rich histone H1 in eukaryote chromatin. The three structural regions of the histone H1 molecule. *European Journal of Biochemistry*, **77**, 45–51.

Hayashi, H., Iwai, K., Johnson, J. D. & Bonner, J. (1977). Pea histones H2A and H2B. *Journal of Biochemistry, Tokyo*, **82**, 503–10.

Hewish, D. R. & Burgoyne, L. A. (1973). Chromatin sub-structure. The digestion of chromatin DNA at regularly spaced sites by a nuclear deoxyribonuclease. *Biochemical and Biophysical Research Communications*, **52**, 504–10.

Hjelm, R. P., Kneale, G. G., Suau, P., Baldwin, J. P., Bradbury, E. M. & Ibel, K. (1977). Small angle neutron scattering studies of chromatin subunits in solution. *Cell*, **10**, 139–51.

Hooper, J. A., Smith, E. L., Summer, K. R. & Chalkley, R. (1973). Amino acid sequence of histone III of the testes of the carp. *Journal of Biological Chemistry*, **248**, 3275–79.

Inglis, R. J., Langan, T. A., Matthews, H. R., Hardie, D. G. & Bradbury, E. M. (1976). Advance of mitosis by histone phospholinase. *Experimental Cell Research*, **97**, 418–25.

Jackson, V., Shires, A., Chalkley, R. & Granner, D. K. (1975). Studies on highly metabolically active acetylation and phosphorylation of histones. *Journal of Biological Chemistry*, **250**, 4856–63.

Jackson, V., Shires, A., Tanphaichitr, N. & Chalkley, R. (1976). Modifications to histones immediately after synthesis. *Journal of Molecular Biology*, **104**, 471–83.

Johnson, E. M., Littau, V. C., Allfrey, V. G., Bradbury, E. M. & Matthews, H. R. (1976). The sub-unit structure of chromatin from *Physarum polycephalum*. *Nucleic Acids Research*, **3**, 3313–29.

Kelly, R. I. (1973). Isolation of a histone IIb1–IIb2 complex. *Biochemical and Biophysical Research Communications*, **54**, 1588–94.

Kornberg, R. (1974). Chromatin structure: a repeating unit of histones and DNA. *Science*, **184**, 868–71.

Kornberg, R. & Thomas, J. O. (1974). Chromatin structure: oligomers of the histones. *Science*, **184**, 865–8.

Lake, R. S. (1973). Further characterization of the F1 histone phosphokinase of metaphase-arrested animal cells. *Journal of Cell Biology*, **58**, 317–33.

Langan, T. A. (1978). Methods for the assessment of site-specific histone phosphorylation. *Methods of Cell Biology*, ed. G. S. Stein & J. Stein, vol. 19, pp. 127–42. New York & London: Academic Press.

Langmore, J. P. & Wooley, J. P. (1975). Chromatin architecture: investigation of a subunit of chromatin by dark field electron microscopy. *Proceedings of the National Academy of Sciences, USA*, **72**, 2691–95.

Leffak, I. M., Grainger, R. & Weintraub, H. (1977). Conservative assembly and segregation of nucleosomal histones. *Cell*, **12**, 837–45.

Levy, W. B., Wong, N. C. W. & Dixon, G. H. (1977). Selective association of the trout-specific H6 protein with chromatin regions susceptible to DNase II: possible location of HMG-T in the spacer region between core nucleosomes. *Proceedings of the National Academy of Sciences, USA*, **74**, 2810–14.

Lohr, D., Cordieu, J., Tatchell, K., Kovak, R. T. & van Holde, K. E. (1977). A comparative subunit structure of HeLa, yeast and chicken erythrocyte chromatin. *Proceedings of the National Academy of Sciences, USA*, **74**, 79–83.

Louie, A. J., Candido, E. P. M. & Dixon, G. H. (1973). Enzymatic modifications and their possible role in regulating the binding of basic proteins to DNA and controlling chromosome structure. *Cold Spring Harbor Symposia in Quantitative Biology*, **38**, pp. 803–19.

Lutter, L. C. (1979). Precise location of DNase I cutting sites in the nucleosome core determined by high resolution gel electrophoresis. *Nucleic Acids Research*, **6**, 41–56.

MacLeod, A. R., Wong, N. C. W. & Dixon, G. H. (1977). The amino acid sequence of trout-testis histone H1. *European Journal of Biochemistry*, **78**, 281–91.

Marushige, Y. & Marushige, K. (1975). Enzymatic unpacking of bull sperm chromatin. *Biochimica et Biophysica Acta*, **403**, 180–91.

Matsumoto, Y., Hasuda, H., Mita, S., Marunouchi, T. & Yamada, M. (1980). Evidence for the involvement of H1 histone phosphorylation in chromosome condensation. *Nature, London*, **284**, 181–3.

Mittermeyer, C., Braun, R. & Rusch, H. P. (1964). RNA synthesis in the mitotic cycle of *Physarum polycephalum. Biochimica et Biophysica Acta,* **91**, 399–405.

Mohberg, J., Babcock, K. L., Haugli, F. B. & Rusch, H. P. (1973). Nuclear DNA content and chromosome numbers in the myxomycete *Physarum polycephalum. Developmental Biology*, **34**, 228–45.

Molgaard, H. V., Matthews, H. R. & Bradbury, E. M. (1976). Organisation of genes for ribosomal RNA in *Physarum polycephalum. European Journal of Biochemistry*, **68**, 541–9.

Morris, R. (1976). Nucleosome structure in *Aspergillus nidulans. Cell*, **8**, 357–63.

Moss, T., Cary, P. D., Abercrombie, B. D., Crane-Robinson, C. & Bradbury, E. M. (1976*a*). A pH-dependent interaction between histone H2A and H2B involving secondary and tertiary folding. *European Journal of Biochemistry*, **71**, 337–50.

Moss, T., Cary, P. D., Crane-Robinson, C. & Bradbury, E. M. (1976*b*). Physical studies on the H3/H4 histone tetramer. *Biochemistry*, **15**, 2261–67.

Moss, T., Stephens, R. M., Crane-Robinson, C. & Bradbury, E. M. (1977). A nucleosome-like structure containing DNA and the arginine-rich histones H3 and H4. *Nucleic Acids Research*, **4**, 2477–85.

Nelson, N., Perry, M. E. & Chalkley, R. (1979). A correlation between nucleosome spacer region susceptibility to DNase I and histone acetylation. *Nucleic Acids Research*, **6**, 561–74.

Newrock, K. M., Alfagame, C. R., Nardi, R. V. & Cohen, L. H. (1977). Histone changes during chromatin remodeling in embryogenesis. *Cold Spring Harbor Symposia in Quantitative Biology*, **42**, 421–31.

Noll, M. & Kornberg, R. (1977). Action of micrococcal nuclease on chromatin and the location of histone H1. *Journal of Molecular Biology*, **109**, 393–404.

Olins, A. L. & Olins, D. E. (1974). Spheroid chromatin units. *Science*, **183**, 330–32.

Panyim, S. & Chalkley, R. (1969). High resolution acrylamide gel electrophoresis of histones. *Archives of Biochemistry and Biophysics*, **130**, 337–46.

Pardon, J. F., Worcester, D. L., Wooley, J. C., Tatchall, K., van Holde, K. E. & Richards, B. M. (1975). Low angle neutron scattering from chromatin subunit particles. *Nucleic Acids Research*, **2**, 2163–76.

Prior, C. P., Cantor, C. R., Johnson, E. M. & Allfrey, V. G. (1980). Incorporation of exogenous pyrene-labeled histone into *Physarum* chromatin: a system for studying changes in nucleosomes assembled *in vivo. Cell*, **20**, 597–608.

Rall, S. C. & Cole, R. D. (1971). Amino acid sequence and sequence variability of the amino-terminal regions of lysine-rich histones. *Journal of Biological Chemistry*, **246**, 7175–90.

Reeves, R. & Jones, A. (1976). Genomic transcriptional activity and the structure of chromatin. *Nature, London*, **260**, 495–500.

Richards, B. M., Pardon, J. F., Lilley, D., Cotter, R. & Wooley, J. C. (1977). The sub-structure of nucleosomes. *Cell Biology International Reports*, **1**, 107–16.

Ruiz-Carrillo, A., Waugh, L. J. & Allfrey, V. G. (1975). Processing of newly synthesized histone molecules. *Science*, **190**, 117–28.

Santiére, P., Tyron, D., Laine, S., Mizon, J., Ruffin, P. & Biserte, G. (1974). Covalent structure of calf thymus alk histone. *European Journal of Biochemistry*, **41**, 563–76.

Sollner-Webb, B., Camerini-Otero, R. D. & Felsenfeld, G. (1976). Chromatin structure as probed by nucleases and proteases. Evidence for the central role of histones H3 and H4. *Cell*, **9**, 179–93.

Strickland, M., Strickland, W. N., Brandt, W. F. & Von Holt, C. (1977*a*). The complete amino-acid sequence of histone H2B(1) from sperm of the sea urchin *Parechinas angulosus*. *European Journal of Biochemistry*, **77**, 263–75.

Strickland, W. N., Strickland, M., Brandt, W. F. & Von Holt, C. (1977*b*). The complete amino acid sequence of histone H2B(2) from sperm of the sea urchin *Parechinas angulosus*. *European Journal of Biochemistry*, **77**, 277–86.

Stuhrmann, H. B. (1975). Small angle scattering of proteins in solution. In *Brookhaven Symposia of Biology*, **27**, vol. 4, ed. B. P. Schoenbora, pp. 3–19. Upton, New York: Brookhaven National Laboratory.

Suau, P., Bradbury, E. M. & Baldwin, J. P. (1979). Higher order structures of chromatin in solution. *European Journal of Biochemistry*, **97**, 593–602.

Suau, P., Kneale, G. G., Braddock, G. W., Baldwin, J. P. & Bradbury, E. M. (1977). A low resolution model for the chromatin core particle by neutron scattering. *Nucleic Acids Research*, **4**, 3769–86.

Sung, M. T. & Dixon, G. H. (1970). Modification of histones during spermiogenesis in trout: a molecular mechanism for altering histone binding to DNA. *Proceedings of the National Academy of Sciences, USA*, **67**, 1616–23.

van Holde, K. E. & Isenberg, I. (1975). Nucleosome chromatin structure. *Accounts of Chemical Research*, **8**, 327–35.

van Holde, P. D., Strickland, W. N., Brandt, W. F. & Von Holt, C. (1979). The complete amino-acid sequence of histone H2B from the mollusc *Patella granatina*. *European Journal of Biochemistry*, **93**, 71–8.

Vidali, G., Boffa, L. & Allfrey, V. G. (1977). Selective release of chromosomal proteins during limited DNase I digestion of avian erythrocyte chromatin. *Cell*, **12**, 408–15.

Vidali, G., Boffa, L. C., Bradbury, E. M. & Allfrey, V. G. (1978). Butyrate suppression of histone deacetylation leads to an accumulation of multi-acetylated forms of histones H3 and H4 and increased DNase I sensitivity of the associated DNA sequences. *Proceedings of the National Academy of Sciences, USA*, **75**, 2239–43.

Vogt, V. & Braun, R. (1976). Structure of ribosomal DNA in *Physarum polycephalum*. *Journal of Molecular Biology*, **106**, 567–587.

Von Holt, C., Strickland, W. N., Brandt, W. F. & Strickland, M. (1979). More histone structures. *FEBS Letters*, **100**, 201–18.

Walker, J. M., Goodwin, G. H. & Johns, E. W. (1979). The primary structure of the nucleosome-associated chromosomal protein HMG 14. *FEBS Letters*, **100**, 394–8.

Weintraub, H. & Groudine, M. (1976). Chromosomal subunits in active genes have an altered conformation. *Science*, **193**, 848–56.

Whitlock, J. & Simpson, R. T. (1977). Localization of the sites along the nucleosome DNA which interact with NH-terminal histone regions. *Journal of Biological Chemistry*, **252**, 6516–20.

Woodcock, L. L. F. (1973). Ultrastructure of inactive chromatin. *Journal of Cell Biology*, **59**, 368a.

Parasexual processes in fungi

C. E. CATEN

Department of Genetics, University of Birmingham, Birmingham B15 2TT, UK

Introduction

The term parasexual was first proposed by Pontecorvo in 1954 for those processes other than standard sexual reproduction which result in recombination of hereditary properties (Pontecorvo, 1954). With an exclusive definition of this type we might expect a variety of genetic mechanisms to be encompassed by the heading 'Parasexual processes', and this is certainly the case in fungi. In eukaryotes, recombination of genetic information originating from different individuals requires four steps: (1) the introduction of the two genomes (or parts of genomes) into the same cell, (2) their association in the same nucleus, (3) crossing over leading to intra-chromosomal recombination and (4) reduction whereby the amount of genetic information per nucleus is reduced to the haploid level. Inter-chromosomal recombination typically occurs during the reduction step through independent separation of non-homologous chromosomes. In the standard sexual cycle, step (1) generally involves the fusion of specialized cells, i.e. plasmogamy, step (2) involves the fusion of complete haploid genomes to give a diploid zygote nucleus, i.e. karyogamy, and steps (3) and (4) are accomplished sequentially during two highly co-ordinated cell divisions, i.e. meiosis. A mechanism of recombination that differs from the sexual cycle in one or more of these respects may therefore be designated as parasexual. It follows that parasexual processes are likely to involve one or more of the following: somatic cells, partial diploid zygotes, crossing-over and reduction during mitosis and the absence of a fixed sequence and fine control over the whole process. Some of these features are reflected in the frequently used terms somatic recombination and mitotic recombination which are considered synonymous with parasexual recombination.

As is clear from the above definition, the title 'Parasexual processes in fungi' encompasses a variety of phenomena about which a large body of information has been accumulated. Various aspects have been previously reviewed (Pontecorvo, 1956; Käfer, 1961; Casselton, 1965a; Roper, 1966; Tinline & MacNeill, 1969; Sermonti, 1969; Caten & Day, 1977). This article aims to bring the more recent developments together and to relate them to previous knowledge of parasexuality. Selection of material is unavoidable and I will concentrate on five aspects: (1) the occurrence of parasexual processes, (2) the standard parasexual cycle, (3) factors affecting the parasexual processes, (4) variation in parasexual processes and (5) parasexual processes in nature. Details of methodology, applications of parasexuality and various phenomena related to parasexuality (e.g. mitotic non-conformity, Nga & Roper, 1969) are not covered. Neither is it possible to give detailed consideration to individual species, with the result that occasional idiosyncrasies go unnoted in the search for generalizations.

Recombination has been investigated by genetical, cytological and, more recently, biochemical and physical methods (Catcheside, 1977; Stahl, 1979). Almost all knowledge of parasexual recombination has been obtained from genetical studies. The failure of cytology to contribute significantly may be attributed to the small size of fungal chromosomes and the rarity of parasexual events.

Occurrence of parasexual processes

Parasexual recombination in fungi was first studied in detail in the 1950s by Pontecorvo and associates in *Aspergillus nidulans* (Pontecorvo *et al.*, 1953; Pontecorvo, 1954, 1956; Pontecorvo & Käfer, 1958). Earlier indications of parasexuality can be found in the reports of the origin of nuclei with recombinant mating type in vegetative mycelia of tetrapolar Basidiomycetes (Brunswick, 1924; Quintanilha, 1939; Papazian, 1950). The methods developed in these initial investigations have been applied to other fungi and a survey of the literature revealed forty species in which there is good evidence for parasexual recombination, although not necessarily for a complete parasexual cycle (Table 1). References to the work involving most of these species are given in the reviews by Esser & Kuenen (1967) and Tinline & MacNeill (1969). Among more recent demonstrations of parasexual recombination, those for *Phycomyces blakesleeanus* (Park, Kenehan & Goodgal, 1968) and *Dictyostelium discoideum* (Katz & Sussman, 1972; Williams, Kessin & Newell, 1974) are noteworthy as the first reports in their respective

Table 1. *Occurrence of parasexual processes*

Acrasiomycetes *Dictyostelium discoideum*	Basidiomycetes *Ustilago* (3 species) *Puccinia* (3 species) *Melampsora lini*
Phycomycetes *Phycomyces blakesleeanus*	*Coprinus* (4 species) *Schizophyllum commune*
Ascomycetes *Aspergillus* (9 species) *Penicillium* (3 species) *Saccharomyces cerevisiae* *Schizosaccharomyces pombe* *Cochliobolus sativus* *Leptosphaeria maculans* *Podospora anserina* *Neurospora crassa*	Deuteromycetes *Fusarium oxysporum* *Verticillium* (2 species) *Cephalosporium* (2 species) *Humicola* sp. *Ascochyta imperfecta* *Pyricularia oryzae*

fungal groups. The list of species in Table 1 is not necessarily complete but it indicates the range of fungi in which parasexual processes have been found. Indeed, failure to detect parasexual recombination where appropriate techniques have been used has seldom been reported (see Strømnaes, Garber & Beraha, 1964 for one case in *Penicillium digitatum*), suggesting that parasexual processes are widespread in fungi. *Neurospora crassa* remains a special case in that extensive searches for somatic diploids and somatic recombinants in heterokaryons have given negative results (Case & Giles, 1962; Roper, 1966; Smith, 1974). Diploid mycelia have been isolated from ascospores, however, and these produce parasexual recombinants by mitotic crossing over and haploidization (Smith, 1974).

The standard parasexual cycle

The sequence of events first established in *Aspergillus nidulans* (Pontecorvo *et al.*, 1953; Pontecorvo & Käfer, 1958; Käfer, 1961) has subsequently been shown to occur in many other species and may be considered as the standard parasexual cycle (Fig. 1). The first step is the bringing together of the nuclei from two haploid strains into a heterokaryon. The two genomes exist side by side in the heterokaryon and, very rarely, they may fuse to form a heterozygous somatic diploid nucleus. This divides mitotically alongside the haploid nuclei in the heterokaryon, establishing a clone of diploid nuclei which may eventually be distributed into conidia or form a discrete sector. From such conidia or sectors a homokaryotic mycelium containing only diploid nuclei, i.e. a diploid strain, can be established. Diploid strains are unstable and produce segregants by mitotic recombination and mitotic chromosomal nondisjunction. The degree of instability varies between species but it is generally possible to maintain the diploid. Mitotic recombination reassociates markers carried on the same chromosome and may or may not precede the reduction step. The latter occurs through a sequential loss of chromosomes by mitotic nondisjunction with intervening aneuploid states. Since the homologue lost at each nondisjunction is determined at random the eventual haploid products carry all possible combinations of chromosomes from the parent haploids. Mitotic recombination is not restricted to diploid nuclei but also occurs between the disomic chromosomes of aneuploids (Upshall *et al.*, 1979).

The outcome of the parasexual cycle is therefore a population of haploid genotypes recombinant for characters of the parent haploids, that is the same as the outcome of the sexual cycle. Relative to the sexual cycle, however, the whole process is unco-ordinated and inefficient, and recovery of a workable population of recombinant haploids is generally dependent upon selection at every stage and upon the fact that aneuploids grow poorly and readily produce vigorous haploid sectors. The four major steps in the parasexual cycle will now be considered in detail.

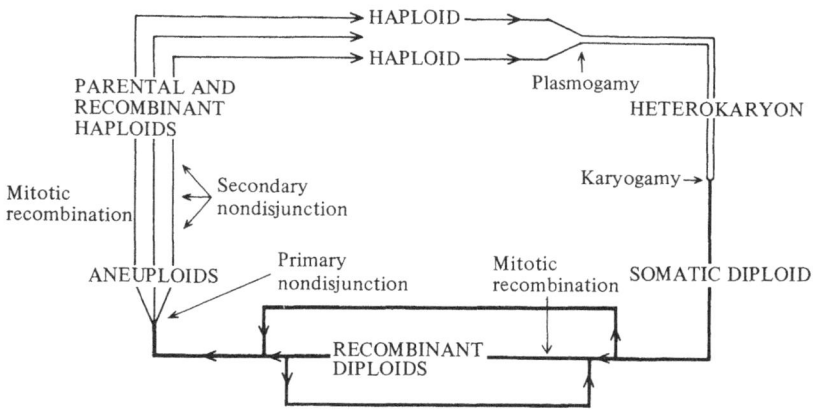

Fig. 1. The standard parasexual cycle. ——, haploid or aneuploid nuclei; ——, diploid nuclei.

Heterokaryon formation

The extent of heterokaryon formation varies greatly between species with the proportion of heterokaryotic cells in a heterokaryotic mycelium ranging from near 100% in many Basidiomycete dikaryons (Raper, 1966) to just a few per cent in certain Ascomycetes and Deuteromycetes (Puhalla & Mayfield, 1974). The degree of nuclear mixing has consequences for the growth of the heterokaryon, for the probability of nuclear fusion producing a heterozygous as opposed to a homozygous diploid and for the development of the heterozygous diploid clone once formed. Little is known, however, about the processes of hyphal anastomosis and nuclear migration which determine the extent of heterokaryosis (Burnett, 1976).

Nuclear fusion

The occurrence of somatic nuclear fusion is generally inferred from the recovery of heterozygous diploid nuclei from heterokaryons (Pontecorvo *et al.*, 1953), although fusion has been observed in a few cases (Parag, 1968; Puhalla & Mayfield, 1974). Estimation of the rate of nuclear fusion is technically difficult and the best approximation is given by the frequency of diploid nuclei in a sample of nuclei from a heterokaryon. This varies widely both within and between species and may range from 10^{-3} to 10^{-7} (e.g. Pontecorvo *et al.*, 1953; Casselton, 1965*b*; Lhoas, 1967; Day & Jones, 1968). While differences in the rate of nuclear fusion undoubtedly contribute to this variation, differences in nuclear mixing in heterokaryons and in the relative division/sporulation rates of haploid and diploid nuclei/hyphae are also important factors.

It is unlikely that fusion in somatic cells occurs only between nuclei of different genotype. Assuming a random process, fusions between nuclei of like genotype would be expected to predominate, considering the associated nature of nuclear distributions within heterokaryons (Atwood & Mukai, 1955; Clutterbuck & Roper, 1966). The resultant homozygous diploid nuclei would be genetically indistinguishable from their haploid predecessors, however, and hence difficult to detect. The 'gigas' forms of *Penicillium notatum* described by Sansome (1946*a*) were probably homozygous diploids produced in this way. Tolmsoff (1972) has proposed that such diploids are regularly involved in the production of microsclerotia, dark mycelium and chlamydospores by *Verticillium* species.

Upshall, Giddings & Mortimore (1977) observed that diploid strains of *Aspergillus nidulans* and *A. terreus* are more sensitive to benomyl than haploid strains and that this provides a simple test for classifying the ploidy of unknown strains. Using this benomyl test, H. Howell (personal communication) searched for diploids among single conidial and single ascospore cultures from homokaryotic haploid strains of *A. nidulans*. Cultures suspected of being diploid from the benomyl test were further checked by measurements of conidial size and by tests for the frequency of recessive mutations to chlorate resistance (Cove, 1976). Standard haploid and diploid strains were included in all tests. Three diploids were recovered from 9265 conidial cultures, giving a frequency of homozygous diploids among conidia of 3.2×10^{-4}. This is approximately one hundred times greater than the frequency of recovery of heterozygous diploids from heterokaryons in the same species (Pontecorvo *et al.*, 1953) and is consistent with the idea that fusions occur more frequently between like than between unlike nuclei. (The possibility that diploid nuclei may also arise through errors of mitosis can not be ignored.) No diploids were found among 9288 ascospore cultures, which is surprising in view of the report (Pritchard, 1954) that ascospores with diploid nuclei are produced in *A. nidulans* with a frequency of the order of 0.1%.

Mitotic recombination

The generally accepted mechanism for intergenic–mitotic recombination is by mitotic crossing-over. This occurs after chromosome replication and involves a reciprocal exchange between two non-sister chromatids (Roper & Pritchard, 1955; Hastie, 1968). Following crossing-over the centromeres divide and separate in typical mitotic manner with the result that either one recombinant and one non-recombinant chromatid pass to each pole (Fig. 2A) or both recombinant chromatids pass to one pole and both non-recombinant chromatids to the other (Fig. 2B). The former segregation pattern produces daughter nuclei that are homozygous for genes distal to the point of exchange while the latter retains heterozygosity for all markers but changes the linkage relationships. In contrast to this process, intragenic–mitotic recombination does not usually yield reciprocal products (Roman, 1956; Putrament, 1964) and is thought to occur through gene conversion. At the molecular level the mechanisms of mitotic crossing-over and mitotic gene conversion are believed to be similar to those occurring during meiosis (Catcheside, 1977).

The spontaneous frequency of mitotic crossing-over varies both between species and between chromosome regions within a species.

Estimates of the frequency per genome per division range from < 0.01 in yeast (Roman, 1956) and *Ustilago maydis* (Holliday, 1961) to > 0.4 in *Verticillium albo-atrum* (Hastie, 1967) and *Neurospora crassa* (Smith, 1974). The frequency in *Aspergillus nidulans* is not less than 0.03 per genome per division (Käfer, 1977). The distribution of mitotic crossing-over within a chromosome complement is best documented for *A. nidulans* in which it differs from that of meiotic crossing-over and is concentrated around the centromeres (Pontecorvo & Käfer, 1958; Käfer, 1977). Frequent if not continuous pairing of homologous chromosomes in somatic diploids is implied by the high rates of mitotic exchange, and this idea is supported by the failure to observe the expected doubling of the number of chromatinic elements in the dividing nuclei of diploid strains (Robinow & Caten, 1969; Day & Jones, 1972). In meiosis the synaptonemal complex is believed to play a part in chromosome pairing, leading to crossing-over (Westergaard & von Wettstein, 1972). Olson & Zimmermann (1978) failed to find synaptonemal complexes in yeast diploids undergoing induced mitotic crossing-over and gene conversion.

Reduction
From the results of detailed genetical analyses of unselected segregants from diploids of *Aspergillus nidulans,* Käfer (1961, 1977)

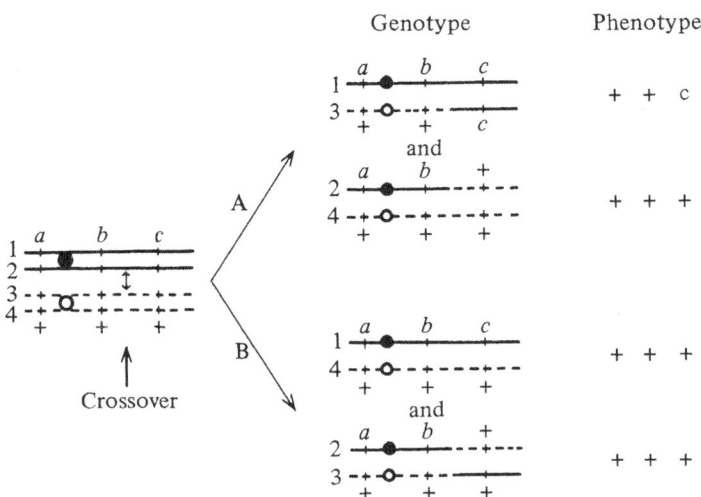

Fig. 2. The genetic consequences of mitotic crossing-over. 1,2,3,4 are the four chromatids of the replicated homologous chromosomes. A and B are alternative patterns of chromatid segregation.

proposed that reduction to the haploid state (haploidization) occurred through sequential chromosome loss by mitotic nondisjunction. Primary mitotic nondisjunction in *Aspergillus nidulans* occurs with a frequency of around 0.02 per division, and produces trisomic and monosomic daughter nuclei (Fig. 3). The fate of these aneuploid products depends upon the occurrence of secondary nondisjunctional events. A second misdistribution involving the same chromosome in the trisomic nucleus will restore the diploid state and in one-third of the cases the product should be homozygous for the chromosome involved; these segregants are referred to as nondisjunctional diploids (Fig. 3A). Secondary nondisjunctions involving the other chromosomes in the monosomic nucleus will give a series of aneuploids of the type 2n − 2, 2n − 3, etc., until the haploid level is reached (Fig. 3B). Under ideal experimental conditions, haploid segregants show independent assortment of the chromosomes originating from the two parents, indicating that the homologue lost at each nondisjunction is determined at random. Direct cytological evidence for Käfer's nondisjunctional model of haploidization has been obtained in *Dictyostelium discoideum*, where nuclei of all classes from 2n + 2 to n + 1 have been observed to originate from diploids (Brody & Williams, 1974).

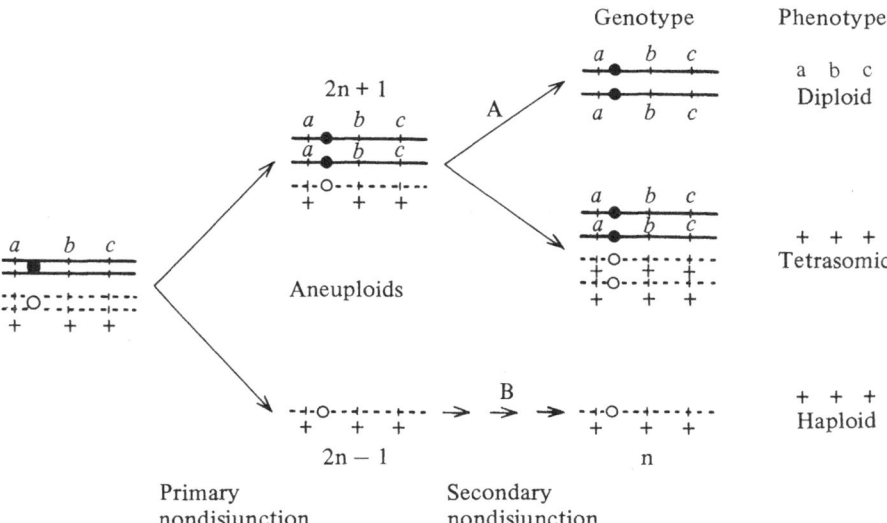

Fig. 3. The genetic consequences of mitotic chromosome nondisjunction. A shows the formation of nondisjunctional diploids; B shows the formation of haploids. The genotypes and phenotypes given are examples of possible products depending upon which homologue fails to divide normally.

As with mitotic recombination there is considerable specific variation in the frequency of haploidization. Diploids of yeast, *Ustilago maydis* and *U. violacea* appear not to haploidize spontaneously (Holliday, 1961; Day & Jones, 1968; Fincham, Day & Radford, 1979). Diploids of *Neurospora crassa,* however, are very unstable and mitotic nondisjunction has been estimated to occur with a frequency of 0.1 per division (Smith, 1974). Nondisjunction occurs at high frequency in *Verticillium* spp. also; after three weeks incubation, 90% of conidia sampled from a diploid may be haploid (Typas & Heale, 1977). It is not possible to determine whether the frequency of secondary nondisjunction is comparable to that of primary nondisjunction since competitive effects prevent reliable estimation.

The relationship between the steps in the parasexual cycle

The steps in the parasexual cycle are obviously related in the sense that the early stages are prerequisites for the later. Within this restriction, however, considerable independence exists and even the sequential dependence does not apply to mitotic recombination and haploidization. Coincidences of two mitotic crossovers or of mitotic crossing-over with nondisjunction occur no more frequently than expected from independent events (Pontecorvo & Käfer, 1958; Käfer, 1977). There is a suggestion, however, in the high frequency of hyperdiploids with two or more extra chromosomes, that nondisjunctional events may tend to be associated (Käfer, 1961). An expected positive correlation between the extent of heterokaryosis and the frequency of heterozygous diploids is probably masked by the greater competitive advantage of diploid nuclei in poor heterokaryons. Thus, in practice, heterozygous diploids may be easier to isolate in species with restricted heterokaryosis, or between mutants that complement poorly in heterokaryons, than in species in which vigorous heterokaryons are formed.

Although within a species mitotic crossing-over and nondisjunction appear to be independent events, their frequencies are associated when comparisons are made between species. The spontaneous rates of these two processes are both low in yeast and *Ustilago* sp., both high in *Neurospora crassa* and *Verticillium*, and both intermediate in *Aspergillus nidulans*. Smith (1974) proposed that mitotic crossing-over and nondisjunction are two consequences of competition between homologues for attachment sites on the nuclear membrane; attachment of both homologues to the same site accomplishes somatic pairing and hence facilitates exchange, whereas failure of one homologue to attach results in nondisjunction. The differences between species in the stability of somatic diploid nuclei may then be explained by differences in the degree of competition for attachment sites; diploids such as yeast have sufficient sites whereas sites are limiting in habitually haploid species such as *N. crassa, Verticillium* and *A. nidulans.*

Factors affecting the parasexual processes

Knowledge of factors affecting a cellular process may help elucidate the underlying mechanism(s). The parasexual processes, especially mitotic recombination and mitotic nondisjunction, are well documented in this respect. However, much of this information derives from the use of fungal parasexual systems as screens for environmental agents with genetic activity (e.g. Morpurgo, 1963; Zimmermann, 1971; Kappas, Georgopoulos & Hastie, 1974) and has not contributed greatly to an understanding of the mechanisms of parasexuality. In reviewing the factors affecting heterokaryosis, nuclear fusion, mitotic recombination and nondisjunction it is convenient to divide them into those that are extrinsic to the organism, i.e. environmental, and those that are intrinsic, i.e. genetic. That environmental and genetic factors frequently interact should not be overlooked, however.

Heterokaryosis

There has been little systematic investigation of environmental agents affecting heterokaryosis although it is generally appreciated that culture conditions such as media and temperature are important (Parmeter, Snyder & Reichle, 1963; Puhalla & Mayfield, 1974; Burnett, 1975). Genetic controls include specific heterokaryon compatibility loci (Garnjobst, 1955; Caten & Jinks, 1966; Esser, 1971; Mylyk, 1976; Dales & Croft, 1980), the mating-type loci in heterothallic fungi (Sansome, 1946b; Gross, 1952; Raper, 1966) and *tol*, a suppressor of the mating-type heterokaryon incompatibility in *Neurospora crassa* (Newmeyer, 1970).

Nuclear fusion

Early investigations with *Aspergillus* spp. indicated that (+) camphor (Pontecorvo *et al.*, 1953) and far ultraviolet (Ishitani, Ikeda & Sakaguchi, 1956) treatment of heterokaryotic cells increased diploid recovery. It was not clear, however, whether this reflected direct induction of nuclear fusion or some secondary selective effect. A. W.

Day & L. L. Day (1974) conducted an extensive survey of environmental factors influencing somatic karyogamy in *Ustilago violacea*. They confirmed the activity of camphor and ultraviolet irradiation and also found that diploid recovery was dependent upon temperature and culture age. However, none of over 40 biologically-active chemicals tested had an effect. In *U. violacea*, ultraviolet irradiation increased the absolute number of diploids recovered, demonstrating that the effect is inductive rather than selective (Day, A. W. & Day, L. L., 1974).

Among genetic factors the mating-type genes are undoubtedly important but their specific role in fusion is difficult to separate from their widespread developmental effects (Caten & Day, 1977; Day, 1979). Certain ultraviolet-sensitive mutants of *Aspergillus nidulans* (Lanier, Tuveson & Lennox, 1968) and *Ustilago violacea* (Day, A. W. & Day, L. L., 1974) are reduced or blocked in somatic diploid recovery, indicating a link between nuclear fusion and DNA repair. Dikaryons of *Schizophyllum commune*, homozygous for the recessive spontaneous mutation *dik*, produce diploid sectors at high frequency (Koltin & Raper, 1968). It is tempting to conclude that in normal dikaryons somatic karyogamy is repressed by the product of the *dik+* allele.

Mitotic recombination

The frequency of mitotic recombination is increased by a variety of physical and chemical treatments including ultraviolet irradiation (Holliday, 1961; Wood & Käfer, 1969; Zimmermann, 1971), ionizing radiation (Zimmermann, 1971), mitomycin C (Holliday, 1964), fluorodeoxyuridine (Esposito & Holliday, 1964; Beccari, Modigliani & Morpurgo, 1967), fluorouracil (Beccari *et al.*, 1967), and various alkylating agents (Morpurgo, 1963; Zimmermann, 1971; Bignami, 1977). Many of these treatments are mutagenic or inhibit DNA synthesis and it is likely that damage to the DNA induces the recombination system(s) (Fabre & Roman, 1977).

Several types of mutation affect mitotic recombination. Some have been directly selected for this function and both decreasing (*rec*) and increasing (*pop*) mutations have been isolated (Parag, Y. & Parag, G., 1975). In other cases mutations selected for a defect in some other aspect of DNA metabolism, e.g. ultraviolet sensitivity (Holliday, 1967; Shanfield & Käfer, 1969; Jansen, 1970) or nuclease deficiency (Holloman & Holliday, 1973), have proved to be altered in mitotic recombination. These mutations may or may not also affect meiotic recombination (Holliday, 1967; Shanfield & Käfer, 1969), implying steps unique to mitotic recombination. However, the precise role in recombination of the genes represented by these mutations is not clear.

Nondisjunction

Morpurgo (1961) and Lhoas (1961) demonstrated that *p*-fluorophenylalanine induced mitotic haploidization in *Aspergillus* spp. and this amino acid analogue has since been shown to be active in other fungi (Gutz, 1966; Day & Jones, 1968; De Bertoldi & Caten, 1975). Subsequent screens for potentially harmful nondisjunctional agents using fungal somatic diploids have identified a wide range of active substances, including arsenate (Van Arkel, 1963), nitrosoguanidine (Shanfield & Käfer, 1971), acridine yellow (Kinghorn & Pateman, 1975), chloral hydrate (Singh & Sinha, 1976), benzimidazole and thiophanate fungicides (Kappas *et al.*, 1974) and various pharmaceutical drugs (Bignami *et al.*, 1974) and pesticides (Bignami, 1977). The mode of action of most of these nondisjunctional agents is unknown but benomyl (a benzimidazol fungicide) interacts with the β-subunit of tubulin, thereby interfering with the polymerization and function of microtubules in the mitotic spindle (Davidse & Flach, 1977; Sheir-Neiss, Lai & Morris, 1978; Gull, pp. 113–32). Mutations in the structural gene for β-tubulin in *A. nidulans* are cross-resistant to *p*-fluorophenylalanine suggesting that this also acts directly on tubulin (Morris & Oakley, 1979).

As regards genetic factors, several observations indicate that compatibility for the mating-type genes induces haploidization of somatic diploids. In *Coprinus lagopus* and *Schizophyllum commune* diploids have been isolated from incompatible heterokaryons but not from compatible dikaryons (Casselton, 1965*b*; Frankel & Ellingboe, 1976). That nuclear fusion occurs in dikaryons is indicated by the frequent production of recombinant haploid nuclei (Parag, 1962). Compatible diploid strains of *S. commune*, selected from basidiospore progenies of common-AB diploids by haploid crosses, were unstable, producing aneuploid and haploid sectors and showing meiotic-like somatic recombination (Frankel & Ellingboe, 1976, 1977). A haploidizing effect of mating type is consistent with the observation that diploid nuclei of *C. lagopus* are stable on their own but haploidize when combined in a dikaryon (Casselton, 1965*b*; Casselton & Lewis, 1966). Clearly genes in addition to mating type are also involved in mitotic chromosome disjunction and Upshall *et al.* (1979) have isolated mutations in *Aspergillus nidulans* which yield and stabilize specific disomic states.

Variation in parasexual processes

Since it was first worked out in *Aspergillus nidulans*, the standard parasexual cycle has been shown to occur in many Ascomycetes, Deuteromycetes and in incompatible heterokaryons of Basidiomycetes. In a number of cases, however, parasexual recombination does not appear to conform to the standard parasexual cycle. While some of these are simply extreme variants of the cycle, the differences in others are such as to suggest that different mechanisms are involved.

Variants of the parasexual cycle

There is considerable variation in the frequency of the primary events in the parasexual cycle and in the duration of the individual steps. In extreme cases one of the steps may be so transient as to regularly escape detection. For example, diploids of a *Humicola* sp. arise directly in mixed cultures of auxotrophic mutants on minimal medium without an intervening heterokaryon (De Bertoldi & Caten, 1975). Presumably nuclear migration did not occur and heterokaryosis was limited to the fusion cells and was insufficient to support a visible heterokaryotic mycelium. Under these conditions there was intense selection for rare heterozygous fusion nuclei that produced vigorous sectors. In *Cephalosporium acremonium*, it is the diploid stage that is transient and apparently bypassed. Platings from heterokaryons onto selective media yield recombinant haploids, but only rarely a segregating colony that could be a diploid or aneuploid (Hamlyn & Ball, 1979).

Non-standard parasexual processes

Investigations of somatic recombination in dikaryons of Basidiomycetes, using either direct recovery of haploids or illegitimate di–mon matings, have suggested the operation of non-standard processes. Three of these will be considered: meiotic-like recombination, specific non-reciprocal transfer of a single gene, and chromosome transfer.

Meiotic-like recombination has been reported in dikaryons and di–mon matings of *Schizophyllum commune* (Parag, 1962; Ellingboe, 1964). In a recent study, Frankel (1979) examined monokaryons developed from platings of macerates of three differently marked dikaryons. No diploids or aneuploids were recovered but overall 28% of unselected haploid monokaryons were of non-parental genotype. These included both inter- and intra-chromosomal recombinants, and the intra-chromosomal frequencies were similar to the meiotic frequencies.

Hence this parasexual process possesses the high frequency of crossing-over and direct reduction typical of meiosis rather than the conservation of linkages and stepwise haploidization of the parasexual cycle.

Ellingboe (1963) reported a di–mon mating in *Schizophyllum commune* where mating-type factors from the nuclei of the dikaryon were inserted into the genome of the monokaryon. Only the mating-type factors were recombined and therefore the process appeared to involve internuclear transfer of single genes (specific factor transfer). A similar case, also in *S. commune* but not involving mating type, has recently been analysed (Leonard, Dick & Gaber, 1978; Leonard, Gaber & Dick, 1978). Dikaryons heteroallelic for the recessive mutation *mnd* produced sectors which had become homoallelic for *mnd* (transformed dikaryons). If this homoallelism resulted from either conventional parasexual recombination or the meiotic-like process, the *mnd* allele should occur in a range of recombinant genotypes with respect to unselected markers. However, resolution of transformed dikaryons by di–mon matings or dedikaryotization yielded only monokaryons of the two parental marker genotypes, although both now carried the *mnd* allele. Reciprocal transformation of the original *mnd* nucleus into *mnd*+ was not observed. The transformed *mnd* allele behaved as a normal chromosomal gene in sexual crosses. Leonard, Gaber & Dick (1978) concluded that a novel process of somatic recombination involving the non-reciprocal internuclear transfer of the *mnd* allele was operating. Recent results suggest that the process involves a whole chromosome segment, as a linked auxotrophic marker is simultaneously transferred (T. J. Leonard, personal communication).

A similar restricted genetic transfer occurs between nuclei in conjugating cells of *Ustilago violacea* (Day, 1978). The plating of conjugating cells carrying complementary auxotrophic mutations onto minimal medium produced, in addition to the normal diploids (Day & Jones, 1968), prototrophic colonies which carried only one mating-type allele and hence were only partially diploid. These partial diploids were not conventional aneuploids since, when grown under non-selective conditions, they reverted to the haploid parental genotypes and no recombinants were formed. Linked markers were always acquired and lost together, and markers on two or more chromosomes could be acquired simultaneously, although at a reduced frequency. Where markers on different chromosomes were co-acquired in this way, they were also generally lost as a group. Day (1978) interpreted these observations in terms of the breakdown of one of the nuclei in the dikaryon followed by the acquisition by the other nucleus (the recipient) of one or more of the released chromosomes either by direct uptake or in the form of an associated micronucleus. In either case the acquired chromosomes do not mix randomly with the recipient genome, as in a conventional aneuploid, but are organized separately such that they can be lost as a whole.

Parasexual processes in nature

There has been considerable speculation over the role of parasexuality in the variability of natural fungal populations (Pontecorvo, 1958; Buxton, 1959; Webster, 1974). Certain asexual fungi show high frequencies of diploid formation, mitotic recombination and haploidization, and this has encouraged the idea that parasexuality has evolved to replace sexuality in these organisms (Pontecorvo, 1958). However, our knowledge of parasexual processes is largely derived from the application of specialized laboratory techniques and it is appropriate to enquire to what extent these processes occur in nature.

Caten & Jinks (1966) examined the evidence for the natural occurrence of heterokaryosis in Ascomycetes and Deuteromycetes, and concluded that while it did occur its frequency and significance had been generally overestimated. That there have been few further reports of natural heterokaryons in these groups (e.g. Christensen, López & Benjamin, 1965; Ming, Lin & Yu, 1966) reinforces this conclusion. The dikaryons of heterothallic Basidiomycetes are invariably heterokaryotic but constitute a special case through their role in the sexual cycle of these fungi. The sexually incompatible heterokaryons of heterothallic Basidiomycetes (Raper, 1966) are more directly equivalent to heterokaryons in other groups but there is no evidence that these are of any significance in nature where lack of a compatible nucleus with which to dikaryotize is rare.

Direct evidence for the natural occurrence of parasexual processes is provided by the isolation of somatic diploid strains. Day & Anagnostakis (1971) found that 0.26% and 0.43% of teliospores from two galls of *Ustilago maydis* produced diploid sporidial colonies. Single diploid strains of *Verticillium dahliae* (Ingram, 1968), *V. albo-atrum* (Typas & Heale, 1976) and *Aspergillus niger* (Nga, Teo & Lim, 1975) have been isolated from nature. In a recent survey, fourteen of 154 independent isolates of *A. nidulans* proved to be somatic diploids (A. Upshall, personal communication; H. Howell, personal communication). These isolates had been in culture for several years and diploidy may have originated in the laboratory but this seems unlikely for all fourteen.

In nature, as in the laboratory, somatic diploids would be expected to undergo spontaneous mitotic recombination and haploidization, but these events would be difficult to observe under natural conditions and have never been reported. Certain fungicides and pesticides have recombinogenic and nondisjunctional effects (Kappas *et al.*, 1974; Bignami, 1977) and will therefore increase the 'natural' activity of the parasexual processes. Somatic recombination has been proposed to explain the appearance of new strains of rust fungi following infection of hosts with mixtures of dikaryons under near-natural conditions (Ellingboe, 1961; Bartos *et al.*, 1969). Somatic diploids have never been observed in these experiments, although they have been isolated from axenic cultures of *Puccinia graminis tritici* (Maclean, Tommerup & Scott, 1974). The apparent transient nature of diploidy and the high frequency with which unselected recombinants are recovered suggest that one of the non-standard parasexual processes observed in dikaryons of *Schizophyllum commune* is operating. However, unlike *S. commune,* the rust experiments involved mixtures of dikaryons and there appear to be no reports of somatic recombinants arising from individual dikaryons. Whether this deficiency means that an interaction between dikaryons is essential for somatic recombination in rusts or is simply the result of insufficient testing or the homozygosity of the individual dikaryons for relevant markers requires clarification.

The significance of the parasexual processes as a source of variation in nature depends not only upon the frequency with which they occur but also on the diversity of the interacting genomes. Heterogenic vegetative incompatibility systems are present in populations of many Ascomycetes and Deuteromycetes and these will block heterokaryon formation and parasexual recombination between strains of different genotype (Caten & Jinks, 1966; Esser & Blaich, 1973; Croft & Jinks, 1977). A similar incompatibility (intraspecific antagonism) occurs both in nature and in the laboratory between dikaryons of wood-decaying Basidiomycetes (Rayner & Todd, 1979) and this will stop nuclear exchange between adjacent mycelia, a situation in which extensive genetic interchange was believed to occur (Raper, 1966; Burnett, 1976). While these vegetative incompatibility systems are widespread they are not necessarily present in all species, neither do they constitute an absolute barrier to heterokaryosis and parasexuality. Somatic diploids have been constructed between unrelated strains of *Fusarium oxysporum* (Buxton,

1956), between unrelated strains of *Verticillium albo-atrum* (Hastie, 1962), between different species of *Verticillium* (Hastie, 1973) and between heterokaryon-incompatible strains of *Aspergillus nidulans* (Dales & Croft, 1977).

The predicted effect of vegetative incompatibility in restricting somatic diploidization to near-isogenic genomes is borne out by the lack of segregation observed on haploidization of the natural diploid of *Verticillium dahliae* (Hastie, 1970) and eight of the fourteen diploids of *Aspergillus nidulans* (H. Howell, personal communication). In contrast, the single natural diploids of *A. niger* (Nga *et al.*, 1975) and of *V. albo-atrum* (Typas & Heale, 1976) and the remaining six diploids of *A. nidulans* (A. Upshall, personal communication) all segregated on haploidization and hence were heterozygous for at least one gene. The heterozygous diploids of *A. nidulans* may have originated through errors of meiosis rather than a parasexual mechanism but this can not be the case for the two asexual species. The heterozygosity of the latter may have arisen after diploidization, or it may reflect the genetic diversity present within heterokaryon-compatibility groups (Croft & Jinks, 1977) or the rare formation of diploids across an incompatibility barrier. Whatever their origin, these natural heterozygous diploids will act as stores of genetic variability and will release this variability by mitotic recombination and haploidization.

In conclusion, it is now clear that parasexual processes occur in nature but the available information is insufficient to permit evaluation of their frequency and significance. The contribution of the parasexual processes to natural variation will be limited by the vegetative-incompatibility systems, which severely restrict the diversity of the interacting genomes. Nevertheless, heterozygous diploids exist in natural populations and there is a need for more work to establish how common they are, and the extent and nature of the variation they carry.

Concluding remarks

It is more than twenty years since G. Pontecorvo, J. A. Roper and E. Käfer elucidated the parasexual cycle in *Aspergillus nidulans* and our understanding of the basic parasexual processes has changed little since then. More recent studies have confirmed the impression, periodically advanced in the literature, that processes other than the standard parasexual cycle also operate but the mechanisms involved remain to be established. Transformation has recently been convincingly demonstrated in fungi (Hinnen *et al.*, 1979; Mishra, 1979), and should be added to the processes described since it comes within the definition of parasexual. Fungal protoplast fusion is another new development which, by making possible heterokaryon formation between strains too distantly related to anastomose naturally, greatly extends the scope for investigations and applications of parasexual interactions (Peberdy, 1980).

In considering the variability of the parasexual processes it is noticeable that the standard parasexual cycle is normally found where compatible mating-type factors are not present. Parasexual recombination occurs in sexually compatible combinations but it appears to involve non-standard mechanisms where diploids cannot be recovered from vegetative cells or are only recovered under special conditions when perhaps the mating-type genes are inactive. Thus although parasexuality is by definition outside the normal sexual interactions, the mating-type genes still have a strong influence upon the mechanisms involved.

Acknowledgements. I am grateful to Dr H. Howell and Dr A. Upshall for permission to refer to their unpublished results. I thank Professor J. L. Jinks, Dr J. Croft and Dr A. Upshall for many valuable discussions on fungal genetics and for their helpful comments on the manuscript.

References

Atwood, K. C. & Mukai, F. (1955). Nuclear distribution in conidia of *Neurospora* heterokaryons. *Genetics*, **40**, 438–43.

Bartos, P., Fleischmann, G., Samborski, D. J. & Shipton, W. A. (1969). Studies on asexual variation in the virulence of oat rust, *Puccinia coronata* f. sp. *avenae* and wheat leaf rust, *Puccinia recondita*. *Canadian Journal of Botany*, **47**, 1383–87.

Beccari, E., Modigliani, P. & Morpurgo, G. (1967). Induction of inter- and intragenic mitotic recombination by fluorodeoxyuridine and fluorouracil in *Aspergillus nidulans*. *Genetics*, **56**, 7–12.

Bignami, M. (1977). Mutagenic and recombinogenic action of pesticides in *Aspergillus nidulans*. *Mutation Research*, **46**, 395–402.

Bignami, M., Morpurgo, G., Pagliani, R., Carere, A., Conti, G. & Di Giuseppe, G. (1974). Non-disjunction and crossing-over induced by pharmaceutical drugs in *Aspergillus nidulans*. *Mutation Research*, **26**, 159–70.

Brody, T. & Williams, K. L. (1974). Cytological analysis of the parasexual cycle in *Dictyostelium discoideum*. *Journal of General Microbiology*, **82**, 371–83.

Brunswick, H. (1924). Untersuchungen über die Geschlechts- und Kernverhältnisse bei der Hymenomyzetengattung, *Coprinus*. *Abhandlungen der Botanisch königlich Goebel*, **5**, 1–152.

Burnett, J. H. (1975). *Mycogenetics*. London: Wiley.

Burnett, J. H. (1976). *Fundamentals of Mycology*, 2nd edn. London: Edward Arnold.

Buxton, E. W. (1956). Heterokaryosis and parasexual recombination in pathogenic strains of *Fusarium oxysporum*. *Journal of General Microbiology*, **15**, 133–9.

Buxton, E. W. (1959). Mechanisms of variation in *Fusarium oxysporum* in relation to host–parasite interactions. In *Plant Pathology: Problems and Progress 1908–1958*, ed. C. S. Holton, G. W. Fischer, R. W. Fulton, H. Hart, S. E. A. McCallan, pp. 183–91. Madison: University of Wisconsin Press.

Case, M. E. & Giles, N. H. (1962). The problem of mitotic recombination in *Neurospora*. *Neurospora Newsletter*, **2**, 6–7.

Casselton, L. A. (1965*a*). Somatic recombination in fungi. *Science Progress, Oxford*, **53**, 107–15.

Casselton, L. A. (1965*b*). The production and behaviour of diploids of *Coprinus lagopus*. *Genetical Research*, **6**, 190–208.

Casselton, L. A. & Lewis, D. (1966). Compatibility and stability of diploids in *Coprinus lagopus*. *Genetical Research*, **8**, 61–72.

Catcheside, D. G. (1977). *The genetics of recombination*. London: Edward Arnold.

Caten, C. E. & Day, A. W. (1977). Diploidy in plant pathogenic fungi. *Annual Review of Phytopathology*, **15**, 295–318.

Caten, C. E. & Jinks, J. L. (1966). Heterokaryosis: its significance in wild homothallic Ascomycetes and Fungi Imperfecti. *Transactions of the British Mycological Society*, **49**, 81–93.

Christensen, C. M., López, F. L. C. & Benjamin, C. R. (1965). A new *Eurotium* from rough rice stored in Mexico. *Mycologia*, **57**, 535–42.

Clutterbuck, A. J. & Roper, J. A. (1966). A direct determination of nuclear distribution in heterokaryons of *Aspergillus nidulans*. *Genetical Research*, **7**, 185–94.

Cove, D. J. (1976). Chlorate toxicity in *Aspergillus nidulans*: the selection and characterisation of chlorate resistant mutants. *Heredity*, **36**, 191–203.

Croft, J. H. & Jinks, J. L. (1977). Aspects of the population genetics of *Aspergillus nidulans*. In *Genetics and Physiology of* Aspergillus, ed. J. E. Smith & J. A. Pateman, pp. 339–360. New York & London: Academic Press.

Dales, R. B. G. & Croft, J. H. (1977). Protoplast fusion and the isolation of heterokaryons and diploids from vegetatively incompatible strains of *Aspergillus nidulans*. *FEMS Microbiology Letters*, **1**, 201–4.

Dales, R. B. G. & Croft, J. H. (1980). Protoplast fusion and the genetical analysis of vegetative incompatibility in *Aspergillus nidulans*. *Advances in Protoplast Research*, ed. L. Ferenczy & G. L. Farkas, pp. 73–84. Proceedings of the 5th International Protoplast Symposium. Budapest: Akadémiai Kiadó.

Davidse, L. C. & Flach, W. (1977). Differential binding of methyl benzimidazol-2-yl carbamate to fungal tubulin as a mechanism of resistance to this antimitotic agent in mutant strains of *Aspergillus nidulans*. *Journal of Cell Biology*, **72**, 174–93.

Day, A. W. (1978). Chromosome transfer in dikaryons of a smut fungus. *Nature, London*, **273**, 753–5.

Day, A. W. (1979). Mating type and morphogenesis in *Ustilago violacea*. *Botanical Gazette*, **140**, 94–101.

Day, A. W. & Day, L. L. (1974). The control of karyogamy in somatic cells of *Ustilago violacea*. *Journal of Cell Science*, **15**, 619–32.

Day, A. W. & Jones, J. K. (1968). The production and characteristics of diploids in *Ustilago violacea*. *Genetical Research*, **11**, 63–81.

Day, A. W. & Jones, J. K. (1972). Somatic nuclear division in the sporidia of *Ustilago violacea*. I. Acetic orcein staining. *Canadian Journal of Microbiology*, **18**, 663–70.

Day, P. R. & Anagnostakis, S. L. (1971). Meiotic products from natural infections of *Ustilago maydis*. *Phytopathology*, 61, 1020–1.

De Bertoldi, M. & Caten, C. E. (1975). Isolation and haploidization of heterozygous diploid strains in a species of *Humicola*. *Journal of General Microbiology*, 91, 63–73.

Ellingboe, A. H. (1961). Somatic recombination in *Puccinia graminis tritici*. *Phytopathology*, 51, 13–15.

Ellingboe, A. H. (1963). Illegitimacy and specific factor transfer in *Schizophyllum commune*. *Proceedings of the National Academy of Sciences, USA*, 49, 286–92.

Ellingboe, A. H. (1964). Somatic recombination in dikaryon K of *Schizophyllum commune*. *Genetics*, 49, 247–51.

Esposito, R. E. & Holliday, R. (1964). The effect of 5-fluorodeoxyuridine on genetic replication and somatic recombination in synchronously dividing cultures of *Ustilago maydis*. *Genetics*, 50, 1009–17.

Esser, K. (1971). Breeding systems in fungi and their significance for genetic recombination. *Molecular and General Genetics*, 110, 86–100.

Esser, K. & Blaich, R. (1973). Heterogenic incompatibility in plants and animals. *Advances in Genetics*, 17, 107–52.

Esser, K. & Kuenen, R. (1967). *Genetics of Fungi*. New York: Springer-Verlag.

Fabre, F. & Roman, H. (1977). Genetic evidence for inducibility of recombination competence in yeast. *Proceedings of the National Academy of Sciences, USA*, 74, 1667–71.

Fincham, J. R. S., Day, P. R. & Radford, A. (1979). *Fungal Genetics*, 4th edn. Oxford: Blackwell.

Frankel, C. (1979). Meiotic-like recombination in vegetative dikaryons of *Schizophyllum commune*. *Genetics*, 92, 1121–6.

Frankel, C. & Ellingboe, A. H. (1976). Isolation and characterisation of compatible diploids of *Schizophyllum commune*. *Molecular and General Genetics*, 148, 225–31.

Frankel, C. & Ellingboe, A. H. (1977). Sexual incompatibility factors and somatic recombination in *Schizophyllum commune*. *Genetics*, 85, 427–37.

Garnjobst, L. (1955). Further analysis of genetic control of heterocaryosis in *Neurospora crassa*. *American Journal of Botany*, 42, 444–8.

Gross, S. R. (1952). Heterokaryosis between opposite mating types in *Neurospora crassa*. *American Journal of Botany*, 39, 574–7.

Gutz, H. (1966). Induction of mitotic segregation with *p*-fluorophenylalanine in *Schizosaccharomyces pombe*. *Journal of Bacteriology*, 92, 1567–8.

Hamlyn, P. F. & Ball, C. (1979). Recombination studies with *Cephalosporium acremonium*. In *Genetics of Industrial Microorganisms*, ed. O. K. Sebek & A. I. Laskin, pp. 185–91. Proceedings of the 3rd International Symposium on Genetics of Industrial Microorganisms. Washington, DC: American Society for Microbiology.

Hastie, A. C. (1962). Genetic recombination in the hop-wilt fungus, *Verticillium albo-atrum*. *Journal of General Microbiology*, 27, 373–82.

Hastie, A. C. (1967). Mitotic recombination in conidiophores of *Verticillium albo-atrum*. *Nature, London*, 214, 249–52.

Hastie, A. C. (1968). Phialide analysis of mitotic recombination in *Verticillium*. *Molecular and General Genetics*, 102, 232–40.

Hastie, A. C. (1970). The genetics of asexual phytopathogenic fungi with special reference to Verticillium. In *Root Disease and Soil-borne Pathogens*, ed. T. A. Toussoun, R. V. Bega & P. E. Nelson, pp. 55–62. Berkeley: University of California Press.

Hastie, A. C. (1973). Hybridization of *Verticillium albo-atrum* and *Verticillium dahliae*. *Transactions of the British Mycological Society*, **60**, 511–23.

Hinnen, A., Hicks, J. B., Ilgen, C. & Fink, G. R. (1979). Yeast transformation: a new approach for the cloning of eucaryotic genes. In *Genetics of Industrial Microorganisms*, ed. O. K. Sebek & A. I. Laskin, pp. 36–43. Proceedings of the 3rd International Symposium on Genetics of Industrial Microorganisms. Washington, DC: American Society for Microbiology.

Holliday, R. (1961). Induced mitotic crossing-over in *Ustilago maydis*. *Genetical Research*, **2**, 231–48.

Holliday, R. (1964). The induction of mitotic recombination by mitomycin C in *Ustilago* and *Saccharomyces*. *Genetics*, **50**, 323–35.

Holliday, R. (1967). Altered recombination frequencies in radiation sensitive strains of *Ustilago maydis*. *Mutation Research*, **4**, 275–88.

Holloman, W. K. & Holliday, R. (1973). Studies of a nuclease from *Ustilago maydis*. I. Purification, properties and implications in recombination of the enzyme. *Journal of Biological Chemistry*, **248**, 8107–13.

Ingram, R. (1968). *Verticillium dahliae* var. *logisporum*, a stable diploid. *Transactions of the British Mycological Society*, **51**, 339–41.

Ishitani, C., Ikeda, Y. & Sakaguchi, K. (1956). Hereditary variation and genetic recombination in Koji-molds (*Aspergillus oryzae* and *Asp. sojae*). VI. Genetic recombination in heterozygous diploids. *Journal of General and Applied Microbiology*, **2**, 401–30.

Jansen, G. J. O. (1970). Abnormal frequencies of spontaneous mitotic recombination in *uvsB* and *uvsC* mutants of *Aspergillus nidulans*. *Mutation Research*, **10**, 33–41.

Käfer, E. (1961). The processes of spontaneous recombination in vegetative nuclei of *Aspergillus nidulans*. *Genetics*, **46**, 1581–609.

Käfer, E. (1977). Meiotic and mitotic recombination in *Aspergillus* and its chromosomal aberrations. *Advances in Genetics*, **19**, 33–131.

Kappas, A., Georgopoulous, S. G. & Hastie, A. C. (1974). On the genetic activity of benzimidazole and thiophanate fungicides on diploid *Aspergillus nidulans*. *Mutation Research*, **26**, 17–27.

Katz, E. R. & Sussman, M. (1972). Parasexual recombination in *Dictyostelium discoideum*: selection of stable diploid heterozygotes and stable haploid segregants. *Proceedings of the National Academy of Sciences, USA*, **69**, 495–8.

Kinghorn, J. R. & Pateman, J. A. (1975). Mutations which affect amino acid transport in *Aspergillus nidulans*. *Journal of General Microbiology*, **86**, 174–84.

Koltin, Y. & Raper, J. R. (1968). Dikaryosis: genetic determination in *Schizophyllum*. *Science*, **160**, 85–6.

Lanier, W. B., Tuveson, R. W. & Lennox, J. E. (1968). A radiation sensitive mutant of *Aspergillus nidulans*. *Mutation Research*, **5**, 23–31.

Leonard, T. J., Dick, S. & Gaber, R. F. (1978). Internuclear genetic transfer in vegetative dikaryons of *Schizophyllum commune*. I. Di–mon mating analysis. *Genetics*, **88**, 13–26.

Leonard, T. J., Gaber, R. F. & Dick, S. (1978). Internuclear genetic transfer in dikaryons of *Schizophyllum commune*. II. Direct recovery and analysis of recombinant nuclei. *Genetics*, **89**, 685–93.

Lhoas, P. (1961). Mitotic haploidization by treatment of *Aspergillus niger* diploids with *p*-fluorophenylalanine. *Nature, London*, **190**, 744.

Lhoas, P. (1967). Genetic analysis by means of the parasexual cycle in *Aspergillus niger*. *Genetical Research*, **10**, 45–61.

Maclean, D. J., Tommerup, I. C. & Scott, K. J. (1974). Genetic status of monokaryotic variants of the wheat stem rust fungus isolated from axenic culture. *Journal of General Microbiology*, **84**, 364–78.

Ming, Y. N., Lin, P. C. & Yu, T. F. (1966). Heterokaryosis in *Fusarium fujikuroi* (Sacc.). *Scientia Sinica*, **15**, 371–8.

Mishra, N. C. (1979). DNA-mediated genetic changes in *Neurospora crassa*. *Journal of General Microbiology*, **113**, 255–9.

Morpurgo, G. (1961). Somatic segregation induced by p-fluorophenylalanine. *Aspergillus Newsletter*, **2**, 10.

Morpurgo, G. (1963). Induction of mitotic crossing-over in *Aspergillus nidulans* by bifunctional alkylating agents. *Genetics*, **48**, 1259–63.

Morris, N. R. & Oakley, C. E. (1979). Evidence that p-fluorophenylalanine has a direct effect on tubulin in *Aspergillus nidulans*. *Journal of General Microbiology*, **114**, 449–54.

Mylyk, O. M. (1976). Heteromorphism for heterokaryon incompatibility genes in natural populations of *Neurospora crassa*. *Genetics*, **83**, 275–84.

Newmeyer, D. (1970). A suppressor of the heterokaryon-incompatibility reaction associated with mating type in *Neurospora crassa*. *Canadian Journal of Genetics and Cytology*, **12**, 914–26.

Nga, B. H. & Roper, J. A. (1969). A system generating spontaneous intra-chromosomal changes at mitosis in *Aspergillus nidulans*. *Genetical Research*, **14**, 63–70.

Nga, B. H., Teo, S.-P. & Lim, G. (1975). The occurrence in nature of a diploid strain of *Aspergillus niger*. *Journal of General Microbiology*, **88**, 364–6.

Olson, L. W. & Zimmermann, F. K. (1978). Mitotic recombination in the absence of synaptonemal complexes in *Saccharomyces cerevisiae*. *Molecular and General Genetics*, **166**, 161–5.

Papazian, H. P. (1950). Physiology of the incompatibility factors in *Schizophyllum commune*. *Botanical Gazette*, **112**, 143–63.

Parag, Y. (1962). Studies in somatic recombination in dikaryons of *Schizophyllum commune*. *Heredity*, **17**, 305–18.

Parag, Y. (1968). Phase-microscopic observations of fusions of nuclei in somatic cells of a heterokaryon of *Schizophyllum commune*. *American Journal of Botany*, **55**, 984–88.

Parag, Y. & Parag, G. (1975). Mutations affecting mitotic recombination frequency in haploids and diploids of the filamentous fungus *Aspergillus nidulans*. *Molecular and General Genetics*, **137**, 109–23.

Park, S., Kenehan, P. & Goodgal, S. (1968). Heterocaryons and recombination in *Phycomyces blakesleeanus*. *Genetics*, **60**, 209–10.

Parameter, J. R., Snyder, W. C. & Reichle, R. E. (1963). Heterokaryosis and variability in plant-pathogenic fungi. *Annual Review of Phytopathology*, **1**, 51–76.

Pederdy, J. F. (1980). Protoplast fusion – a new approach to interspecies genetic manipulation and breeding in fungi. In *Advances in Protoplast Research*, pp. 63–72. Proceedings of the 5th International Protoplast Symposium. Budapest: Akadémiai Kiadó.

Pontecorvo, G. (1954). Mitotic recombination in the genetic system of filamentous fungi. *Caryologia, Florence*, **6**, Suppl., 192–200.

Pontecorvo, G. (1956). The parasexual cycle. *Annual Review of Microbiology*, **10**, 393–400.

Pontecorvo, G. (1958). *Trends in genetic analysis*. New York: Columbia University Press.

Pontecorvo, G. & Käfer, E. (1958). Genetic analysis based on mitotic recombination. *Advances in Genetics*, **9**, 71–104.

Pontecorvo, G., Roper, J. A., Hemmons, L. M., Macdonald, K. D. & Bufton, A. W. J. (1953). The genetics of *Aspergillus nidulans*. *Advances in Genetics*, **5**, 141–238.

Pritchard, R. H. (1954). Ascospores with diploid nuclei in *Aspergillus nidulans*. *Caryologia, Florence*, **6** Suppl., 1117.

Puhalla, J. E. & Mayfield, J. E. (1974). The mechanism of heterokaryotic growth in *Verticillium dahliae*. *Genetics*, **76**, 411–22.

Putrament, A. (1964). Mitotic recombination in the *pabal* cistron of *Aspergillus nidulans*. *Genetical Research* **5**, 316–27.

Quintanilha, A. (1939). Étude génétique du phénomène de Buller. *Boletim da Sociedade Broteriana*, **13**, 425–86.

Raper, J. R. (1966). *Genetics of Sexuality in Higher Fungi*. New York: Ronald Press.

Rayner, A. D. M. & Todd, N. K. (1979). Population and community structure and dynamics of fungi in decaying wood. *Advances in Botanical Research*, **7**, 333–420.

Robinow, C. F. & Caten, C. E. (1969). Mitosis in *Aspergillus nidulans*. *Journal of Cell Science*, **5**, 403–31.

Roman, H. (1956). Studies of gene mutation in *Saccharomyces*. *Cold Spring Harbor Symposia on Quantitative Biology*, **21**, 175–83.

Roper, J. A. (1966). The parasexual cycle. In *The Fungi* (ed. G. C. Ainsworth & A. S. Sussman), vol. 2, pp. 589–617. New York & London: Academic Press.

Roper, J. A. & Pritchard, R. H. (1955). The recovery of the complementary products of mitotic crossing over. *Nature, London*, **175**, 639.

Sansome, E. R. (1946*a*). Induction of 'gigas' forms of *Penicillium notatum* with camphor vapour. *Nature, London*, **157**, 843–4.

Sansome, E. R. (1946*b*). Heterokaryosis, mating-type factors and sexual reproduction in *Neurospora*. *Bulletin of the Torrey Botanical Club*, **73**, 397–409.

Sermonti, G. (1969). *Genetics of Antibiotic-producing Microorganisms*. London: Wiley Interscience.

Shanfield, B. & Käfer, E. (1969). UV-sensitivity mutants increasing mitotic crossing-over in *Aspergillus nidulans*. *Mutation Research*, **7**, 485–7.

Shanfield, B. & Käfer, E. (1971). Chemical induction of mitotic recombination in *Aspergillus nidulans*. *Genetics* **67**, 209–19.

Sheir-Neiss, G., Lai, M. H. & Morris, N. R. (1978). Identification of a gene for β-tubulin in *Aspergillus nidulans*. *Cell*, **15**, 639–47.

Singh, M. & Sinha, U. (1976). Chloral hydrate induced haploidization in *Aspergillus nidulans*. *Experientia*, **32**, 1144–5.

Smith, D. A. (1974). Unstable diploids of *Neurospora* and a model for their somatic behaviour. *Genetics*, **76**, 1–17.

Stahl, F. W. (1979). *Genetic Recombination: thinking about it in Phage and Fungi*. San Francisco: W. H. Freeman.

Strømnaes, Ø., Garber, E. D. & Beraha, L. (1964). Genetics of phytopathogenic fungi. IX. Heterocaryosis and the parasexual cycle in *Penicillium italicum* and *Penicillium digitatum*. *Canadian Journal of Botany*, **42**, 423–7.

Tinline, R. D. & MacNeill, B. H. (1969). Parasexuality in plant pathogenic fungi. *Annual Review of Phytopathology*, **7**, 147–70.

Tolmsoff, W. J. (1972). Diploidization and heritable gene repression–derepression as major sources for variability in morphology, metabolism and pathogenicity of *Verticillium* species. *Phytopathology*, **62**, 407–13.

Typas, M. A. & Heale, J. B. (1976). Acriflavine-induced hyaline variants of *Verticillium albo-atrum* and *V. dahliae*. *Transactions of the British Mycological Society*, **66**, 15–25.

Typas, M. A. & Heale, J. B. (1977). Analysis of ploidy levels in strains of *Verticillium* using a coulter counter. *Journal of General Microbiology*, **101**, 177–80.

Upshall, A., Giddings, B. & Mortimore, I. D. (1977). The use of benlate for distinguishing between haploid and diploid strains of *Aspergillus nidulans* and *Aspergillus terreus. Journal of General Microbiology*, **100**, 413–18.

Upshall, A., Giddings, B., Teow, S. C. & Mortimore, I. D. (1979). Novel methods of genetic analysis in fungi. In *Genetics of Industrial Microorganisms*, ed. O. K. Sebek & A. I. Laskin, pp. 197–204. Proceedings of the 3rd International Symposium on Genetics of Industrial Microorganisms. Washington, DC: American Society for Microbiology.

Van Arkel, G. A. (1963). Sodium arsenate as an inducer of somatic reduction. *Aspergillus Newsletter*, **4**, 9.

Webster, R. K. (1974). Recent advances in the genetics of plant pathogenic fungi. *Annual Review of Phytopathology*, **12**, 331–53.

Westergaard, M. & von Wettstein, D. (1972). The synaptonemal complex. *Annual Review of Genetics*, **6**, 71–110.

Williams, K. L., Kessin, R. H. & Newell, P. C. (1974). Parasexual genetics in *Dictyostelium discoideum:* mitotic analysis of acriflavin resistance and growth in axenic medium. *Journal of General Microbiology*, **84**, 59–69.

Wood, S. & Käfer, E. (1969). Effects of ultraviolet irradiation on heterozygous diploids of *Aspergillus nidulans*. I. UV-induced mitotic crossing-over. *Genetics*, **62**, 507–18.

Zimmermann, F. K. (1971). Induction of mitotic gene conversion by mutagens. *Mutation Research,* **11**, 327–37.

Regular and aberrant segregation at meiosis

B . C . LAMB

Department of Botany, Imperial College, London SW7 2BB, UK

Introduction

This review will mainly concentrate on various types of aberrant segregation because most work on regular segregation is older and well established. In a cross between a fungus with a wild-type allele (+) and a mutant (*m*) strain, regular Mendelian segregation gives a ratio of 1+ : 1*m* amongst random meiotic products, 2+ : 2*m* in tetrads and 4+ : 4*m* in octads, where the four meiotic products have become eight by a mitosis giving four pairs of sister spores. Several different types and causes of aberrant segregation at meiosis in fungi will be considered here, including polarized segregation, nondisjunction, chromosome aberrations and gene conversion. Other causes of aberrant ratios include gene interaction, polyploidy and polysomy, somatic crossing-over, aberrant nuclear behaviour and spontaneous mutation: they are described in the classic review by Emerson (1956) with reference to *Saccharomyces*.

Gene conversion is a most important kind of aberrant segregation and one major section here will be on how its study in fungi can be used to elicit molecular details of meiotic recombination. The other major section will be on the implications of gene conversion for the population genetics and evolution of fungi. Gene conversion was defined by Lindegren (1955) as: 'the interaction, occurring at meiosis, between the dominant and recessive alleles in a heterozygote resulting in the transformation of one or more dominant alleles into the corresponding recessive allele, or vice versa. Gene conversion is essentially a directed mutation occurring at meiosis as a result of the effect of homologous alleles upon each other; it does not occur (or is not apparent) at meiosis of homozygous diploids.' As it is now known that conversion can occur

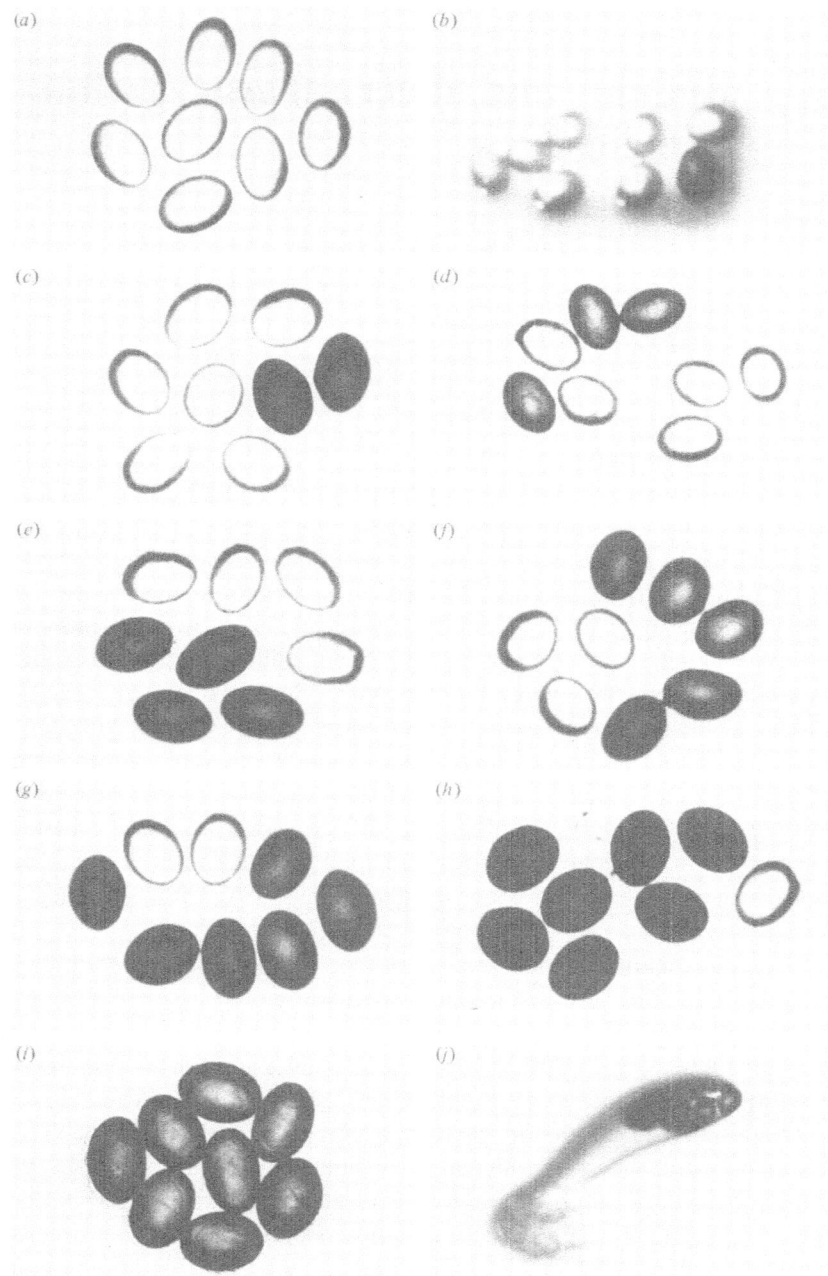

Fig. 1. *Ascobolus immersus*, + × *w* crosses. (*a–i*) unordered octads
from the cross 1WT3, + × EC11 *w* – 78, 7, – showing wider ratio (*a*,

at mitosis, though very rarely, I prefer to define gene conversion as: a process of chemical interaction between alleles, usually at meiosis, where one allele can convert another to being of its own kind in a heterozygote. For example, at meiosis in an Aa ascus, A may convert an a allele to A, affecting the expected $2A : 2a$ ratio in the meiotic tetrad. So gene conversion can cause aberrant segregation ratios such as $3A : 1a$ or $1A : 3a$, or in octads, ratios such as $6 : 2, 2 : 6, 5 : 3, 3 : 5, 7 : 1$, etc. Figs. 1 and 2 show normal ($4+ : 4$ ascospore colour mutant) and aberrant segregation ratios in octads from *Ascobolus immersus*, *Sordaria fimicola* and *Neurospora crassa*. For reviews of the mechanisms of crossing-over and gene conversion, see Fogel *et al.* (1979), Catcheside (1977), Pukkila (1977), Meselson & Radding (1975).

There are three main uses of regular segregation, as detected by studying the products of meiosis either as random spores or by tetrad analysis: (*a*) mapping genes or centromeres, (*b*) studying aspects of recombination such as whether it is reciprocal or not, and (*c*) detecting chromosome and chromatid interference. Crosses involving regular segregation at one locus are made to establish that a particular character difference really is controlled by a pair of Mendelian alleles at one locus, rather than environmentally or polygenically. Crosses involving segregation at two loci are used for determining recombination frequencies and hence linkage relations, while three-factor crosses are used to establish gene order for three linked loci and to investigate chromosome (chiasma position) interference between crossovers. Unordered tetrads from dihybrid crosses give information on linkage relations from the relative frequencies of parental ditype, non-parental ditype and tetratype tetrads, and on whether recombination is reciprocal or not. Unordered tetrads from crosses with three linked markers give

---

b, *h*, *i*) and narrower ratio (*c*, *d*, *f*, *g*) gene conversion classes and a normal 4 : 4 segregation (*e*), with all possible ratios from 0 : 8 (*a*) to 8 : 0 (*i*). + spores are red, $w - 78^-$ spores are white. (*j*) shows an intact 8 : 0 ascus from the cross V46 − 2R, + × 549 $w - 78$, −; in this + × $w - 78$ cross, finding an 8 : 0 segregation in an intact ascus with eight good spores shows that the wider ratio is not a result of false clustering of unrelated spores from different asci in dehisced octads, and has not arisen by nondisjunction. With aberrant segregation octads, back-crosses from germinated spores confirmed normal segregation ($4+ : 4−$) for mating type and confirmed spore genotypes, showing that these aberrant ratios came from gene conversion, not false clustering, additional nuclear divisions, aneuploidy, etc. The ascospores range from about 50 to 60 μm long and 30 to 35 μm wide.

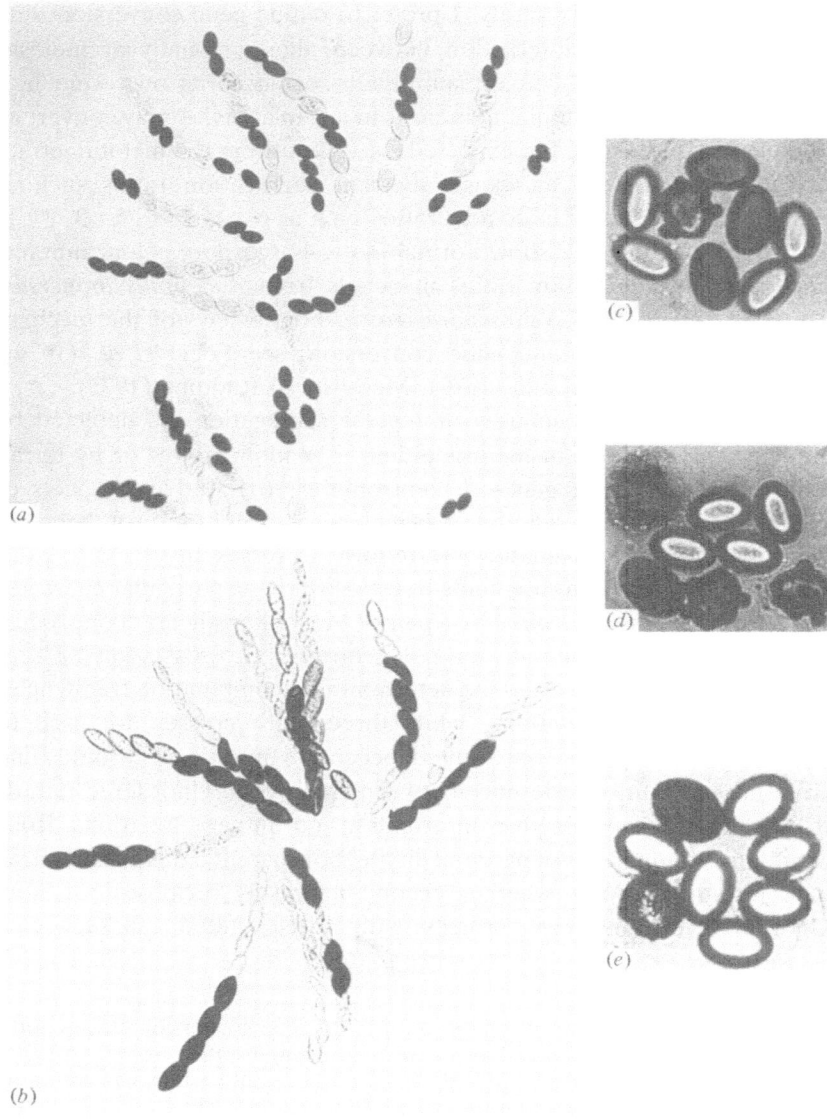

Fig. 2 (*a*) Ordered octads from the cross C7+ × C7 *hyaline* in *Sordaria fimicola*. Wild-type spores are black, *h* are hyaline, and the apical pores are peripheral to each group of asci. Most asci show regular 4+ : 4*h* segregations of the six ordered types I to VI (see Fig. 3) with a high frequency of second division segregation (III to VI) classes, showing that the *h* locus is far from its centromere. The third ascus from the right shows a 3+ : 5*h* gene conversion ratio. Ascospores are about 20 – 23 μm long. (*b*) Ordered octads from the cross A4+*a* ×

information on chromatid (strand) interference between crossovers. The distance between a locus and its centromere can be determined in ordered tetrads or octads in monohybrid crosses, and by the method of Whitehouse (1957) one can calculate centromere distances from unordered tetrads for a set of three unlinked loci. Details of these techniques based on regular segregation at meiosis are given by Fincham, Day & Radford (1979), Catcheside (1977) and Burnett (1975). The aspects of the mechanisms of regular segregation to be considered here are the role of the synaptonemal complex in meiotic chromosomal pairing and recombination, and the function of recombination nodules.

Polarized segregation

Meiosis in linear asci, such as those giving ordered octads in *Neurospora crassa,* provides ideal conditions for studying whether the two members of a homologous pair of chromosomes truly segregate at random with respect to the two poles of a spindle, as these asci have an identifiable apex and base. The results of segregation for a pair of alleles can be observed directly in intact asci by using an autonomously controlled ascospore pigmentation marker, such as wild-type (black) and *asco* (very pale) in *N. crassa* (Fig. 2b). The six possible ordered ascus classes for such a pair of alleles are shown in Fig. 3.

Polarized segregation was defined as 'the non-random segregation of the members of a pair of homologous chromosomes, or of a pair of daughter chromosomes, with respect to the two poles of a spindle in meiosis' (Lamb, 1966). It was first reported in Ascomycetes by Catcheside (1944) from data of Zickler on *Bombardia lunata,* and from *Neurospora crassa* by Nakamura (1961) and McNelly (1962). It is possible to study separately whether segregation is polarized at the first

---

asco33A in *Neurospora crassa.* Wild-type spores are black, *asco* are very pale; apical pores are peripheral to the group of asci. All asci show regular $4+$: 4 *asco* segregation (one ascus is immature). There is a low frequency of second division segregation as *asco* is not a long way from its centromere. Ascospores are about 25–31 μm long. $(c - e)$ Unordered octads from the cross $w - 78^-, gr - 3^+ \times w - 78^+, gr - 3^-$ in *Ascobolus immersus* (Ghikas, 1978), where $w - 78^+$ is epistatic to $gr - 3$; $w - 78^+, gr^+$ spores are red, smooth; $w - 78^+, gr^-$ spores are red, granular, and $w - 78^-, gr^+$ and $w - 78^-, gr^-$ spores are white, smooth. (c), (d) and (e) have normal segregation for $gr - 3$; (c) and (d) have aberrrant 4 : 4 for $w - 78$ and (e) has 2 : 6 (2) for $w - 78$ (see Ghikas & Lamb, 1977, and the present text). The ascospores are about 50–60 μm long.

division spindle (by comparing classes I and II), at the second division upper spindle (by comparing III plus V with IV plus VI) and at the second division lower spindle (by comparing IV plus V with III plus VI). In the works quoted, biased segregation – with the chromosome or chromatid bearing the wild type appearing to segregate more often to the upper than to the lower spindle – was usually found at the first division spindle and the second division lower spindle, but less often at the second division upper spindle.

Catcheside (1944) suggested that there was a gradient of some diffusible substance in the áscus to which the chromosomes responded at segregation. The phenomenon was investigated by Lamb (1966) for several loci in *Sordaria fimicola* (Fig. 2*a*) and *Neurospora crassa* (Fig. 2*b*), using crosses with a fairly high degree of synchrony in perithecial development to determine whether polarized segregation arose at meiosis, or whether it was an artifact arising post-meiotically from a differential maturation and bursting of the six ascal classes. Segregation patterns were scored from perithecia at widely differing stages of maturity. Significant polarized segregation was not found for any of the

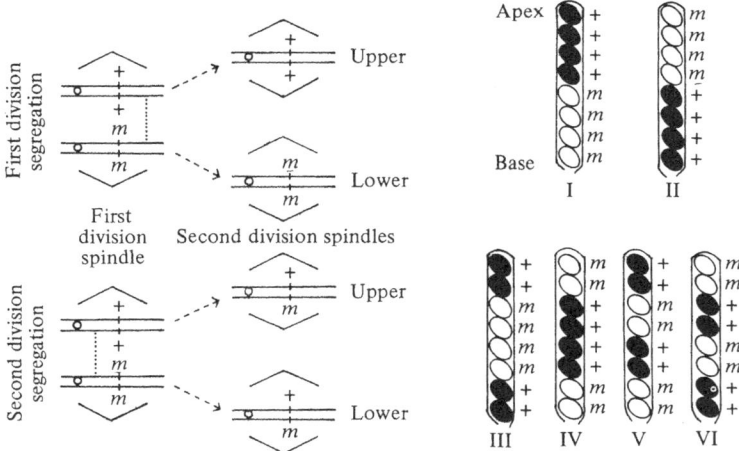

Fig. 3. The origin of the six ordered ascus classes for a single pair of alleles, in octads. +, wild-type (e.g. black ascospore), *m*, mutant (e.g. colourless ascospore). I and II are first division segregation classes arising if there is no crossover between the locus and the centromere (which is shown as a circle between two sister chromatids). III, IV, V and VI are second division segregation classes and arise after a single crossover (dotted line) between the locus and the centromere. Each chromatid is shown as a single line, with a shallow 'V' indicating a spindle pole.

three spindles when little or no dehiscence had occurred from a cross, but apparent polarized segregation became readily manifest as dehiscence of mature asci proceeded. This showed that the differences in frequency between the six ascal classes, that had previously been interpreted as showing polarized segregation at meiosis, were actually due to a post-meiotic differential bursting of asci. Analysis of bias in terms of segregation at the three meiotic spindles therefore had to be replaced by analysis in terms of individual ascus classes, where class II matured and burst before class I, and class VI before III, before IV, before V. Bursting was studied separately from the usually concurrent pigmentation of immature asci by comparing results from asci mounted in water with those from asci mounted in the 2 M sucrose solution used throughout the main experiments: the differential bursting produced by mounting asci in water gave similar polarized segregation results to those from natural dehiscence with time. An hypothesis to account for differential maturation of the six ordered classes was advanced, based on superior metabolic activity in the vicinity of wild-type nuclei, especially in the transport of nutrients in through the base of the ascus (Lamb, 1966).

It seems likely that cases of apparent polarized segregation described in other Ascomycetes are also due to post-meiotic phenomena, giving similar biases in ascal segregation class frequencies.

Meiotic nondisjunction and aneuploidy

Fungal tetrads or octads, ordered or unordered, are ideal for studies of meiotic chromosomal nondisjunction and its production of aneuploids, which have chromosome numbers other than whole multiples of the haploid set. In the three species chiefly studied for aneuploidy, its main origin is from nondisjunction at meiosis in *Neurospora crassa,* nondisjunction in diploid mitosis in *Aspergillus nidulans,* and triploid meiosis in *Saccharomyces cerevisiae* (references in Fincham *et al.,* 1979). Yeast seems unusually tolerant of additional chromosomes, and segregation ratios from tetrads from crosses of haploid mutants to disomic (n + 1) wild-type strains (especially to multiple disomics) have been used by Mortimer & Hawthorne (1973) for rapid mapping of new mutations to particular linkage groups.

Disomic (n + 1) spores can either result from an extra chromosome replication or from nondisjunction: in the latter case an octad should also contain two or four spores with n − 1 chromosomes, depending on whether the irregular segregation occurred at the second or first division of meiosis, with these spores usually being visibly aborted. A useful system for studying these events has been described by Bond (1976, 1979) for *Sordaria brevicollis*. Two complementing alleles at the *buff* ascospore colour locus were crossed in repulsion; black wild-type spores from recombination or from reversion were not found, so black spores were deduced to be disomic, with complementation in the *trans* arrangement, and nullisomic (n − 1) spores were visibly abortive. Nondisjunction for the *buff* chromosome occurred with a frequency of about 4.3×10^{-4} at both the first and second divisions of meiosis, calculated from the frequencies of asci with black spores and abortive spores. Asci with from two to eight black spores and no abortive spores also occurred, but the frequency of additional chromosome replication – a likely cause of these asci – could not be directly calculated. The classic work of Pittenger and others on disomic 'pseudowild types' from ascospores of *Neurospora crassa* showed disomic frequencies for any one chromosome of about 1×10^{-3}, though with some variation; some 'pseudowild types' arose by nondisjunction, as shown by abortive spores in the same asci, while others probably arose by additional chromosomal replication as the other spores were all viable (references in Fincham *et al.*, 1979).

Chromosome aberrations

The visual inspection of ascospores and asci has provided rapid, effective techniques for detecting and characterizing chromosome arrangements, especially in *Neurospora*. In their important review, Perkins & Barry (1977) reported evidence for chromosome rearrangements from spore abortion in *Saccharomyces cerevisiae, Sordaria macrocarpa* and *Neurospora crassa,* and from phenomena such as meiotic and mitotic linkage, pachytene pairing, meiotic bridges and other cytological sources, and unstable bisexual progeny, in *Ascobolus immersus, Aspergillus nidulans, Coprinus radiatus, Sordaria brevicollis, Sordaria fimicola, Neurospora crassa, Cochliobolus heterostrophus* and *Phytophthora.*

The kinds of chromosome aberrations reported from fungi include non-reciprocal and reciprocal translocations, deletions, duplications (mainly or entirely non-tandem) and inversions (mainly or entirely pericentric). In *Neurospora*, non-deficient ascospores (even those that include duplications) are normally black (B), while deficiency ascospores are usually non-black (W). Perkins (1974) and Perkins & Barry (1977) have extensively shown that the frequencies of various unordered octad segregation classes (8B : 0W; 6B : 2W; 4B : 4W; 2B : 6W; 0B : 8W) from meiosis with a heterozygous rearrangement are diagnostic for the type of rearrangement and can provide information on the positions of the break points.

Gene conversion, crossing-over, chromosome pairing and the synaptonemal complex

A correlation between the occurrence of gene conversion and of crossing-over was found by Kitani, Olive & El-Ani (1962) in *Sordaria fimicola*, and various later studies on gene conversion in fungi have provided important information on recombination mechanisms (see Catcheside, 1977; Fogel *et al*, 1979). For example, Lamb (1977) showed how gene conversion could be used to study the topography and pairing relationships of the four chromatids of a bivalent at the time of crossing-over and hybrid DNA formation, the lengths of intimately paired segments, the frequency of intimate pairing at particular sites, and aspects of synaptonemal complex (SC) structure related to DNA pairing and exchange.

Chromatid pairing arrangements

Some possible chromatid arrangements during zygotene and pachytene pairing are shown in Fig. 4, with the chromatids in transverse section. With a linear order of chromatids in a bivalent at any point (Fig. 4a), two non-sister chromatids are adjacent at any point but intimate pairing for crossing-over and conversion only occurs occasionally (e.g. at Fig. 4a, i and v) along the generally paired section of chromatids. With this *linear* arrangement – as in the SC diagram of Moens (1974), Fig. 4(a), and as shown diagrammatically in most text-books – only two non-sister chromatids could pair at any one point. Thus if chromatids 2 and 3 are paired at a point, 1 and 4 could not also be paired there, though they could be paired further along the bivalent.

In the *compact* arrangement (Fig. 4b) it is possible to have no (iii), one (i), or both (ii) pairs of non-sister chromatids paired at any one point. The compact arrangement might, at any point, allow one chromatid to pair only or preferentially with one non-sister chromatid (e.g. 1 with 3 only in Fig. 4b, i), or with equal ease with either (but not both) non-sister chromatid (Fig. 4b, iv). The *linear/compact* diagrams (Fig. 4c) are based on the pairing hypothesis of von Wettstein (1971), especially his Fig. 6. In Fig. 4(c) i, before close pairing, each pair of sister chromatids is associated with one lateral component of the SC, but during close pairing (Fig. 4c, ii and iii), only one of each pair of sister chromatids is joined to the lateral element at any point; so only one pair of non-sister chromatids can pair at any point, although different chromatids can pair at different points (Fig. 4c, ii and iii). The *linear/compact* arrangement therefore has the pairing possibilities of the linear, not the compact, arrangement.

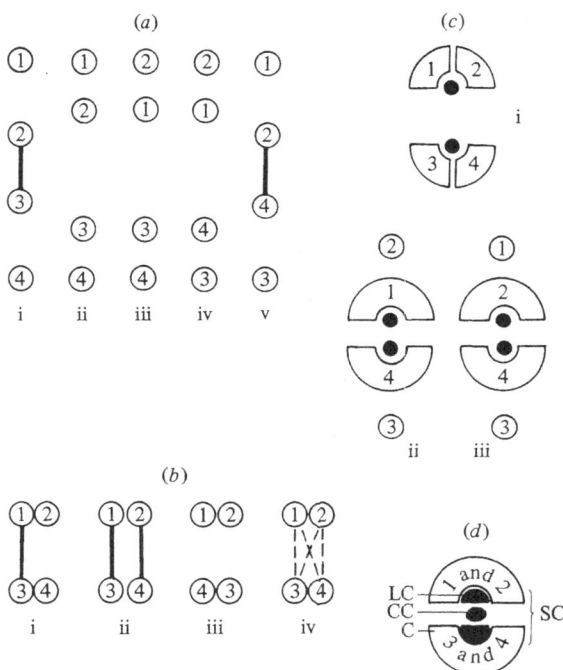

Fig. 4 (*a–c*) Possible chromatid arrangements in transverse section. The two pairs of sister chromatids are 1 and 2, and 3 and 4, respectively. Bold lines indicate intimate pairing; dashed lines show potential pairing. (*a*) *Linear* arrangements: i, 2 and 3 paired, so 1 and 4 cannot; ii, 2 and 3 adjacent but not intimately paired; iii–v, other possibilities. (*b*) *Compact* arrangements: i, 1 and 3 pair, 2 and 4 could pair but have not; ii, intimate pairing between both pairs of chromatids; iii, no pairing; iv, potential pairing. (*c*) *Linear/compact* arrangements: i, pairs of sister chromatids with a lateral component (dark circle) during approximate alignment of chromatids; ii, precise pairing of only two non-sister chromatids at any point, depending on which sister chromatid attaches to the lateral component at that point; iii, another arrangement. (*d*) Diagram of pachytene pairing based on electron micrographs of *Neottiella* (Figs. 1–8, Westergaard & von Wettstein, 1970). LC, lateral component; CC central component; C, chromosome, with chromatin of the two sister chromatids closely associated; SC synaptonemal complex.

In electron micrographs, the two chromatids of a chromosome are seldom distinguishable from each other during pachytene pairing. This was shown for the fungus *Neottiella* by Westergaard & von Wettstein (1970), and Fig. 4(*d*) here, of pachytene pairing, is based on their Figs. 1–8. Although the three pairing arrangements described above have not been distinguished microscopically, gene conversion makes it possible to distinguish between the compact arrangement, which permits both pairs of non-sister chromatids to be paired at one point (as in Fig. 4*b*, ii), and the other two arrangements which, in their basic forms, do not.

The pairing of two non-sister chromatids to give hybrid DNA can give the 'narrower ratio' conversion classes in $+ \times m$ crosses: 6 : 2, 2 : 6, 5 : 3 and 3 : 5 (all written $+ : m$), and these are consistent with compact, linear, or compact/linear arrangements. 'Wider ratio' octads (Lamb & Wickramaratne, 1973), 8 : 0, 0 : 8, 7 : 1 and 1 : 7, and 'unique' narrower ratio classes, 6 : 2(2), 2 : 6(2), 5 : 3(3), 3 : 5(3) and 4 : 4(4) (where the number in parentheses is the number of pairs of non-identical sister-spores) are not consistent with linear or compact/linear arrangements because these segregation classes require both pairs of non-sister chromatids in a bivalent to pair at exactly the same point, as is only possible with the compact arrangement. Genuine wider ratio and/or 'unique' narrower ratio classes have been proved to occur in *Neurospora crassa, Sordaria fimicola, S. brevicollis, Saccharomyces cerevisiae* and both European and American strains of *Ascobolus immersus* (see Lamb & Wickramaratne, 1973; Lamb, 1977; Ghikas & Lamb, 1977; Yu-Sun, Wickramaratne & Whitehouse, 1977; Fogel *et al.*, 1979, and also Fig. 1*a*, *b*, *h–j*; Fig. 2*e*).

The fungal conversion data thus show that both pairs of chromatids can be involved in hybrid DNA formation (and have therefore paired intimately) at exactly the same point in a bivalent, consistent with the compact arrangement of chromatids at pachytene. It might be argued that all three arrangements could occur, with corresponding-site events (those taking place in *both* pairs of non-sister chromatids in a bivalent at corresponding sites) being possible only when compact arrangements occur. This is readily tested from data on corresponding-site interference, which was defined by Lamb & Wickramaratne (1973) as: 'interference between the two pairs of non-sister chromatids of a bivalent in hybrid DNA formation at exactly corresponding sites'. If all three arrangements contribute to narrower ratios such as 6 : 2 but only compact ones contribute to wider ratios such as 8 : 0, then far fewer wider ratio octads will occur than calculated from narrower ratio frequencies, giving strong positive corresponding-site interference. The best data are from *Ascobolus* (see Lamb, 1977; Ghikas & Lamb, 1977) and *Sordaria brevicollis* (Yu-Sun *et al.*, 1977) and in no case is there strong positive corresponding-site inference, with this type of interference being absent or slight, with slight negative interference being more common than slight positive interference. This suggests that the compact pairing arrangement is the usual one, though occasional pairing by other methods is not excluded.

The extent of intimate pairing lengths

By comparing simultaneous conversions (co-conversion) at two linked sites in dihybrid crosses with independent events at each site, one can study co-hybrid DNA formation and/or co-correction of mispaired bases in hybrid DNA at the two sites. As the DNA length undergoing co-conversion can be limited by both the length of hybrid DNA and by the length of the excision track, co-conversion data provide only minimum estimates of molecular pairing lengths. The fungal co-conversion data from yeast, *Neurospora crassa, Schizosaccharomyces pombe* and *Sordaria fimicola* (see Lamb, 1977, and Fincham *et al.*, 1979, for references), indicate that the average lengths of converted segments of DNA are from about a hundred to over a thousand nucleotides, sometimes extending over large fractions of a gene or more than one gene. This is strong evidence against all models of the SC which involve severe restrictions on lengths of DNA that can be continuously paired: details are given by Lamb (1977).

The frequency and location of intimate pairing at particular sites

Gene conversion is easily the best method for determining how often a given region of DNA is involved in molecular-level pairing between homologous chromosomes in meiotic prophase, though conversion frequencies underestimate pairing because some events of hybrid DNA formation do not give detected conversion, e.g. those giving 'correction 4 : 4' octads. The frequency distributions of conversion frequencies for mutants in data from *Ascobolus*, yeast and *Sordaria* are shown in Fig. 5. More limited data from *Neurospora* and *Bombardia lunata* have conversion frequencies around 1% (Catcheside, 1977); Boone & Keitt (1956) found a value of 14.6% for a mutation in *Venturia inequalis*, and quite a few values of over 30% have been reported from European strains of *Ascobolus* (e.g. Paquette & Rossignol, 1978). Most conversion frequencies from fungi are in the range 0.2–10%, but the appreciable number of values over 20% shows that molecular-level meiotic pairing at some particular sites must occur in at least one-fifth to one-third (or more, as these are minimum estimates) of all meioses. This compares with electron micrograph data quoted by Stern, Westergaard & von Wettstein (1975) that a DNA strand parallel to the lateral component of the SC in *Neurospora, Drosophila* and *Zea* would represent only 0.3%, 0.2% and 0.015%, respectively, of their genomic DNA. If only such a small proportion of total DNA can be linearly matched along the axis of the SC in the central region, yet hybrid DNA can form very frequently at particular sites, then either detailed DNA matching must be able to extend outside the central regions of the SC, or there must be very selective incorporation of particular DNA segments into the central region. The existence of very high conversion frequencies eliminates all pairing hypotheses (such as that of Stern *et al.*, 1975) which postulate that only a very small proportion of DNA is paired at each meiosis, with random incorporation of lengths of DNA into the paired regions. An alternative hypothesis, that certain 'pairing' regions of DNA are usually included in matched regions in the SC, and that the bulk of DNA is regularly not intimately paired, is very unlikely because conversion frequencies of zero should then predominate; however, they are actually rare.

It can therefore be concluded that nearly all loci are included in intimately paired meiotic DNA, at least at a low frequency, and an appreciable proportion of sites is included at moderate to high frequencies. The easiest way to reconcile data on conversion and co-conversion frequencies with possible pairing lengths in the SC is to suggest that DNA pairing could start in the central region of the SC where parts of non-sister chromatids from homologues are brought into specific alignment. But once started, DNA pairing spreads outwards into the main chromatin regions, allowing a much larger proportion of DNA to be intimately paired than just the equivalent of the SC length. Callan & Pearce (1979) also concluded that recombination takes place in regions outside the synaptonemal complex, using quite different evidence (from interlocking bivalents in the newt, *Triturus*).

Recombination nodules

Recombination nodules were described in the fruit fly *Drosophila* by Carpenter (1975, 1979) as: 'dense, spherical structures located adjacent to and between the chromatin of the two homologues, above and adjacent to the central element of the synaptonemal complex'. The total numbers of nodules per nucleus and their locations along the euchromatic portion of the bivalent arms corresponded quite closely to the numbers and locations of genetically detected exchange events, suggesting that the recombination nodule performs a role in the recombination process. Carpenter (1979) described two kinds of nodule in *Drosophila*, with different shape, distribution and timing. She presented considerable evidence that the spherical nodules originally described occurred at the sites of reciprocal meiotic crossing-over, and suggested that the ellipsoidal nodules correlated with sites of gene conversion, though the evidence for that was less strong.

Recombination nodules have now been reported from a wide range of organisms, including fungi, algae, nematodes and humans (references in Carpenter, 1979). There is marked variability between species or groups in morphology (fungal nodules are typically broad ellipsoids), in position relative to the SC (fungal ones are within the central space and include or replace the central element, unlike those in rat and *Drosophila* which are external to the SC itself), and in time of occurrence (in *Sordaria* and *Neurospora* they are seen from zygotene to diplotene, but in *Drosophila* they only occur in mid-pachytene). The distinctive position of fungal nodules led to them being called 'central component nodes' but 'nodules' can also be used (see Gillies, 1979; Beckett, pp. 37–61 and Zickler pp. 63–83).

In fungi, studies comparing nodes in reconstructions of serially sectioned zygotene and pachytene nuclei with linkage maps and recombination data have shown that the number of nodules agrees well with the number of crossovers in *Neurospora* (Gillies, 1979), *Sordaria macrocarpa* (Zickler, 1977) and yeast (Byers & Goetsch, 1975). The distribution of nodules along chromosomes and the spacing between them also show correlations with crossover localization and chromosome (chiasma-position) interference. The evidence from fungi and *Drosophila* is therefore in good agreement that at least some types of recombination nodule observed in the electron microscope are correlated with crossing-over. Gillies (1979) found no obvious separation of *Neurospora* nodules into two discrete populations for shape, and there was no conclusive evidence there for two functionally different types as was suggested by Carpenter (1979) for *Drosophila*.

How gene conversion can change allele frequencies in populations: its evolutionary significance

If gene conversion were equally frequent in both directions (+ to *m* and *m* to +), it would have no effect on allele frequencies. In fungal tetrads and octads it is easy to determine the direction of conversion from the aberrant segregation ratios, and disparity in the direction of conversion is very common. Let *b* be the frequency of a particular allele (say *A* for *A/a*, or + for +/*m*) in the products of meiotic tetrads or octads with aberrant segregation ratios, so

$$b = \frac{(4:0 \times 4) + (3:1 \times 3) + (1:3 \times 1) + (0:4 \times 0)}{4 \times \text{Total number of aberrant ratio tetrads}},$$

or

$$b = \frac{\begin{array}{c}(8:0 \times 8) + (7:1 \times 7) + (6:2 \times 6) + (5:3 \times 5) + \\ (3:5 \times 3) + (2:6 \times 2) + (1:7 \times 1) + (0:8 \times 0)\end{array}}{8 \times \text{Total number of aberrant ratio octads}}.$$

With no disparity in direction of conversion, $b = 0.5$; with more conversion to *A* than to *a*, $b > 0.5$; with more conversion to *a* than *A*, $b < 0.5$. Fig. 6 shows the frequency distribution of *b* for sets of data from *Ascobolus, Saccharomyces, Sordaria fimicola* and *S. brevicollis*, with clear disparity in all sets, right up to the maximum disparity values expected ($b = 0.25$ or $b = 0.75$, with nearly all conversion asci being 2 : 6 or 6 : 2 respectively).

The effects of conversion on allele frequencies will increase with increasing disparity (i.e. increasing departure of *b*, upwards or downwards, from $b = 0.5$) and with increasing conversion frequency, which will be represented (as a fraction) by *c*, where

$$c = \frac{\text{Number of tetrads or octads with aberrant segregation ratios}}{\text{Total number of normal and aberrant ratio tetrads or octads}}.$$

The frequency distributions for *c* in various fungi are given in Fig. 5. The force of gene conversion on allele frequencies can be represented as *y*, where $y = c(b - 0.5)$. In theory *y* has a possible range of +0.5 to −0.5, but with *c* usually being in the range 0 to 0.3 (sometimes to 0.4) and *b* usually being in the range 0.25 to 0.75, *y* in practice will usually be within the range +0.1 to −0.1.

Conversion will only alter allele frequencies if occurring in a meiosis heterozygous for a character: this is true whether the main vegetative stage is haploid or diploid. Let the frequency of allele *A* in a population be *p* and the frequency of *a* be *q*, so $p + q = 1$, and the frequency of *A* in meiotic spores from *Aa* meioses will be $cb + 0.5(1 - c)$, which is $0.5 + y$, where *y* is positive for $b > 0.5$ and negative for $b < 0.5$ (and *a* in meiotic products from *Aa* has frequency $0.5 - y$). One can therefore produce equations for p from populations with any degree of non-random, or random mating and/or differential fertility of the three diploid meiotic genotypes. For typical Hardy-Weinberg conditions, with random mating, no differential fertility or selection, allele frequencies in successive generations are related by:

$$P_{n+1} = p_n^2 + 2p_n (1 - p_n) (0.5 + y),\tag{1}$$

when p_{n+1} is the generation after n. By the method of Gutz & Leslie

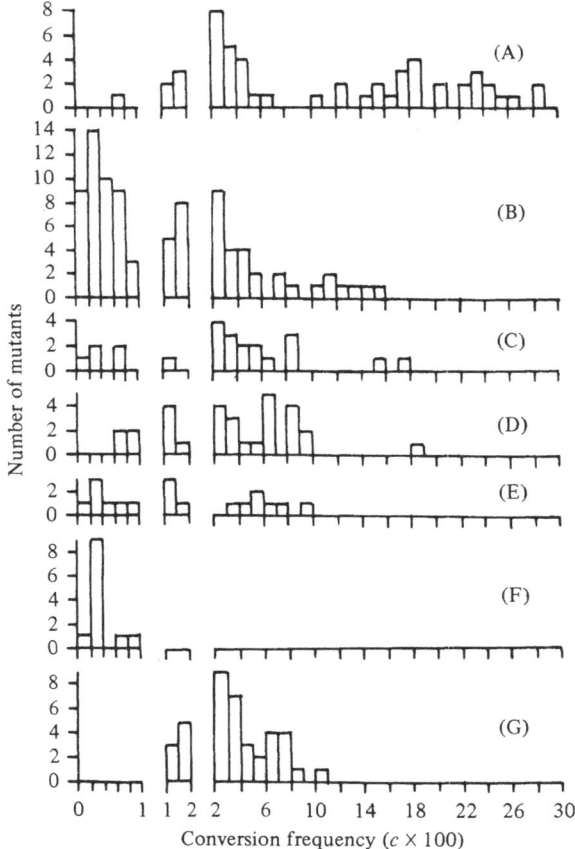

Fig. 5. Frequency distributions for conversion frequencies from a range of fungi. (A) *Ascobolus immersus* (Leblon, 1972); (B) *Ascobolus immersus* (Ghikas, 1978; Lamb & Ghikas, 1979); (C) *Ascobolus immersus* (Yu-Sun, 1966); (D) *Saccharomyces cerevisiae* (Fogel *et al.*, 1979); (E) *Saccharomyces cerevisiae* (Fogel *et al.*, 1971); (F) *Sordaria fimicola* (Kitani & Olive, 1967; Lamb, 1969); (G) *Sordaria brevicollis* (Yu-Sun *et al.*, 1977).

(1976) who were pioneers in this field, one can show that if the change in p in one generation is small in relation to p, then by calculus:

$$p_n = \frac{(p_0/q_0) \cdot e^{2yn}}{1 + (p_0/q_0) \cdot e^{2yn}} \tag{2}$$

where p_n is the frequency of A in the nth generation and p_0 and q_0 are the allele frequencies in the initial population. If there is non-random

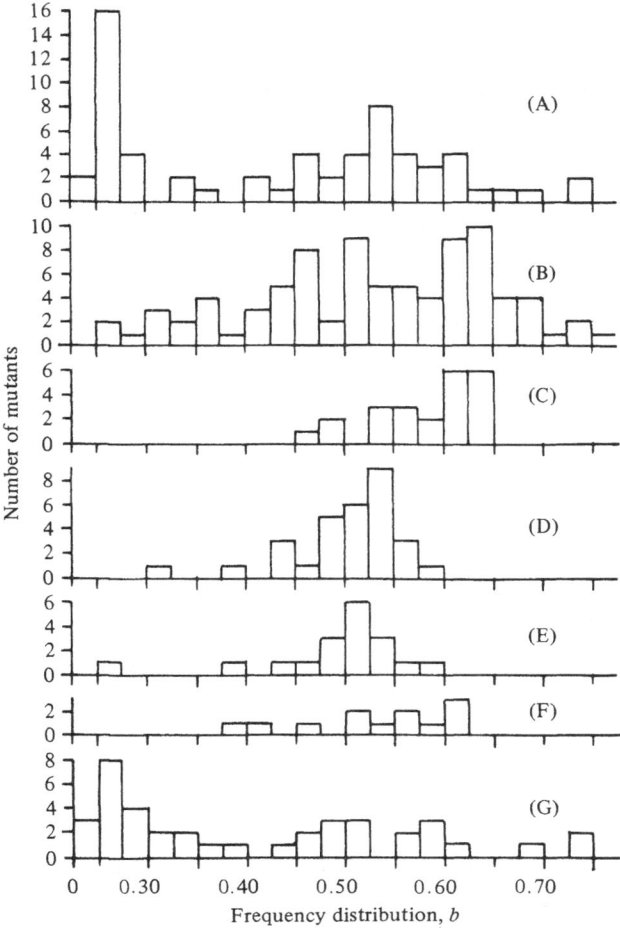

Fig. 6. Frequency distributions, from a range of fungi, for b, the frequency of a particular allele (wild-type or A) in the products of meiotic tetrads or octads with aberrant segregation ratios. $b > 0.5$ shows conversion disparity in favour of wild-type or A; $b < 0.5$ shows disparity in favour of mutant or a. Symbols A to G refer to the same fungi as in Fig. 5. See text for equations involving b.

mating in the direction of outbreeding, conversion will change alleles faster than in Equations (1) or (2), as heterozygous meioses will be more frequent than expected from Hardy-Weinberg proportions, while preferential inbreeding (unlikely in many fungi because of incompatibility systems) will decrease the rate of change due to conversion.

With the above conditions of random mating, and no selection, together with no fertility differences, no appreciable mutation or migration, then any non-zero value of y should eventually lead to complete fixation (allele frequency $= 1.0$) of the allele favoured by the disparity in conversion direction. Except for random fluctuations due to genetic drift, which should be slight in a large population, the changes in allele frequency with the number of generations should follow Equation (1) or, less accurately Equation (2), to fixation. Fig. 7 shows the calculated changes in gene frequency for several fungal mutants, with experimentally determined values of c and b, but with initial allele frequencies set arbitrarily for illustration; Equation (1) was used repetitively. The change in allele frequency is greatest when $p = q = 0.5$ under random mating, as heterozygotes are then most frequent. Fungal polymorphisms should only be retained indefinitely if y is zero, which occurs if c is zero and/or b is exactly 0.5, as for case (v) in Fig. 7, where $b = 0.5$ in the original yeast data. Fig. 7 gives an indication of how fast gene conversion can change allele frequencies. Calculations based on Equation (1) show that for values of c and b commonly found in fungi, the time taken to change allele frequencies from 0.05 to 0.95 or vice versa vary from less than 100 generations to a few thousand generations. It is thus clear that for selectively neutral alleles, gene conversion is a very powerful force in evolution, giving fixation for the allele favoured by the disparity in conversion direction.

If one considers a situation with conversion and mutation, but no selection, one can show that the equilibrium situation is:

$$\hat{p} = \frac{2y - v - \mu \pm \sqrt{(v + \mu - 2y)^2 + 8vy,}}{4y} \tag{3}$$

where A mutates to a with frequency μ, a mutates to A with frequency v, and the circumflex indicates the equilibrium frequency. With y having typical absolute values of 10^{-2} and 10^{-4} and mutation frequencies being in the range 10^{-5} to 10^{-8}, p and q will in most cases depend far more on conversion parameters than on mutation values. Calculations from Equation (3) show that typical fungal positive and negative values of y have profound effects on p over all normal ranges of μ and v.

If one considers selection as well as conversion and mutation, the relevant equilibrium equations then differ for alleles with dominance, recessiveness or no dominance, and whether the main vegetative stage is haploid, diploid, or haploid but capable of heterokaryosis, which gives the possibility of dominance effects in 'heterozygous' heterokaryons, although the nuclei are haploid. For diploids, the main equations are as follows.

For a deleterious recessive, selection coefficient s against aa,

$$\hat{q}^4 \ (2sy) - \hat{q}^3 \ (2sy + vs + \mu s + s) - \hat{q}^2$$
$$(2y - s - \mu s) + \hat{q} \ (\mu + v + 2y) - \mu = 0. \qquad (4)$$

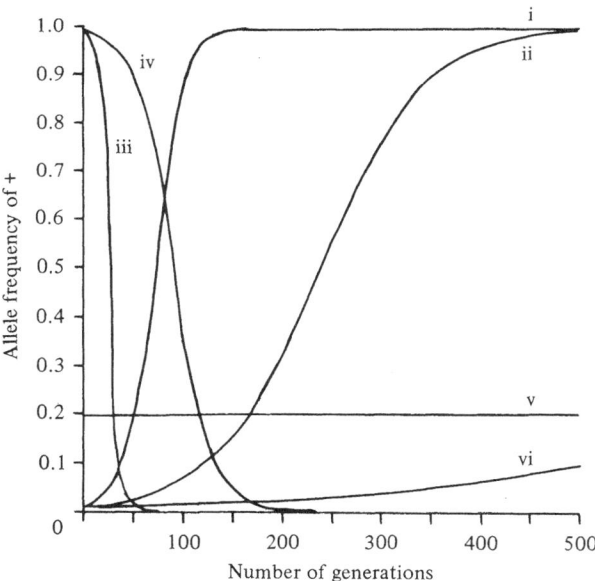

Fig. 7. Changes in allele frequency with number of generations, from gene conversion in the absence of appreciable mutation or selection, calculated using Equation (91) repetitively; random mating is assumed. i, *Ascobolus immersus* (S. Helmi, personal communication), +, $P,S \times w - 78$, P,S, $c = 0.27528$, $b = 0.62389$, $y = 0.0341$; ii, *Ascobolus immersus* (S. Helmi, personal communication), +, $P,nS \times w - 78$, P,nS, $c = 0.11538$, $b = 0.58408$, $y = 0.0097$; iii, *Ascobolus immersus* (Girard & Rossignol, 1974), b_2 allele 130, $c = 0.39296$, $b = 0.23059$, $y = -0.1059$; iv, *Sordaria brevicollis* (Yu-Sun et al., 1977), YS50, $c = 0.1030$, $b = 0.2480$, $y = -0.0260$; v, *Saccharomyces cerevisiae* (Fogel et al., 1971), *ura* $1 - 1$, $c = 0.0359$, $b = 0.5000$, $y = 0.0000$; vi, *Saccharomyces cerevisiae* (Fogel et al., 1971), *arg* $4 - 1$, $c = 0.0594$, $b = 0.5395$, $y = 0.0024$. Initial allele frequencies are chosen for purposes of illustration.

For a deleterious allele with complete dominance, selection coefficient s against AA and Aa,

$$\hat{p}^4 (2sy) + \hat{p}^3 (s + sv + s\mu - 6sy) +$$
$$\hat{p}^2 (2y + 4sy - 2s - 3sv - 2s\mu) +$$
$$\hat{p} (s + v + \mu + 2sv - 2y) - v = 0. \tag{5}$$

For a deleterious allele, a, with no dominance, selection coefficient s for Aa, $2s$ for AA,

$$\hat{q}^3 (4sy) - \hat{q}^2 (2s\mu + 2sv + s + 2y + 4sy) +$$
$$\hat{q} (\mu + v + 2 s\mu + s + 2y) - \mu = 0. \tag{6}$$

Calculations using these equations for diploids show that for alleles with low selection coefficients, conversion often has very large effects on their equilibrium frequencies and may lead to fixation. Where selection coefficients are higher, conversion has major effects on the frequencies of recessive alleles, intermediate effects on alleles with no dominance, and lesser effects on fully dominant alleles. For a haploid where a is deleterious, with selection coefficient s against a,

$$\hat{q}^3 (sy) - \hat{q}^2 (2y + s + \mu s + vs + 2sy) +$$
$$\hat{q} (\mu + v + 2y + s + s\mu) - \mu = 0. \tag{7}$$

Under these conditions, conversion has major effects on allele frequencies when selection coefficients are low, but lesser effects when selection coefficients are high. The situations in fungi with both diploid and haploid vegetative stages, and in haploids that can form heterokaryons, are more complicated mathematically. Also, in many fungi there are problems as to what constitutes an individual or a population. Some difficulties of fungal population genetics were considered by Burnett (1975). The molecular factors controlling disparity and meiotic gene conversion frequencies are not fully understood. The effects of conversion on allele frequencies will be considered in more detail elsewhere (B. C. Lamb & S. Helmi, unpublished), with fuller details of the above equations and with specimen calculations.

In conclusion, gene conversion must be an important force in fungal evolution, but in a non-adaptive way, changing gene frequencies in the direction dictated by the disparity in direction of gene conversion. On its own, conversion should often cause fixation to one allele, and in conjunction with mutation and selection, various equilibria may be set up between these forces, as in Equations (3) to (7).

Acknowledgements. It is a pleasure to thank Mr Shaker Helmi for stimulating discussions on the role of gene conversion in populations. I am grateful to Dr Lewis Frost, of Bristol University, for originally introducing me to the mysteries and delights of regular and aberrant meiotic segregation in fungi.

References

Bond, D. J. (1976). A system for the study of meiotic non-disjunction using *Sordaria brevicollis*. *Mutation Research*, **37**, 213–20.

Bond, D. J. (1979). The origin and chemical induction of meiotic aneuploidy in *Sordaria brevicollis*. *Heredity*, **43**, 151–2.

Boone, D. N. & Keitt, G. W. (1956). *Venturia inequalis* (Cke) Wint. VIII. Inheritance of color mutant characters. *American Journal of Botany*, **43**, 226–33.

Burnett, J. H. (1975). *Mycogenetics*. London: Wiley.

Byers, B. & Goetsch, L. (1975). Electron microscope observations on the meiotic karyotype of diploid and tetraploid *Saccharomyces cerevisiae*. *Proceedings of the National Academy of Sciences, USA*, **72**, 5056–60.

Callan, H. G. & Pearce, S. M. (1979). An experimental analysis of bivalent interlocking in spermatocytes of the newt *Triturus vulgaris*. *Journal of Cell Science*, **37**, 125–41.

Carpenter, A. T. C. (1975). Electron microscopy of meiosis in *Drosophila melanogaster* females. II. The recombination nodule – a recombination-associated structure at pachytene? *Proceedings of the National Academy of Sciences, USA*, **72**, 3186–9.

Carpenter, A. T. C. (1979). Synaptonemal complex and recombination nodules in wild-type *Drosophila melanogaster* females. *Genetics*, **92**, 511–41.

Catcheside, D. G. (1944). Polarized segregation in an Ascomycete. *Annals of Botany* N.S. **8**, 119–30.

Catcheside, D. G. (1977). *The Genetics of Recombination*. London: Edward Arnold.

Emerson, S. (1956). Notes on the identification of different causes of aberrant tetrad ratios in *Saccharomyces*. *Compte rendu des travaux du laboratoire de Carlsberg, Série physiologique*, **26**, 71–86.

Fincham, J. R. S., Day, P. R. & Radford, A. (1979). *Fungal Genetics*. 4th edn. Oxford: Blackwell.

Fogel, S., Hurst, D. D. & Mortimer, R. K. (1971). Gene conversion in unselected tetrads from multipoint crosses. In *Stadler Genetics Symposia*, vols. 1 and 2, ed. G. Kimber & G. P. Rédei, pp. 89–110. Missouri: Columbia University Press.

Fogel, S., Mortimer, R. K., Lusnak, K. & Tavares, K. (1979). Meiotic gene conversion – a signal of the basic recombination event in yeast. *Cold Spring Harbor Symposia on Quantitative Biology*, **43**, 1325–41.

Ghikas, A. (1978). Recombination mechanisms and the controls of gene conversion in *Ascobolus immersus*. PhD Thesis, Imperial College, London.

Ghikas, A. & Lamb, B. C. (1977). The detection, in unordered octads, of 6+ : 2m and 2+ : 6m ratios with postmeiotic segregation, and of aberrant 4 : 4s, and their use in corresponding-site interference studies. *Genetical Research*, **29**, 267–78.

Gillies, C. B. (1979). The relationship between synaptonemal complexes, recombination nodules and crossing over in *Neurospora crassa* bivalents and translocation quadrivalents. *Genetics*, **91**, 1–17.

Girard, J. & Rossignol, J-L. (1974). The suppression of gene conversion and intragenic crossing over in *Ascobolus immersus:* evidence for modifiers acting in the heterozygous state. *Genetics*, **76**, 221–43.

Gutz, H. & Leslie, J. F. (1976). Gene conversion: a hitherto overlooked parameter in population genetics. *Genetics*, **83**, 861–6.

Kitani, Y. & Olive, L. S. (1967). Genetics of *Sordaria fimicola*. VI. Gene conversion at the *g* locus in mutant × wild-type crosses. *Genetics*, **57**, 767–82.

Kitani, Y., Olive, L. S. & El-Ani, A. S. (1962). Genetics of *Sordaria fimicola*. V. Aberrant segregation at the *g* locus. *American Journal of Botany*, **49**, 697–706.

Lamb, B. C. (1966). Polarized segregation in Ascomycetes and the differential bursting of asci. *Genetical Research*, **7**, 325–34.

Lamb, B. C. (1969). Related and unrelated changes in conversion and recombination frequencies with temperature in *Sordaria fimicola*, and their relevance to hybrid-DNA models of recombination. *Genetics*, **62**, 67–78.

Lamb, B. C. (1977). The use of gene conversion to study synaptonemal complex structure and molecular details of chromatid pairing in meiosis. *Molecular and General Genetics*, **157**, 31–7.

Lamb, B. C. & Ghikas, A. (1979). The intergradation, genetic interchangeability and interpretation of gene conversion spectrum types. *Genetics*, **92**, 49–65.

Lamb, B. C. & Wickramaratne, M. R. T. (1973). Corresponding-site interference, synaptonemal complex structure, and $8+ : 0m$ and $7+ : 1m$ octads from wild-type × mutant crosses of *Ascobolus immersus*. *Genetical Research*, **22**, 113–24.

Leblon, G. (1972). Mechanism of gene conversion in *Ascobolus immersus*. I. Existence of a correlation between the origin of mutants induced by different mutagens and their conversion spectrum. *Molecular and General Genetics*, **115**, 36–48.

Lindegren, C. C. (1955). Non-Mendelian segregation in a single tetrad of *Saccharomyces* ascribed to gene conversion. *Science*, **121**, 605–7.

McNelly, C. A. (1962). Studies on sexual reproduction and the frequency of recombination over a range of temperatures in *Neurospora crassa*. PhD Thesis, University of Bristol.

Meselson, M. S. & Radding, C. M. (1975). A general model for genetic recombination. *Proceedings of the National Academy of Sciences, USA*, **72**, 358–61.

Moens, P. B. (1974). Coincidence of modified crossover distribution with modified synaptonemal complexes. In *Mechanisms in Recombination*, ed. R. F. Grell, pp. 377–83. New York: Plenum Press.

Mortimer, R. K. & Hawthorne, D. C. (1973). Genetic mapping in *Saccharomyces*. IV. Mapping of temperature-sensitive genes and use of disomic strains in localizing genes. *Genetics*, **74**, 33–54.

Nakamura, K. (1961). Preferential segregation in linkage group V of *Neurospora crassa*. *Genetics*, **46**, 887.

Paquette, N. & Rossignol, J-L. (1978). Gene conversion spectrum of 15 mutants giving post-meiotic segregation in the b_2 locus of *Ascobolus immersus*. *Molecular and General Genetics*, **163**, 313–26.

Perkins, D. D. (1974). The manifestation of chromosome rearrangements in unordered asci of *Neurospora*. *Genetics*, **77**, 459–89.

Perkins, D. D. & Barry, E. G. (1977). The cytogenetics of *Neurospora*. *Advances in Genetics*, **19**, 133–285.

Pukkila, P. J. (1977). Biochemical analysis of genetic recombination in Eukaryotes. *Heredity*, **39**, 193–217.

Stern, H., Westergaard, M. & von Wettstein, D. (1975). Presynaptic events in meiocytes of *Lilium longiflorum* and their relation to crossingover: a preselection hypothesis. *Proceedings of the National Academy of Sciences, USA*, **72**, 961–5.

von Wettstein, D. (1971). The synaptonemal complex and four-strand crossing over. *Proceedings of the National Academy of Sciences, USA*, **68**, 851–5.

Westergaard, M. & von Wettstein, D. (1970). Studies on the mechanism of crossing-over. IV. The molecular organisation of the synaptonemal complex in *Neottiella* (Cooke) Saccardo. *Compte rendu des travaux du laboratoire de Carlsberg, Série physiologique,* **26**, 71–86.

Whitehouse, H. L. K. (1957). Mapping chromosome centromeres from tetratype frequencies. *Journal of Genetics*, **55**, 348–60.

Yu-Sun, C. C. C. (1966). Linkage groups in *Ascobolus immersus*. *Genetica*, **37**, 569–80.

Yu-Sun, C. C. C., Wickramaratne, M. R. T. & Whitehouse, H. L. K. (1977). Mutagen specificity in conversion pattern in *Sordaria brevicollis*. *Genetical Research*, **29**, 65–81.

Zickler, D. (1977). Development of the synaptonemal complex and the 'recombination nodules' during meiotic prophase in the seven bivalents of the fungus *Sordaria macrospora* Auersw. *Chromosoma, Berlin*, **61**, 289–316.

Somatic incompatibility in fungi and Myxomycetes

E. BIRGITTE LANE

Department of Zoology, University College, Gower Street, London WC1E 6BT, UK

Introduction

The ability to distinguish self from non-self is not a prerogative of a well-defined immune system, but is clearly present in fungi and Myxomycetes, and a study of incompatibility mechanisms in these simpler organisms may help to shed light on the origin of the complexity of the immune response in man. 'Incompatibility' in fungi refers to the restrictions that prevent two genotypes from coexisting in the same cytoplasm. Fungi have attracted attention in this respect because cell fusion can take place at several points in the life cycle of these organisms, in contrast to the situation in higher animals where it is usually limited to gamete fusion in sexual reproduction.

Esser divides the phenomenon into *homogenic* and *heterogenic* incompatibility, according to whether it acts against identical or non-identical genotypes (Esser & Blaich, 1973). When acting on different stages of the life cycle, homogenic and heterogenic incompatibility lead to very different effects.

Homogenic incompatibility influences sexual reproduction, limiting fusion to cells of dissimilar mating types or sexes. If one accepts the value to a species of sexual reproduction for genetic recombination, then the advantage of homogenic incompatibility at this stage is self-evident as it ensures a degree of outbreeding. *Heterogenic incompatibility* has also been described as influencing sexual fusion, but strictly speaking this is verging on speciation, and thus will not be discussed here; the term is more usually applied to somatic cell fusion interactions.

Although incompatibility has been repeatedly observed in a wide variety of fungi and Myxomycetes, the mechanisms affecting the phenomenon have only been examined in a few cases. Wherever such data are available, the genetic control of somatic incompatibility appears to be by polygenic, biallelic systems, whereas homogenic incompatibility is governed by mating type genes, two or more alleles at one or two loci.

Upon contact between incompatible cells, fusion is either avoided, or it takes place with subsequent cytoplasmic mixing, and a post-fusion incompatibility reaction is initiated which is directed at (*a*) limiting further cytoplasmic mixing and (*b*) the destruction of 'contaminated' mixed material. Cases of cell wall fusion without plasmogamy as in *Thanatephorus* (Flentje & Stretton, 1964) have been reported, but their significance is not clear. A 'non-fusion' reaction is the most widespread response to incompatibility and is obviously the most efficient, but negative interactions are difficult to analyse, and the study of post-fusion incompatibility reactions may tell us more about *how* and *why* incompatibility mechanisms operate. Clark & Collins (1973) have suggested in the context of Myxomycetes that some apparent non-fusion may be the result of localized and rapidly terminated cytoplasmic interactions, but so far gene loci governing fusion appear to be distinct from those governing post-fusion incompatibility reactions.

Information regarding the genetics of mating-type interactions is now becoming substantial, as are reports on the immense variety of devices and strategies shown by fungi to achieve fertilization. This literature has been reviewed by Carlile & Gooday (1978) and will not be discussed in detail here, other than to mention the curious and possibly important association that exists between sexual fertility and somatic incompatibility in *Neurospora crassa* (Newmeyer, Howe & Galeazzi, 1973) and in *Podospora anserina* (Esser & Blaich, 1973; Boucherie, Bégueret & Bernet, 1976), the two species about which most genetical data exist. The following discussion will therefore specifically concern heterogenic incompatibility. From a consideration of what is seen to happen when incompatible cells fuse, and of the circumstances in which somatic incompatibility is naturally found, an attempt will be made to reconcile the existing information with suggestions as to the possible biological advantages that may have promoted the evolution of heterogenic somatic incompatibility.

Observations on fungal incompatibility reactions

Somatic fusion, as distinct from sexual fusion, appears to be rare in the lower fungi (Griffin & Perrin, 1960); all reports of somatic incompatibility reactions are from observations of Basidiomycetes, Ascomycetes and the related fungi imperfecti. Such reports include extensive work on *Aspergillus nidulans* and other *Aspergillus* species by Caten, Jinks and coworkers (see Croft & Jinks, 1977); on *Podospora anserina* by Esser's group (see Esser, 1974), Bernet, Bégueret and colleagues (Delettre, Boucherie & Bernet, 1978); and on *Neurospora crassa* by Wilson and colleagues, and Newmeyer (see Carlile & Gooday, 1978). Various studies have been made of other species, including *Thanatephorus cucumeris* (Flentje & Stretton, 1964; McKenzie *et al.*, 1969), *Endothia parasitica* (e.g. Anagnostakis, 1977), and *Coriolus versicolor* among other wood-rotting Basidiomycetes (Rayner & Todd, 1979; Todd & Rayner, 1979). Most of this literature has been reviewed by Carlile & Gooday (1978); few reports include morphological observation of the interactions.

The macroscopic manifestation of an incompatibility reaction between neighbouring vegetative fungal mycelia is the appearance between the colonies of 'barrages' or demarcation zones, which are highly pigmented and consist of hyphal debris. Interaction takes place between hyphae from the two colonies: cell wall contact leads to cell wall fusion and degradation, followed by plasmalemmal fusion and subsequent cytoplasmic mixing; if the strains concerned are incompatible this leads to vacuolization and localized cell death, as described in *Neurospora crassa* by Garnjobst & Wilson (1956). Rayner & Todd (1979) have described the reaction between strains of *Coriolus versicolor,* in which the fusion takes place not between hyphal tips but between side branches, and characteristically swollen and distorted cells are produced in the fusion region. They suggest that the distortion is due to 'uncontrolled lysis' by the disruption of lysogenic vesicles present in the tips for normal hyphal growth (Todd & Rayner, 1979). The main hyphal branches involved on either side then die back to leave 'ghosts' in the barrage zone, where pigment finally accumulates. Barrett & Uscuplic (1971) have shown the interaction in *Polyporus schweinitzii* as twisted knots of darkened hyphae, with an electron micrograph to demonstrate the highly vacuolated cytoplasmic remnants containing pigmented debris.

Williams & Wilson (1966, 1968) have attempted to observe this reaction at greater resolution; their experimental material was *Neurospora crassa*, and instead of allowing fusion to occur randomly they initiated the reaction by microinjecting heterogenic cytoplasmic extracts (Wilson, Garnjobst & Tatum, 1961). Septal plugs were rapidly formed which sealed off the affected cell from the rest of the mycelium, and the contaminated cell then degenerated with characteristic vacuolation and abnormal membrane formations. The authors felt that membranes were probably the primary target in this interaction, and analysis of cytoplasmic extracts indicated that the active substance was proteinaceous. Biochemical analyses of incompatibility reactions in *Podospora anserina* have revealed the appearance of novel proteins in the cytoplasm of cells as they undergo an incompatibility reaction (Blaich & Esser, 1971; Bégueret, 1972), which appear to be proteolytic enzymes. That these proteases are directly involved in the incompatibility reaction (Bégueret & Bernet, 1973) is supported by the demonstration that mutations which suppress the reaction are involved in the production of these enzymes (Boucherie *et al.*, 1976).

There appears to be no further information on the structural sequence of events during these reactions, presumably because of the difficulty in obtaining sufficient quantities of material when one depends on individual hyphal fusion or microinjection. However, a temperature-sensitive non-allelic system has been found in *Podospora anserina* (Labarère, 1973) and another one in *Thanatephorus cucumeris* (McKenzie *et al.*, 1969); either or both of these systems could probably be exploited to give useful ultrastructural information.

Observations on Myxomycete incompatibility reactions

Two groups have looked at somatic incompatibility in Myxomycetes in some detail: Collins' group have studied 'fast cytotoxic reactions' in *Didymium iridis*, while Carlile and coworkers have looked at 'slow lethal reactions' in *Physarum polycephalum*. Incompatibility reactions have also been observed in *Badhamia utricularis* (Carlile, 1974) and in *Physarum cinereum* (Clark, 1977), and Schrauwen (1979) has studied the effect of metabolic inhibitors on the reaction in *P. polycephalum*.

The vegetative phase of the Myxomycete is a mobile, saprophytic plasmodium; millions of diploid nuclei within a single cell membrane undergo synchronous mitosis every 8–9 h. The cytoplasm is differentiated into a reticulate system of 'veins' through which shuttle streaming mixes all cytoplasmic constituents throughout the whole cell. Sporulation of the plasmodium gives rise to haploid amoebae of the two parental mating types: fusion of unlike mating types, of which there are many, initiates plasmodium formation (see review by Collins, 1979). In an encounter between two plasmodia of the same species, plasmalemmal fusion and subsequent cytoplasmic mixing depends upon identity at specific fusion loci, which are distinct from the loci that govern the outcome of a post-fusion reaction.

Compared with fungal hyphal fusion, the incompatibility reaction in plasmodia can be 'expanded' in both space and time – it can last between 30 s (fast cytotoxic reaction) and 6 h (slow lethal reaction) – and is thus good material for structural and biochemical analysis of these reactions.

Observing that the magnitude of a reaction appeared to be inversely related to the number of dissimilar incompatibility loci, Clark & Collins (1973) supposed that the greater genetic dissimilarity leads to a faster termination of the cytoplasmic contact, with less mixing and thus less damage. These fast cytotoxic reactions involve a transient cytoplasmic contact between two plasmodia and result in a small 'clear zone' of lysed material which is left behind as both plasmodia withdraw. Upadhyaya & Ling (1976) examined these 'clear zones' by electron microscopy and showed that a rather thick membranous barrier forms through the cytoplasm, which seals off the degenerating 'clear zone' from the rest of the plasmodia. There appeared to be a lot of lipid droplets in the area, but in the short time of the reaction (< 5 min) there was no evidence of a specific target organelle.

The slow lethal reaction in *Physarum polycephalum* has yielded more information. First observed by Carlile & Dee (1967), the cytoplasmic reaction following fusion was sometimes so extensive as to annihilate both participating plasmodia. Non-identical strains will fuse provided they are at least homogenic at the fusion loci, but a macroscopic post-fusion reaction takes place about 5 h later if the strains differ by at least one of the *let* (lethal reaction) loci, of which three have been identified so far in *P. polycephalum* (Carlile, 1976). Surviving material retracts from the reaction area leaving a trail of debris shown to contain discarded nuclei of the 'sensitive' strain (Border & Carlile, 1974), and this plasmodium invariably expresses the dominant incompatibility phenotype in subsequent fusions (Carlile, 1972), such that in a given pair of fusing plasmodia the 'killer' strain and the 'sensitive' strain can be predicted from their genotypes.

Using a pair of closely related strains that were identical at the fusion loci and differed at one of the *let* loci, this post-fusion reaction was examined by electron microscopy (Lane & Carlile, 1979). The fate of nuclei of 'killer' and 'sensitive' strain plasmodia was followed by autoradiography after [³H]thymidine labelling of one of the strains prior to fusion. Changes were seen in the plasmodia as early as 2 h after fusion, well before a reaction was visible to the naked eye at 5–6 h. It was clear that sensitive strain nuclei were being selectively eliminated from the cytoplasm, becoming enclosed in vacuoles and ejected on to the dorsal or invaginated plasmalemmal surfaces. Internal damage to nuclei was also seen, but the extent of this damage did not appear to be correlated with nuclear elimination, which suggested the two processes were effected by separate mechanisms. The nuclear damage took the form of loosening and expansion of the nucleolus, and its segregation into localized clumps of fibrillar and granular components, whilst chromatin clumping, especially adjacent to the nuclear envelope, was increasingly prominent. Chromatin clumping may be non-specific, but

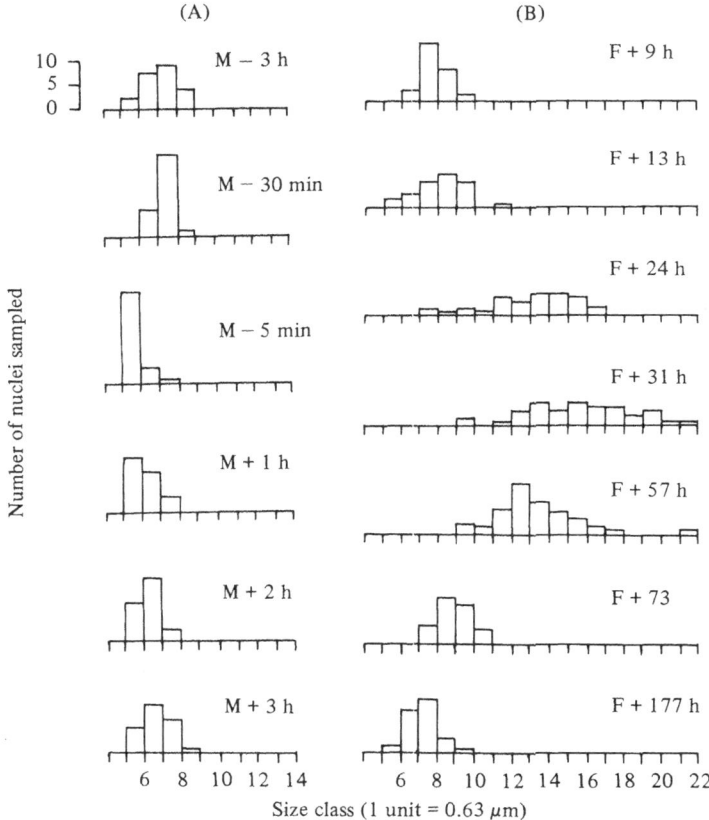

Fig. 1. Relative diameters of nuclei in smear samples taken at different times (A) from a control plasmodium through mitosis, and (B) from a heterokaryotic plasmodium produced by fusion of strains differing at one *let* locus, sampled during the recovery period following an incompatibility reaction. M, mitosis; F, plasmodial fusion.

nucleolar segregation is characteristic of specific inhibition of nucleic acid synthesis (e.g. Bernhard, 1971), and requires continued metabolic activity (Monneron, 1971). Vacuoles to eliminate unwanted nuclei have been seen in sporulating Myxomycetes (Bechtel, 1976) but nucleolar segregation without the influence of drugs appears to be unique to the incompatibility reaction. Concomitant with this nuclear destruction and elimination, another phenomenon was observed in the fused plasmodia, that of extensive nuclear fusion. Nuclei of all combinations of like and unlike types were seen to fuse, giving rise to enlarged nuclei; by 24 h after the reaction, normal nuclei were hard to locate, although by eye the plasmodium appeared to be recovered from the reaction by 12 h after fusion. Nuclear sizes were measured from smears taken from plasmodia at different times after the reaction (Fig. 1) and from these it was found that mean diameters continued to increase for a couple of days but had returned to normal by five days. Figs. 2–5 show the condition of Myxomycete plasmodial nuclei before, during and after a somatic incompatibility reaction induced by fusing two strains of *Physarum polycephalum* which differed from each other at one of the *let* loci; the 'killer' strain has the dominant *let* allele. Nuclear debris was still seen at 24 h after fusion, and throughout the recovery period enlarged nuclei were observed with characteristic nuclear envelope crenellations and associated dense material, which it is tempting to interpret as the shedding of excess chromatin. Whether such a process may be selective cannot be determined at present, since the specificity of the tritium label is lost as degenerate nuclei are recycled by the plasmodium. Taken together with the rather irregular decrease in nuclear diameters from the smear preparations, the data suggest that at least some of the large nuclei can return to normal, and raise the strong possibility of some degree of parasexual recombination having taken place as a result of the induced heterokaryosis. The fusing of *homologous* as well as heterologous nuclei implies that mixing the two plasmodia has disrupted a control mechanism that normally operates to keep the nuclei in a plasmodium separate from each other.

Effector mechanisms for incompatibility reactions

Some comment should be made regarding differences and similarities between heterogenic post-fusion reactions in fungi and Myxomycetes, although it is recognized that the data are still scanty. The most obvious similarity is the immediate action the organism takes to seal off the foreign material, with septal plugs in fungal hyphae, and

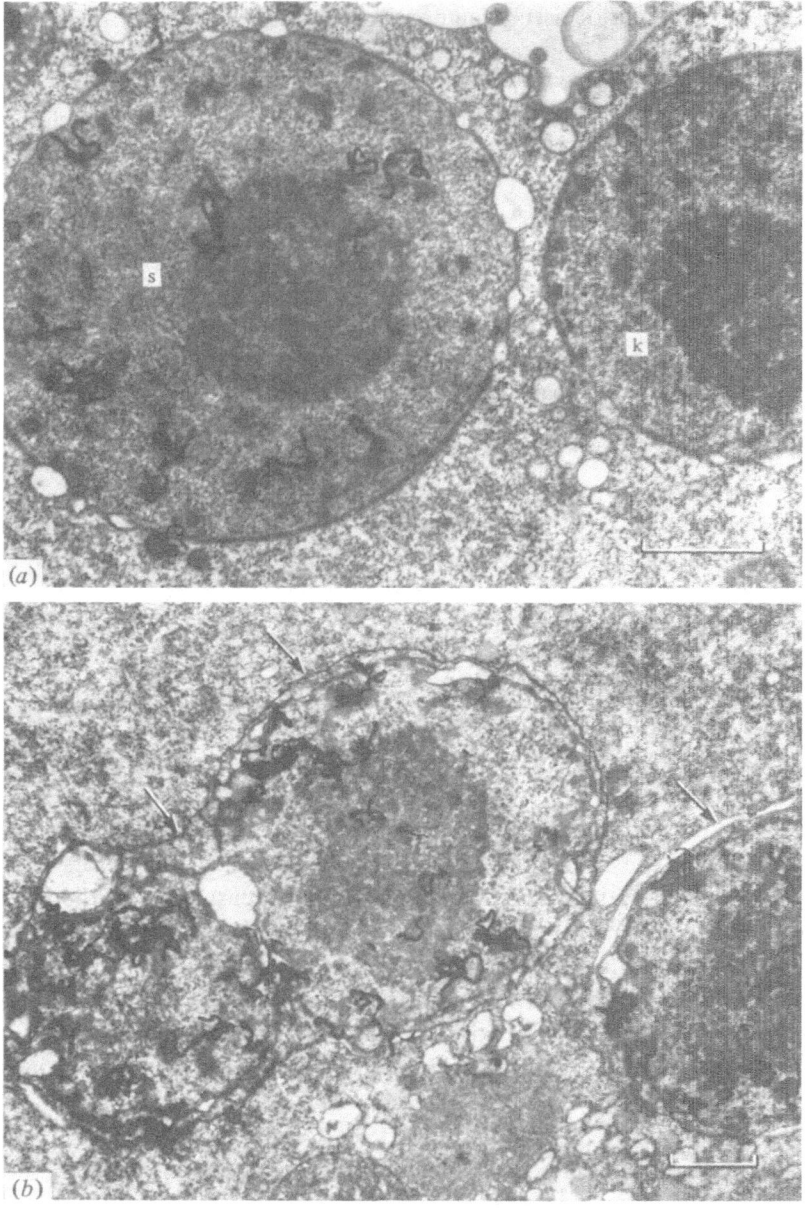

Fig. 2 (a) 'Killer' strain (k, unlabelled) and 'sensitive' strain (s, labelled) nuclei 1½ h after plasmodial fusion. The only changes visible so far are nuclear envelope dilations on all nuclei, associated with increased production of rough endoplasmic reticulum. (b) 3 h after fusion widespread karyogamy is seen, shown here between two membranes in the Myxomycete cytoplasm which enclose just the foreign nuclei plus adherent cytoplasm once the foreign cytoplasmic contents have become well dispersed. Nothing indicates the specific involvement of mitochondria. In the sealed-off areas the foreign material is destroyed.

In *Podospora anserina* several proteases appear in conjunction with an incompatibility reaction, and incompatibility-induced lysis can be prevented by a protease inhibitor (Delettre *et al.*, 1978). The dependence of the *Physarum polycephalum* reaction on RNA and protein synthesis *de novo* has been demonstrated by Schrauwen (1979): specific inhibitors are ineffective after $2\frac{1}{2}$ h and 4 h respectively, in preventing a reaction that becomes visible by 6 h after fusion.

It has been suggested that membranes are the primary reaction target, at least in *Neurospora crassa*; membranous fragments were described in the reaction zone of *Didymium*, but in *Physarum polycephalum* there was no reason to believe that membranes were selectively attacked. In the *P. polycephalum* reaction, a new membrane (not, apparently, the product of multiple vesicle fusion) appeared specifically around the target 'sensitive' nuclei; nuclear envelope dilations were seen on all nuclei in the very early stages of the reaction, and later predominantly on 'killer' nuclei, but these were thought to be contributing to the increase in rough endoplasmic reticulum production which was also seen at the time.

The apparent specificity of the slow lethal reaction for nuclei has two components. The vacuole formation may well have been directed against a cytoplasmic rather than nuclear factor in view of the parallels in other species mentioned; there was always a variable amount of cytoplasm included in each ejection vacuole. The intranuclear damage, however, does appear to be specific, and resembles the effects produced in mammalian cells by RNA (or DNA) synthesis inhibition. Degeneration following a complete metabolic shut-down causes nucleoli to shrink intact rather than to disassemble. In *Physarum polycephalum*, RNA synthesis drops steeply just before the visible lysis (Schrauwen, 1979), at a stage which corresponds to the electron microscopic appearance of nuclear damage (Lane & Carlile, 1979). If 'sensitive' DNA is indeed a specific target, this attack could be mediated by an enzymic restriction–modification system such as are known in bacteria. Restriction–

homologous 'sensitive' strain nuclei, which are at the same time becoming enclosed by membrane vacuoles (arrows) prior to elimination from the plasmodium. Scale bars: 1 μm.

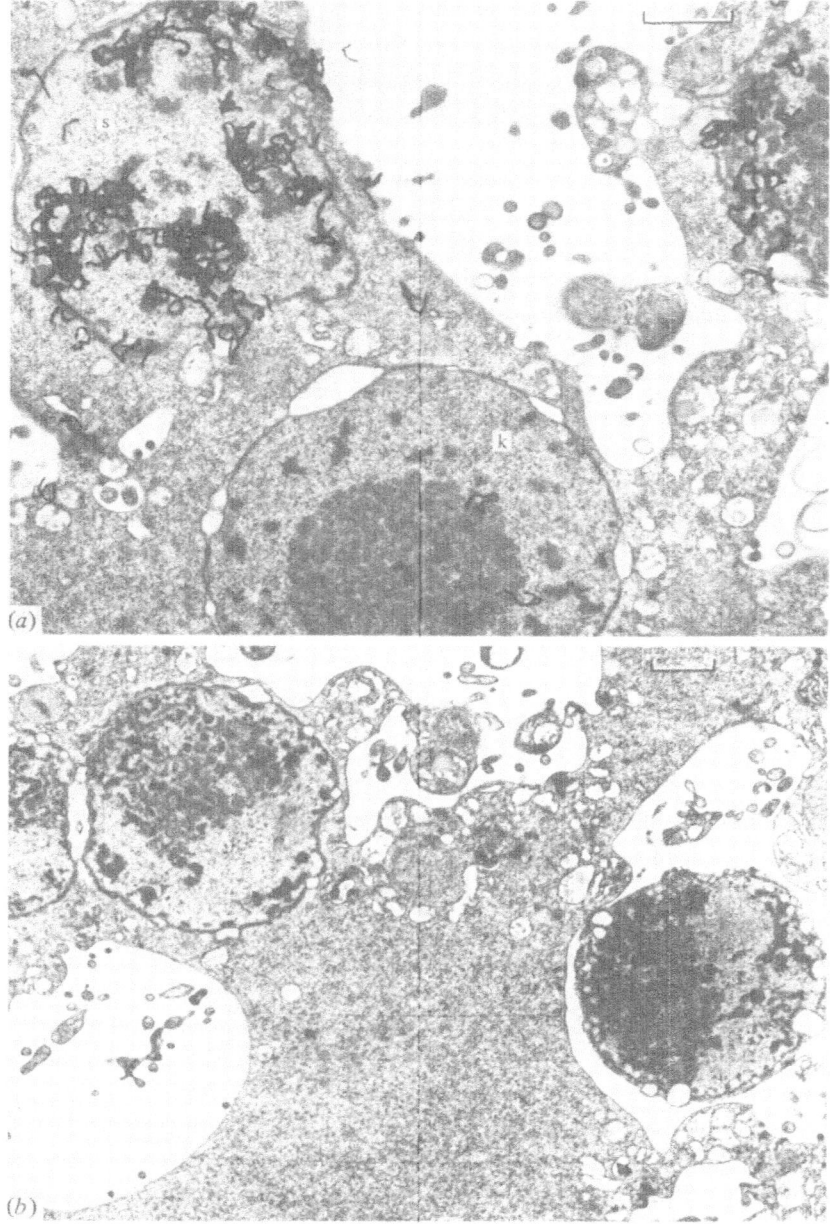

Fig. 3 (*a*) Selective destruction of 'sensitive' (s, labelled) nuclei is well under way by 2½ h after the fusion of incompatible plasmodia, whilst 'killer' nuclei (k) maintain their firm, spherical shape and still show nuclear envelope dilations. (*b*) 'Sensitive' strain nuclei 2½ h after modification has been shown in eukaryote chloroplast DNA (e.g. Royer & Sager, 1979) and suspected but unproven in other eukaryote systems. However, such specific chromatin damage has not been reported in other observations of incompatibility reactions; although this may just be due to lack of information it may also indicate that this damage is not actually central to the post-fusion reaction. The intranuclear damage in *P. polycephalum* did not appear to be dependent on enclosure in elimination vacuoles, nor the vacuoles dependent on nuclear damage. There may be another similarity here with *Podospora anserina*. In an induced reaction the protease activity can be blocked, thus preventing cell lysis; but there is also a complete cessation of RNA synthesis which appears to take effect independently in spite of the protease inhibitor (Delettre *et al.*, 1978).

The observation of conditions in which a reaction can be avoided in *Physarum polycephalum* may also support such a mechanism. If the fusion between *P. polycephalum* strains is initiated during mitosis of the 'killer' strain, no reaction is seen and the heterokaryon appears healthy and is maintained for several days (although reversion to 'killer' phenotype does finally take place after a week or so), and during this time it behaves as a 'neutral' with respect to fusion reactions (Lane & Carlile, 1979). At mitosis RNA synthesis is virtually zero and all metabolic processes are severely curtailed. It is clear that the reaction does require RNA synthesis, but during this hiatus the 'sensitive' strain acquires some measure of protection against subsequent 'killer' metabolism. This aspect of the reaction could also be explained by modification of the 'sensitive' DNA to protect it from subsequent nuclease attack.

Occurrence of incompatibility reactions in nature

Nearly all the work cited above was undertaken in laboratory conditions, and the likelihood of post-fusion incompatibility reactions taking place in nature should be assessed if we are to understand their importance.

The striking incompatibility reaction in *Physarum polycephalum* described above takes place on a nutrient agar between large active plasmodia; an examination of the interaction between strains on water agar, i.e. under starvation conditions, revealed no reaction to the naked

plasmodial fusion showing disruption and partial segregation of nucleolar components; one nucleus has already been ejected into a channel (not processed for autoradiography). Scale bars: 1 μm.

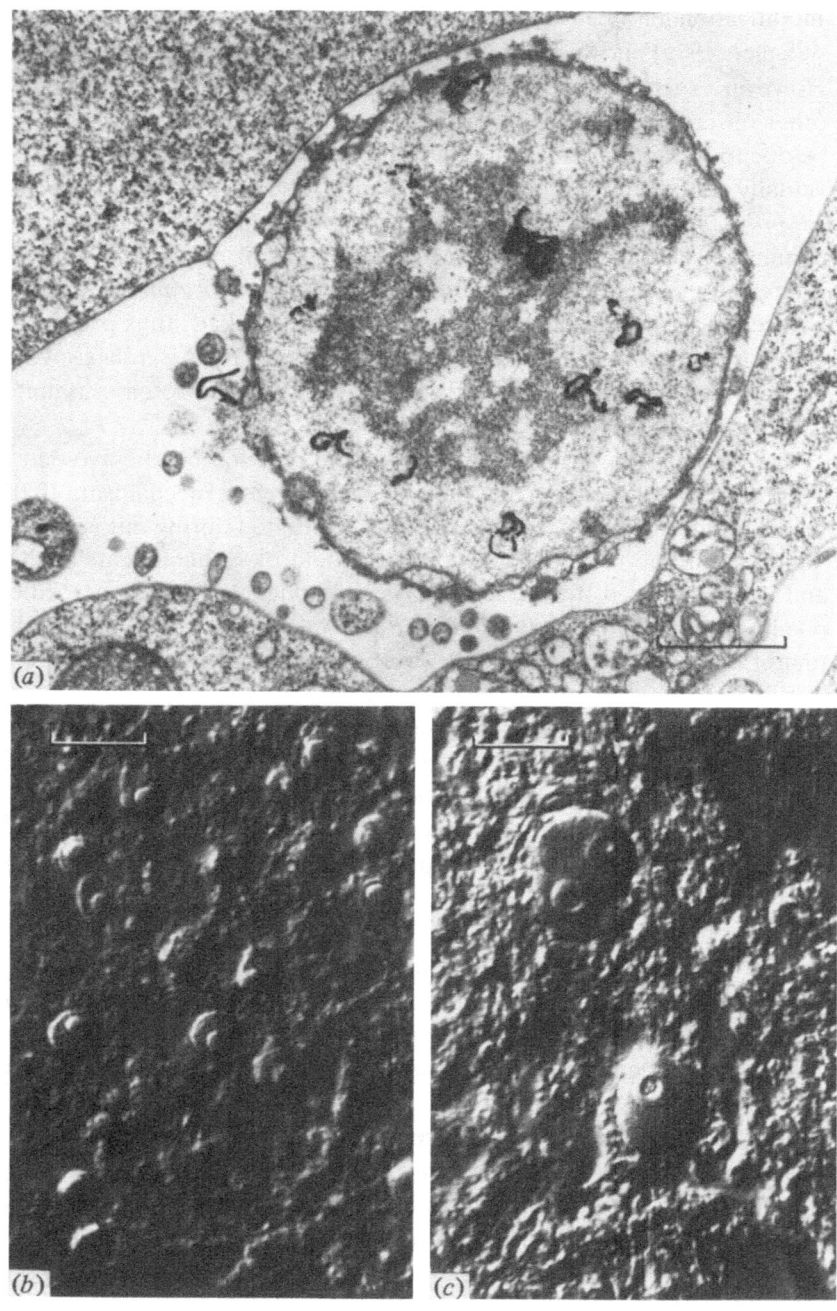

(a)

(b)

(c)

eye, although the same genetic change took place with the 'sensitive' strain being lost (Carlile, 1972). Since plasmodia do not migrate unless food is scarce (Knowles & Carlile, 1978), and so would only encounter another individual under starvation conditions, the 'water agar reaction' is probably the more natural form of the reaction.

Caten & Jinks (1966) among others have drawn attention to the fact that heterokaryon formation in the laboratory between strains paired under forcing conditions would usually be prevented by incompatibility reactions in nature. Vegetative incompatibility between wild isolates has been studied in *Endothia parasitica* (Anagnostakis, 1977), where over two dozen incompatibility groups have now been found, in *Aspergillus nidulans* (see Croft & Jinks, 1977) where the total has reached 19, and also in *Coriolus versicolor* and other wood-rotting Basidiomycetes by Rayner & Todd (1979). The Basidiomycete work is particularly interesting since it indicates that in the wild, incompatibility-induced 'barrages' or demarcation zones may be the barrier which keeps interbreeding individuals distinct from one another, so that in fungal populations at least it is clear that somatic incompatibility is essential to preserve genetic heterogeneity. The actual frequency of balanced heterokaryons in wild fungi may lie somewhere between Caten & Jinks' (1966) caution and Pontecorvo's (1946) suggestion of their widespread importance. The 'dual phenomenon', which is the breakdown of wild isolates of some fungi to two distinct phenotypes when transferred to laboratory culture, does suggest that heterokaryons may be encouraged and maintained by pressures in their natural habitat (Carlile, 1979).

It is also worth considering the likelihood of any genetic similarity between potential fusing individuals in the wild. Since there is a high probability of close genetic relatedness between near neighbours arising from limited dispersal, a mechanism evolved to prevent fusion with non-self must have the capability of recognizing quite small genetic differences. To prevent mixing between two given individuals, it is clear that a single gene difference can be sufficient; to reduce the probability of compatible sibling fusion to near zero, however, requires a number of such genes. Heterozygosity at one gene locus gives a 50% chance of

Fig. 4 (*a*) Nucleus of 'sensitive' strain being broken down in a plasmodial channel after elimination from the cytoplasm, 3 h after fusion. Scale bar: 1 μm. (*b*, *c*) Smear preparations photographed with Nomarski optics, (*b*) from a control plasmodium approaching prophase in the nuclear cycle, and (*c*) from a mixed plasmodium 43 h after fusion of two strains differing at a *let* locus; the nucleoli with vacuoles are characteristic. Scale bars: 10μm.

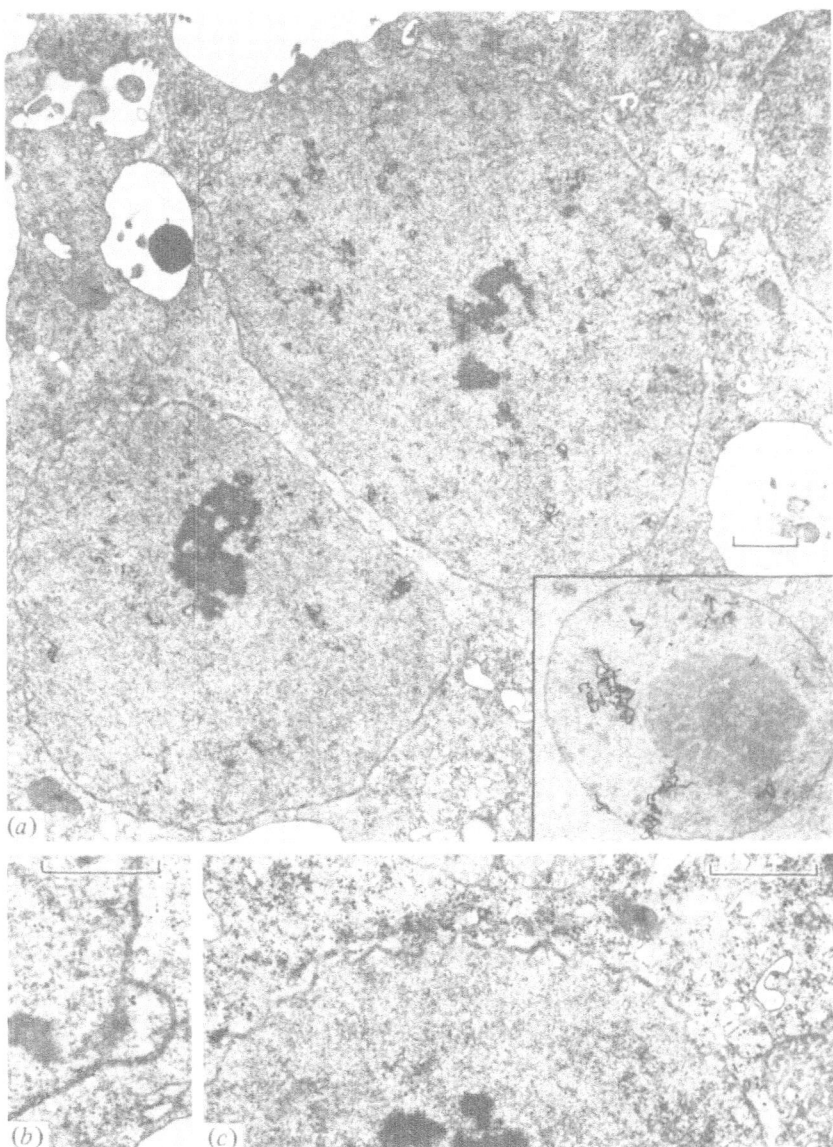

Fig. 5 (*a*) Section through enlarged nuclei seen 24 h after an incompatible plasmodial fusion. *Inset*: normal nucleus, prior to reaction, at the same magnification. (*b*, *c*) Nuclear envelope configurations seen during the recovery phase (here at 48 h) following an incompatibility reaction. Scale bars: 1 μm.

identity between any two F1 siblings, whilst 10 segregating genes for example could reduce that probability to less than 0.1%. A disadvantage in sibling fusion would thus select for multiple incompatibility loci as opposed to multiple alleles at one locus, although the latter system would be adequate, and is frequently used, to promote sexual outcrossing. Many fungi and Myxomycetes have multiallelic mating-type genes but the numerous somatic incompatibility genes identified so far all appear to exhibit simple dominance.

Somatic incompatibility and genetic integrity

The somatic post-fusion reaction is not a direct, passive expression of the incompatibility between two mixed cells, but rather is a specifically evolved preventive mechanism to stop such mixing. This can be seen if the organism's barrier can be bypassed in producing a heterokaryon, which is then fairly normal although sometimes less vigorous, as seen in protoplast fusion between strains of *Aspergillus nidulans* (Dales & Croft, 1977), in the partially diploid strains of *Neurospora crassa* heterozygous for incompatibility genes (Perkins, 1975), and in Myxomycetes fused under certain conditions (Jeffery & Rusch, 1974; Lane & Carlile, 1979).

Heterologous cell fusion usually leads to nuclear fusion. Although homologous nuclear fusion might take place with a very low frequency (Kerr, 1970), it does not appear to be stimulated by homologous cell fusion either in fungi and Myxomycetes or in mammalian cells (viz. multinucleate syncytia of muscle cells). Heterologous cell fusion does, however, lead to karyogamy between sexual gametes in the same species and interspecifically between somatic cells in fungi (Kevei & Peberdy, 1977) and higher organisms. Heterogenic somatic fusion within a species can also stimulate karyogamy, as described above in *Physarum polycephalum* where it leads to the incorporation of foreign DNA which would otherwise have been rejected in the incompatibility reaction. In Basidiomycetes such fusions have been shown to lead to somatic recombination which is often quite selective (see Ellingboe, Prud'homme in Esser & Raper, 1965). Intraspecific heterokaryosis is tolerated in Basidiomycetes as a component of sexual reproduction; in secondary homothallic heterokaryons some specialization to avoid incompatibility has been detected in the tolerance genes of *Neurospora tetrasperma* (Newmeyer, 1970; Metzenberg & Ahlgren, 1973). Yet the coexistence of two genotypes within the same cytoplasm can and does lead to genetic interchange even without karyogamy. This has been clearly demonstrated by the example of the transfer of mound growth mutant genes between intact unfused nuclei in *Schizophyllum commune* (Leonard, Gaber & Dick, 1978). It is therefore not surprising to see that most if not all heterokaryons are genetically unstable for some time after their formation; once two nuclei are together in the same cytoplasm there is no longer so effective a barrier between them to prevent exchange of genetic material.

For any given gene, the stratagem of genetic recombination through sexual reproduction is always a gamble, since the outcome may either increase or decrease the fitness of their genotype. Some fungi have lost sexual recombination from their life cycle, presumably because the advantages no longer outweighed the risks for a species evolved in a restricted habitat. For species that do undergo sexual reproduction, however, the genetic recombination events must be separated by periods of growth and differentiation for any advantage gained by the recombination to be exploited; for natural selection to act. Not only could genetic alteration during the growth phase be anything from disruptive to disastrous, but it could neutralize the achievements of sexual recombination. In fungi the individuals most likely to be involved in attempted asexual fusions would probably be sibling organisms whose genes had just been resegregated by meiosis so that such fusions would negate sexual recombination. As Rayner & Todd (1979) showed in Basidiomycetes, the absence of fusion barriers could lead directly to a loss of individuality and thus of genetical heterogeneity; if as Buller (1931) suggested such co-operation might be necessary for sporulation, a stratagem of pooling resources with a neighbouring dissimilar, and therefore competing, genotype would surely be evolutionary suicide.

If somatic incompatibility had really evolved in an important functional association with sexual reproduction, then one might expect to see some evidence of genetic integration of the two phenomena. In the two systems wherein genetic control of incompatibility has received most attention there is evidence of pleiotropic genes which simultaneously affect somatic incompatibility and sexual fertility. In *Neurospora crassa*, heterogeneity at the mating-type locus allows sexual fusion but simultaneously enforces somatic incompatibility, and all attempts to resolve this function into more than one gene within the locus have failed (Newmeyer *et al.*, 1973). The mating-type locus is, however, only one of ten identified so far which affect somatic incompatibility in this species, and Williams & Wilson (p. 241) were studying two other ones. In *Podospora anserina*, the study of certain mutants that suppress a non-allelic somatic incompatibility system and simultaneously prevent normal sexual differentiation (Bernet, 1971) has led to the conclusion that the genes in question are involved in the production of proteases important for both the incompatibility reaction and female sexual development (Boucherie *et al.*, 1976). While somatic incompatibility between strains does not in itself result in sexual infertility (Butcher, 1968) direct evidence for a genetic link between the two phenomena is still lacking for other fungal species.

Finally, if somatic incompatibility between similar genotypes is as primitive in origin as sexual recombination, there should be signs of it in many different groups of organisms. The observation of the parallels between fungal incompatibility and the immune system has been made before, and Esser & Blaich (1973) have suggested that the similarity has arisen by convergent evolution. But if the two both have their origin in a mechanism for conserving sexual recombination then this might help to explain some of the more perverse features of the immune response. A histocompatibility recognition mechanism is a more basic immunological capability than immunoglobulin production: graft rejection is seen even in coelenterates (Theodor, 1970). It has always been difficult to explain in evolutionary terms why the immune system is apparently better adapted for graft rejection than for preventing the spread of infectious agents. The trigger for antibody production is in fact slightly altered self: the antigen has to be presented in association with a self determinant on the macrophage surface for lymphocytes to be activated. The defensive role against infection may turn out to be a secondary development; it is not immediately clear that the suggested primary requirement is actually still present in higher animals. Protection against infection is a role frequently assigned to the incompatibility mechanisms in fungi (Esser & Blaich, 1973), but whilst no doubt hindering the spread of agents such as viruses through a population, post-fusion heterogenic incompatibility is by no means sufficient to prevent it (Caten, 1972; Anagnostakis & Day, 1980). On the other hand, evidence is accumulating for an association between sex and the major histocompatibility locus complex in the mouse (the only species subject to such genetical studies), both functionally in affecting the animals' breeding preferences (Andrews & Boyse, 1978) and genetically as seen by the segregation of the T locus (Klein, 1975). Another association is that of the H-Y antigen, which is conserved through many classes of vertebrates and gives rise to heterosexual histoincompatibility in inbred strains; this antigen is a product of the gene which apparently determines heterogametic sex, e.g. male in man, female in birds (Wachtel *et al.*, 1975).

Acknowledgements. The author wishes to thank Dr Michael Carlile for helpful discussions during the preparation of this manuscript, and the Science Research Council for financial support of part of this work.

References

Anagnostakis, S. L. (1977). Vegetative incompatibility in *Endothia parasitica*. *Experimental Mycology*, **1**, 306–16.

Anagnostakis, S. L. & Day, P. R. (1980). Hypovirulence conversion in *Endothia parasitica*. *Phytopathology*, **69**, 1226–92.

Andrews, P. W. & Boyse, E. A. (1978). Mapping of an H-2-linked gene that influences mating preference in mice. *Immunogenetics*, **6**, 265–8.

Barrett, D. K. & Uscuplic, M. (1971). The field distribution of interacting strains of *Polyporus schweinitzii* and their origin. *New Phytologist*, **70**, 581–98.

Bechtel, D. B. (1976). Nuclear degeneration in the myxomycete *Physarella oblonga*. *Protoplasma*, **90**, 179–87.

Bégueret, J. (1972). Protoplasmic incompatibility: possible involvement of proteolytic enzymes. *Nature New Biology*, **235**, 56–8.

Bégueret, J. & Bernet, J. (1973). Proteolytic enzymes and protoplasmic incompatibility in the fungus *Podospora anserina*. *Nature New Biology*, **243**, 94–6.

Bernet, J. (1971). Sur un cas de suppression de l'incompatibilité cellulaire chez *Podospora anserina*. *Compte rendu hebdomadaire des séances de l'Académie des sciences*, **273**, 1330–3.

Bernhard, W. (1971). Drug-induced changes in the interphase nucleus. *Advances in Cytopharmacology*, **1**, 49–67.

Blaich, R. & Esser, K. (1971). The incompatibility relationship between geographical races of *Podospora anserina*. V. Biochemical characterization of heterogenic incompatibility on cellular level. *Molecular and General Genetics*, **111**, 265–72.

Border, D. J. & Carlile, M. J. (1974). Somatic incompatibility following plasmodial fusion between strains of the myxomycete *Physarum polycephalum*: the effect on their nuclei. *Journal of General Microbiology*, **85**, 211–19.

Boucherie, H., Bégueret, J. & Bernet, J. (1976). The molecular mechanism of protoplasmic incompatibility and its relationship to the formation of protoperithecia in *Podospora anserina*. *Journal of General Microbiology*, **92**, 59–66.

Buller, A. H. R. (1931). *Researches on Fungi*, vol. 4. London: Longmans.

Butcher, A. C. (1968). The relationship between sexual outcrossing and heterokaryon incompatibility in *Aspergillus nidulans*. *Heredity*, **23**, 443–52.

Carlile, M. J. (1972). The lethal interaction following plasmodial fusion between two strains of the myxomycete *Physarum polycephalum*. *Journal of General Microbiology*, **71**, 581–90.

Carlile, M. J. (1974). Incompatibility in the myxomycete *Badhamia utricularis*. *Transactions of the British Mycological Society*, **62**, 401–2.

Carlile, M. J. (1976). The genetic basis of the incompatibility reaction following plasmodial fusion between two strains of the myxomycete *Physarum polycephalum*. *Journal of General Microbiology*, **93**, 371–6.

Carlile, M. J. (1979). Bacterial, fungal and slime mould colonies. In *Biology and Systematics of Colonial Organisms,* ed. G. Larwood & B. R. Rosen, pp. 3–27. Systematics Association Special Volume 11. London & New York: Academic Press.

Carlile, M. J. & Dee, J. (1967). Plasmodial fusion and lethal interaction between strains in a myxomycete. *Nature, London,* **215,** 832–4.

Carlile, M. J. & Gooday, G. W. (1978). Cell fusion in myxomycetes and fungi. In *Membrane Fusion,* ed. G. Poste & G. L. Nicholson, pp. 219–65. Amsterdam & New York: Elsevier/North-Holland Biomedical Press.

Caten, C. E. (1972). Vegetative incompatibility and cytoplasmic infection in fungi. *Journal of General Microbiology,* **72,** 221–9.

Caten, C. E. & Jinks, J. L. (1966). Heterokaryosis: its significance in wild homothallic Ascomycetes and Fungi imperfecti. *Transactions of the British Mycological Society,* **49,** 81–93.

Clark, J. (1977). Plasmodial incompatibility reactions in the slime mold *Physarum cinereum. Mycologia,* **69,** 46–52.

Clark, J. & Collins, O. R. (1973). Directional cytotoxic reactions between incompatible plasmodia of *Didymium iridis. Genetics,* **73,** 247–57.

Collins, O. R. (1979). Myxomycete biosystematics: some recent developments and future research opportunities. *Botanical Reviews,* **45,** 145–201.

Croft, J. H. & Jinks, J. L. (1977). Aspects of the population genetics of *Aspergillus nidulans.* In *Genetics and Physiology of Aspergillus,* ed. J. E. Smith & J. A. Pateman, pp. 339–60. New York & London: Academic Press.

Dales, R. B. G. & Croft, J. H. (1977). Protoplast fusion and the isolation of heterokaryons and diploids from vegetatively incompatible strains of *Aspergillus nidulans. FEMS Microbiology Letters,* **1,** 201–3.

Delettre, Y. M., Boucherie, H. & Bernet, J. (1978). Protoplasmic incompatibility and cell lysis in *P. anserina:* effect of β-phenyl pyruvic acid. *Biochemie und Physiologie der Pflanzen,* **172,** 27–34.

Esser, K. (1974) *Podospora anserina.* In *Handbook of Genetics,* ed. R. C. King, pp. 531–51. New York: Van Nostrand-Reinhold.

Esser, K. & Blaich, R. (1973). Heterogenic-incompatibility in plants and animals. *Advances in Genetics,* **17,** 107–52.

Esser, K. & Raper, J. R. (eds.) (1965). *Incompatibility in Fungi.* Berlin & New York: Springer-Verlag.

Flentje, N. T. & Stretton, H. M. (1964). Mechanisms of variation in *Thanatephorus cucumeris* and *T. practicolus. Australian Journal of Biological Sciences,* **17,** 686–704.

Garnjobst, L. & Wilson, J. F. (1956). Heterokaryosis and protoplasmic incompatibility in *Neurospora crassa. Proceedings of the National Academy of Sciences, USA,* **42,** 613–18.

Griffin, D. M. & Perrin, H. N. (1960). Anastomoses in Phycomycetes. *Nature, London,* **187,** 1039–40.

Jeffery, W. R. & Rusch, H. P. (1974). Induction of somatic fusion and heterokaryosis in two incompatible strains of *Physarum polycephalum. Developmental Biology,* **39,** 331–5.

Kerr, S. J. (1970). Nuclear size in plasmodia of the true slime mold *Didymium nigripes. Journal of General Microbiology,* **63,** 347–56.

Kevei, F. & Peberdy, J. F. (1977). Interspecific hybridization between *Aspergillus nidulans* and *Aspergillus rugulosus* by fusion of somatic protoplasts. *Journal of General Microbiology,* **102,** 255–62.

Klein, J. (1975). *Biology of the Mouse histocompatibility-2 Complex.* Berlin: Springer-Verlag.

Knowles, D. J. C. & Carlile, M. J. (1978). Growth and migration of plasmodia of the myxomycete *Physarum polycephalum*: the effect of carbohydrates, including agar. *Journal of General Microbiology*, **108**, 9–15.

Labarère, J. (1973). Propriétés d'un système d'incompatibilité chez le champignon *P. anserina* et intérêt de ce système pour l'étude d'incompatibilité. *Compte rendu hebdomadaire des séances de l'Académie des Sciences, Série D*, **276**, 1301–4.

Lane, E. B. & Carlile, M. J. (1979). Post-fusion somatic incompatibility in plasmodia of *Physarum polycephalum*. *Journal of Cell Science*, **35**, 339–54.

Leonard, J. J., Gaber, R. F. & Dicks, S. (1978). Internuclear genetic transfer in dikaryons of *Schizophyllum commune*. II. Direct recovery and analysis of recombinant nuclei. *Genetics*, **89**, 685–93.

McKenzie, A. R., Flentje, N. T., Stretton, H. M. & Mayo, M. J. (1969). Heterokaryon formation and genetic recombination within one isolate of *Thanatephorus cucumeris*. *Australian Journal of Biological Sciences*, **22**, 895–904.

Metzenberg, R. L. & Ahlgren, S. K. (1973). Behaviour of *Neurospora tetrasperma* mating-type genes introgressed into *N. crassa*. *Canadian Journal of Genetics and Cytology*, **15**, 571–6.

Monneron, A. (1971). Action of some drugs on liver nuclei and polysomes. *Advances in Cytopharmacology*, **1**, 131–44.

Newmeyer, D. (1970). A suppressor of the heterokaryon-incompatibility associated with mating type on *Neurospora crassa*. *Canadian Journal of Genetics and Cytology*, **12**, 914–26.

Newmeyer, D., Howe, H. B. & Galeazzi, D. R. (1973). A search for complexity at the mating-type locus of *Neurospora crassa*. *Canadian Journal of Genetics and Cytology*, **15**, 577–85.

Perkins, D. D. (1975). Use of duplication-generating rearrangements for studying heterokaryon incompatibility genes in *Neurospora*. *Genetics*, **80**, 87–105.

Pontecorvo, G. C. (1946). Genetic systems based on heterokaryosis. *Cold Spring Harbor Symposia on Quantitative Biology*, **11**, 193–201.

Rayner, A. D. M. & Todd, N. K. (1979). Population and community structure and dynamics of fungi in decaying wood. *Advances in Botanical Research*, **7**, 333–420.

Royer, H.-D. & Sager, R. (1979). Methylation of chloroplast DNAs in the life cycle of *Chlamydomonas*. *Proceedings of the National Academy of Sciences, USA*, **76**, 5794–8.

Schrauwen, J. A. M. (1979). Post-fusion incompatibility in *Physarum polycephalum*. *Archives of Microbiology*, **122**, 1–7.

Todd, N. K. & Rayner, A. D. M. (1979). Fungal individualism. *Science Progress, Oxford*, **66**, 331–53.

Theodor, J. L. (1970). Distinction between 'self' and 'not-self' in lower invertebrates. *Nature, London*, **227**, 690–2.

Upadhyaya, K. C. & Ling, H. (1976). Ultrastructure of clear zones exhibiting cytoplasmic incompatibility following somatic cell fusion in the myxomycete *Didymium iridis*. *Indian Journal of Experimental Biology*, **14**, 652–58.

Wachtel, S. S., Ohno, S., Koo, G. C. & Boyse, E. A. (1975). Possible role for H-Y antigen in the primary determination of sex. *Nature, London*, **257**, 235–6.

Williams, C. A. & Wilson, J. F. (1966). Cytoplasmic incompatibility reactions in *Neurospora crassa*. *Annals of the New York Academy of Science*, **129**, 853–63.

Williams, C. A. & Wildon, J. F. (1968). Factors of cytoplasmic incompatibility. *Neurospora Newsletter*, **13** (June), 12.

Wilson, J. F., Garnjobst, L. & Tatum, E. L. (1961). Heterocaryon incompatibility in *Neurospora crassa* – micro-injection studies. *American Journal of Botany*, **48**, 299–305.

The effects of ultraviolet radiation on the nuclei of yeasts and filamentous fungi

B. W. BAINBRIDGE

Microbiology Department, Queen Elizabeth College, Campden Hill Road, London W8 7AH, UK

Introduction

The ultraviolet (UV) lamp has been an extremely popular experimental tool in molecular and microbial genetics partly due to its convenience in use but also because of its relative safety in comparison with ionizing radiation and chemical mutagens. The most biologically active wavelengths are 254 nm, usually referred to as far UV, and 350 nm which is known as near UV or black light. Far UV is known to affect survival, mutation and recombination, and near UV has similar effects in the presence of a photosensitizer. Near UV can also increase survival when used after irradiation with far UV, a process known as photoreactivation. Background material and techniques will be found in Jagger (1967, 1977) and Smith (1977*a*). Experiments in which yeast cells or fungal spores are exposed to UV are, however, deceptively simple, and in practice the response of a cell to UV-induced damage is complex and not completely understood. The major photoproducts appear to be pyrimidine dimers, but a complex of factors influence the level of survival, mutation and recombination of damaged cells. Models for the repair of DNA have been based on extensive information from bacterial systems for which a range of UV-sensitive and recombination-deficient mutants are available (Hanawalt *et al.*, 1979). Less complete data are available from yeast and filamentous fungal systems, but in general these models have proved useful. The major organisms that have been studied are *Saccharomyces cerevisiae*, *Schizosaccharomyces pombe*, *Neurospora crassa*, *Aspergillus nidulans*, *Ustilago maydis* and *Schizophyllum commune* (see review articles of genetic aspects in King (1974), also Catcheside (1974) and Fincham, Day & Radford (1979)). Few cytological studies have been made of the influence of UV on nuclear and chromosome structure and this article will concentrate on biochemical, physiological and genetic approaches.

Biochemical effects of ultraviolet radiation
Detection of pyrimidine dimers
The analysis of UV-induced photoproducts in the DNA of yeasts and fungi has been hampered by technical limitations. Firstly it has proved impossible to use a specific label for DNA that is comparable with the use of radioactive thymidine in bacteria. This is due to the absence of thymidine kinase in fungi (Grivell & Jackson, 1968). Specific labelling of DNA has been achieved for certain strains of *Saccharomyces* by using thymidine monophosphate (Wickner, 1974), but this has not been used extensively. Radioactive precursors are used which label both DNA *and* RNA, and methods for analysing DNA must normally allow for the complete removal of contaminating RNA. Allowance must also be made for the presence of very active nucleases which rapidly degrade DNA. A further complication is that mitochondrial DNA may be 10–20% of the total DNA and it may be necessary to separate this from genomic DNA by density gradient centrifugation (Williamson, Moustacchi & Fennell, 1971). Another major problem is the difficulty in removing the fungal wall rapidly. Protoplasts of yeast cells can be made with relative ease, but many fungi may require a minimum digestion time of one to two hours to produce a reasonable yield of protoplasts. During this time, repair of UV-induced lesions may have occurred and it is difficult to study the initial effects of the treatment. Alternative methods of disruption involved crushing, grinding or extrusion at low temperatures in the presence of DNase inhibitors. Heating within 5 °C of the T_m can also be used to inactivate DNases. Morris (1978) has published a technique for obtaining high molecular weight DNA from *Aspergillus nidulans* and intact nuclei can be obtained from the same organism by blending mycelium in the presence of liquid nitrogen (Gealt, Sheir-Neiss & Morris, 1976), but this is not a very convenient technique for large numbers of radioactive samples.

It is generally accepted that the main photoproducts in DNA are pyrimidine dimers which are formed between adjacent pyrimidine residues in the same single strand of DNA. They have been detected directly in *Ustilago* and *Saccharomyces* (Unrau *et al.*, 1973), *Schizosaccharomyces* (Fabre & Moustacchi, 1973), *Neurospora* (Worthy & Epler, 1973) and in *Aspergillus nidulans* (B. W. Bainbridge & K. van de Vate, unpublished, Fig. 1). Unrau *et al.* (1973) have also reported that the frequency of the pyrimidine dimers, cytosine–cytosine ĈC, cytosine–thymine ĈT, and thymine–thymine T̂T, was 1:1:2. These results are based on the hydrolysis of DNA followed by separation of bases by paper chromatography and they are complicated by the presence of radioactive material in the same region as the dimers (Fig. 1). In spite of this, the techniques have been successfully applied by a number of

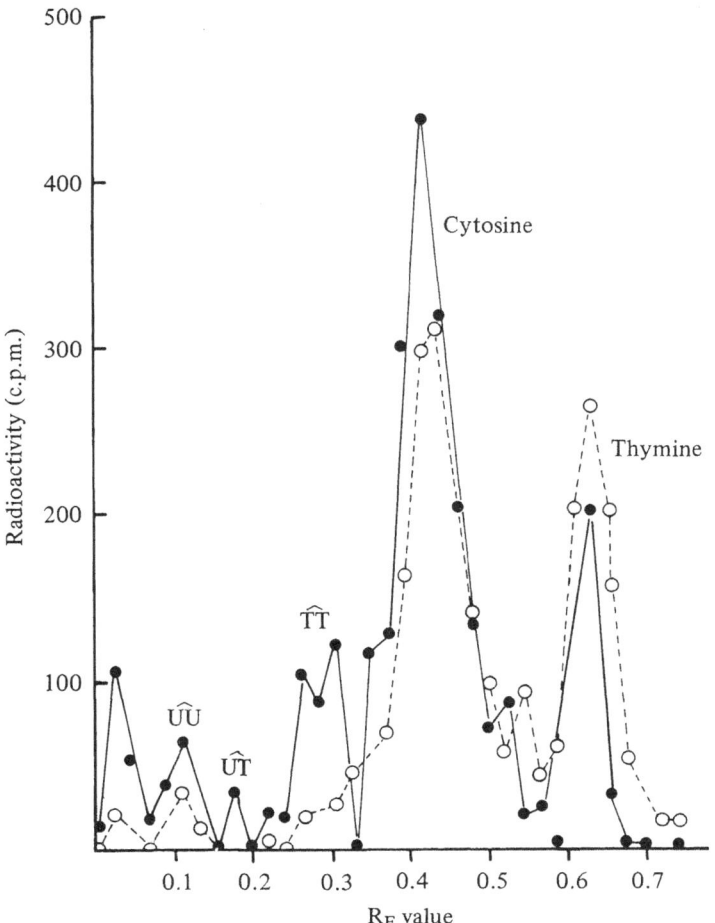

Fig. 1. Detection of pyrimidine dimers in far UV-irradiated DNA of *A. nidulans*. Mycelium was labelled with [5 – ^{3}H]uridine for 16 h at 37 °C and irradiation dose was 600 J m^{-2}. Mycelium was ground with sand and powdered solid carbon dioxide. DNA was extracted by a modified Marmur (1961) technique and hydrolysed by the method of Unrau *et al.* (1972). (●) irradiated culture, (○) unirradiated control.

groups to analyse the frequency of dimers and their removal under a variety of conditions.

A more recent technique, first used in bacteria, avoids some of the technical pitfalls mentioned by using a dimer-specific endonuclease to study the frequency of dimers (Ganesan, 1974). The enzyme makes a single-strand break at each dimer, which can be detected by a reduction in molecular weight as measured on alkaline sucrose-density gradients. Reynolds (1978) has applied this technique to *Saccharomyces cerevisiae*.

Repair processes

The presence of pyrimidine dimers in DNA causes a number of problems for the cell. The coding capacity of the DNA is altered and, although DNA can be replicated, single-strand gaps are left opposite to the dimers. A high frequency of single-strand gaps is lethal unless repair processes can restore the double-strand molecule. At least four repair processes for UV damage are known in bacteria, and there is some evidence for similar processes in yeasts and fungi. *Photoreactivation* involves the enzymatic splitting of dimers mediated by near UV and this has been shown in *Saccharomyces* (Parry & Cox, 1968), *Neurospora* (Schroeder, 1975), *Aspergillus* (Fortuin, 1971*a*) and *Schizophyllum* (Hundert *et al.*, 1978). *Excision repair* involves the removal of dimers and adjacent bases followed by repair synthesis to fill the gap. This process is not light-dependent and is therefore an example of *dark repair*. This process has been directly demonstrated in *Saccharomyces cerevisiae*, *Neurospora crassa* and *Ustilago* (Unrau *et al.*, 1972; Worthy & Epler, 1973; Unrau, 1975). Also included under dark repair is a process that can occur after the DNA has been replicated: *post-replication repair* or daughter-strand gap repair. In bacteria it is known that DNA-containing dimers can be replicated to leave single-strand gaps opposite each dimer. Intact molecules are then salvaged by a recombination-like process. The lesion itself is therefore not repaired, but the cell manages to survive by making one functional DNA molecule. Direct biochemical evidence for the operation of the pathway in fungi is lacking, although there is abundant evidence that UV can induce recombination and that UV-sensitive mutants have altered recombination (see below). Ferguson & Cox (1980), however, have obtained evidence that in yeast the onset of DNA synthesis is to some extent dependent on the repair of dimers. The process thought to be responsible for UV-induced mutation is that of *error-prone* or *SOS* repair (Witkin, 1976). This process is inducible by UV irradiation in bacteria, but again direct biochemical evidence for a similar process in fungi is lacking. There is growing evidence for UV-inducible processes in fungi which will be discussed later.

Photosensitization to near ultraviolet radiation

As already mentioned, near UV in the presence of photosensitizers such as 8-methoxypsoralen can cause killing and mutation comparable to that of far UV. The psoralen molecules, under the influence of near UV, interact with DNA to produce monofunctional and bifunctional adducts (cross links) with pyrimidines (Scott & Anderson, 1971; reviewed by Scott, Pathak & Mohn, 1976). Repair of this damage appears to follow similar pathways to those for repair of far UV damage.

Other photoproducts

Most research effort has been directed towards understanding the effects of pyrimidine dimers but it should be realized that UV irradiation can also produce other photoproducts (Smith, 1977*b*). During replication and transcription of DNA, hydration of pyrimidines may occur to form, for example, dihydrocytosine. This has been shown *in vitro* to alter the coding properties of polycytidylic acid. Near UV has been shown in bacteria to produce growth delay and a complete, but temporary cessation in RNA synthesis. This is thought to occur as a consequence of cross links formed between 4-thiouracil and cytidine in tRNA (Smith, 1977*c*). Amino acids such as tryptophan can absorb UV to generate peroxide radicals which can damage DNA (Hanawalt *et al.*, 1979). On the whole this area has received little attention, but it seems likely that the influence of UV on the cytoplasm may reveal unexpected effects.

Physiological effects of ultraviolet radiation

The technical problems associated with the analysis of UV-induced photoproducts in DNA have meant that many workers have concentrated on the physiological rather than the biochemical effects of UV radiations. The easiest parameter to study is that of survival, and this has produced useful indirect evidence for the operation of repair processes. Information can be obtained from the shape of survival curves and also from the influence of pre- and post-UV treatments, such as liquid holding, repair inhibitors and photoreactivation. The most popular systems have been the survival of yeast cells or fungal spores, particularly spores that have a single nucleus as in *Aspergillus nidulans*.

Irradiation of germinating spores

Interesting information can also be obtained by irradiating germinating spores (Jansen, 1970*a*; Fortuin, 1971*b*). An experiment of this type is shown in Fig. 2 (B. W. Bainbridge & C. Britnell, unpublished). Spores were irradiated, in a soft agar overlay, with 120 J m^{-2} of far UV and the initial survival, after incubation for up to 150 min, was between 96% and 100%, but at 180 min and 210 min the survival had decreased to 80%. Biochemical data for cultures with comparable germination times (Bainbridge, 1971) have shown that DNA synthesis started at 150 min and was complete by 300 min. This would suggest that *Aspergillus nidulans* spores are sensitive to UV irradiation during

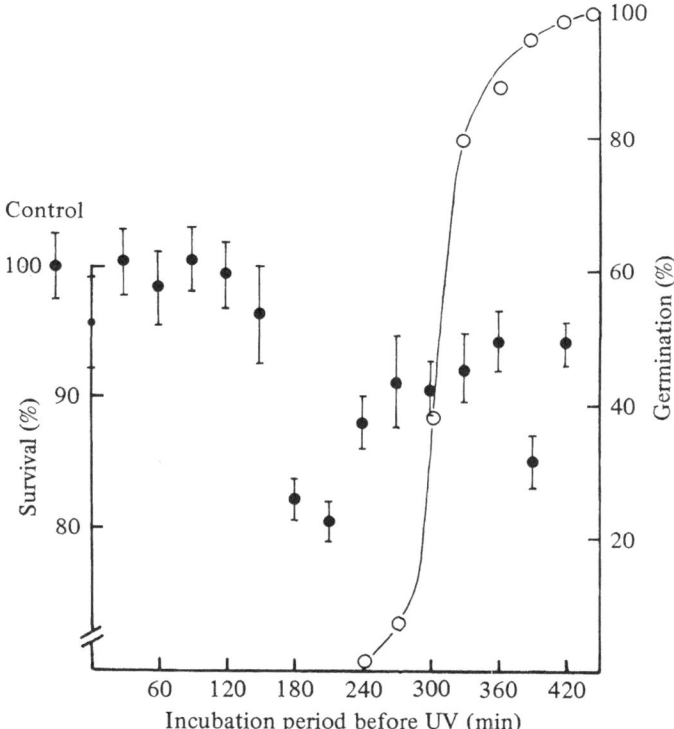

Fig. 2. Influence of far UV on survival of germinating spores of *A. nidulans*. Spores were germinated in an overlay of 0.6% agar on complete medium. At various times plates were exposed to 120 J m^{-2} far UV and incubated at 37 °C. Bars indicate standard errors of the mean values which were based on a minimum count of 1200 colonies. (●) % survival of spores, (○) % germination of spores in parallel cultures.

the early S phase. Synchrony of spore germination is only partial so that maximum sensitivity may occur during the late G1 phase as well as in the S phase. The influence of far UV has also been studied on spores growing on cellophane (Fig. 3, B. W. Bainbridge, unpublished). All spores in early germination, which were 7.5 μm in diameter or less (including germ tube), showed a delay in growth lasting for 40 min which occurred immediately on irradiating with a low dose of far UV (10 J m^{-2}). In the case of spores that were between 7.5 μm and 25 μm in diameter half of them continued growing without any delay (Fig. 4). Again this tends to suggest that there is a phase in early germination, probably the S phase or late G1, which is particularly sensitive to UV damage.

Work on synchronized cultures of *Saccharomyces cerevisiae* (Chanet, Williamson & Moustacchi, 1973; Chanet, Heude & Moustacchi, 1974)

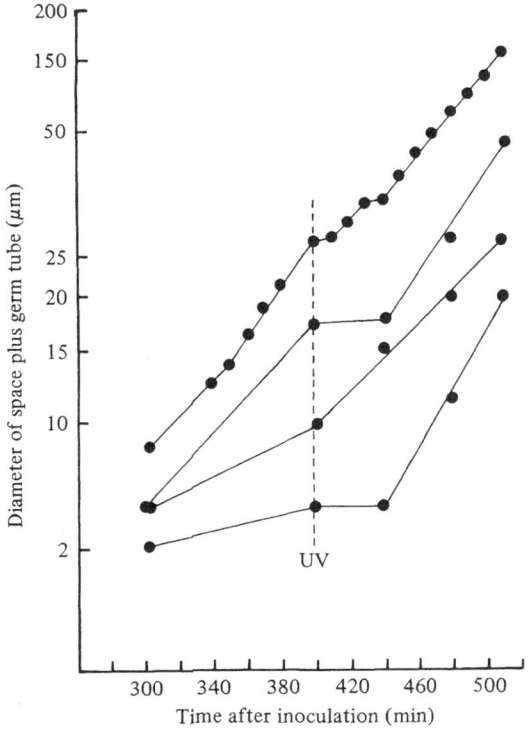

Fig. 3. Influence of far UV on growth of individual spores of *A. nidulans* growing on cellophane. Spores were germinated and then irradiated with 10 J m^{-2} far UV at 399 min after incubation. Growth was analysed from time-lapse photographs taken every 10 min.

has shown that the G1-phase cells were more radiosensitive and the G2-phase cells were more radioresistant. Davies, Tippins & Parry (1978) have reported maximum sensitivity in the G1/early S phase although the yeast cells here were collected from gradients and irradiated in the non-growing state. Survival potential therefore appears to depend on the stage of DNA replication and whether or not there are daughter DNA strands present.

Irradiation of hyphae

Systems comparable with synchronized yeast cells have not been obtained for filamentous fungi although naturally occurring synchrony of mitosis and septation are known in individual hyphae of *Aspergillus nidulans* (Clutterbuck, 1970; Bainbridge, 1976; Fiddy &

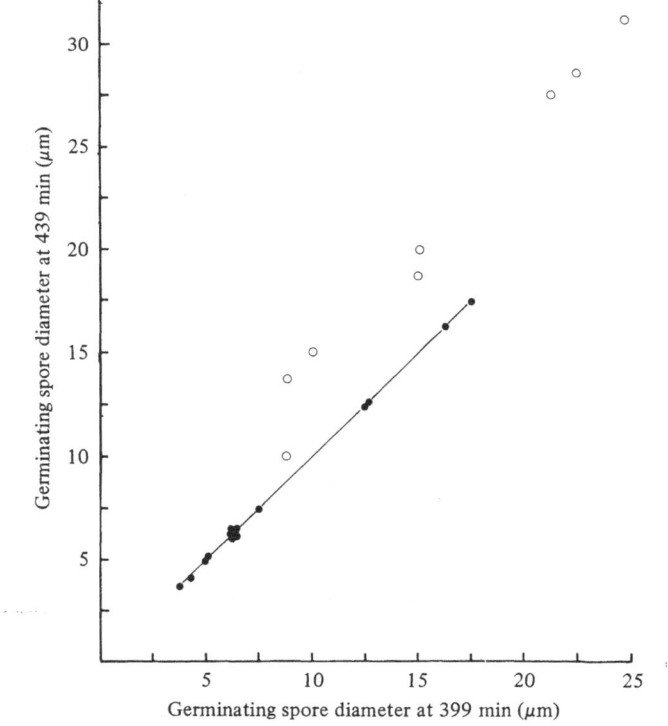

Fig. 4. Summary graph of effect of far UV on germinating spores shown in Fig. 3. Spore diameters include germ tubes. Spores were irradiated at 399 min and diameters recorded at this time and at 439 min. (●) spores in which no growth occurred, (○) spores which showed growth.

Trinci, 1976). This has led to the recognition of the so-called 'duplication cycle' which has some analogies with the cell cycle of unicellular organisms. When an apical cell has reached its maximum length, mitosis occurs in a synchronous wave, and this is followed by separation which results in an apical cell half the length of the original apical cell. It follows from this that the length of an apical cell can be used to estimate the stage reached in the duplication cycle. As mitosis and septation occur more or less synchronously, it appears likely that DNA synthesis is also synchronized. Irradiation of hyphae growing on cellophane has shown that growth ceases immediately and that hyphae responded in two ways: some hyphae recover and continue growing whereas others vacuolate and die (Fig. 5; D. Hogden & B. W. Bainbridge, unpublished). Measurements of apical cell lengths before irradiation have

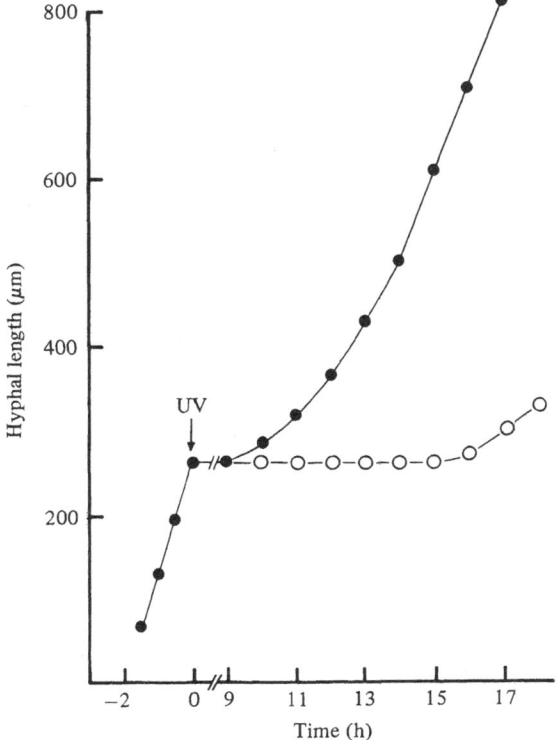

Fig. 5. Influence of far UV on the growth of mature hyphae of *A. nidulans*. Hyphae were grown on cellophane at 37 °C on complete medium and irradiated with 240 J m^{-2} at time 0. (●) hypha that recovered after 9 h lag, (○) hypha that vacuolated and died.

indicated that it is hyphae which are in the last third of the duplication cycle that are most likely to vacuolate (Fig. 6). Identical results were obtained for diploid and haploid strains. These results support the findings from the previous section that the UV-sensitive period is late G1/S phase, and it would also follow that G2 is very short in these hyphae. Results for germination have also suggested that G2 is short under these conditions (Bainbridge, 1971).

It is interesting that growth should cease immediately on irradiation as this implies a direct effect on wall and macro-molecular synthesis. Inhibition of DNA and RNA synthesis is understandable, but the cessation of tip growth perhaps implies that UV can also interfere with precursor levels or energy supplies, perhaps by alteration of ATP or NAD(P) levels.

Genetic effects

Far UV and near UV in the presence of photosensitizers have long been known to induce mutation. However, a wide range of other genetic effects are known and these are listed in Table 1. It is generally assumed that pyrimidine dimers in DNA are responsible for most of these effects but direct proof for this is lacking. The consequences of induced mutation or recombination can usually not be detected until several days after the event, when colonies appear on plates. However, Holliday (1971) has devised an ingenious biochemical method for detecting recombination. Diploids of *Ustilago maydis* were made heteroallelic for *nar*1, nitrate reductase, and homoallelic for *nir*1–1, nitrite reductase. Recombination between *nar*1 alleles gave *nar*1[+] which

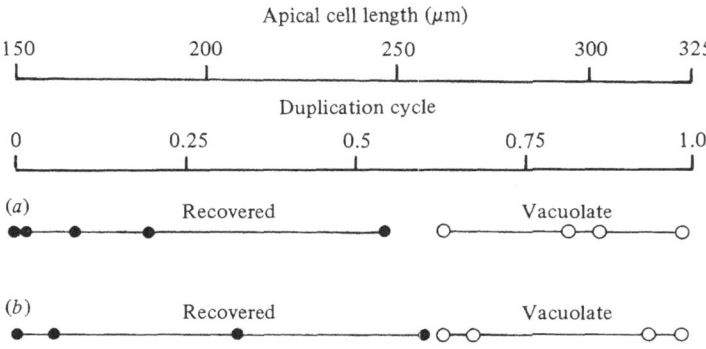

Fig. 6. Influence of apical cell length at irradiation on subsequent recovery or vacuolation of hyphae (conditions as for Fig. 5). (*a*) haploid strain, (*b*) diploid strain.

Table 1. *Genetic effects of ultraviolet radiation on Fungi*[a]

Process affected	System studied
Mutation	Morphological mutants; reversion of nutritional mutants; two-way selection, e.g. fluoracetate resistance/acetate utilization
Recombination	Intra- and intergenic effects in diploids and disomics; mitotic crossing-over; gene conversion
Chromosomal aberrations	Translocations; inversions; deletions; aneuploidy; nondisjunction

[a] General references will be found in the text.

coded for an active enzyme. This could convert nitrate to nitrite which was then assayed. After irradiation of the diploid, recombinants appeared after 4 h although recombination was blocked with higher doses (Holliday, 1974).

Space does not permit a detailed review of the large number of genetic studies in this area but mention should be made of the induction of chromosome aberrations. In *Aspergillus nidulans*, Käfer (1977) has reported 15–25% translocations at survival levels of 2–5%, and in *Neurospora crassa*, 10% translocations were found at survival levels of 50% (D. D. Perkins & A. Radford, unpublished, quoted in Fincham *et al.*, 1974).

Influences of gene mutations on survival

As in bacterial systems, one of the most productive areas of research has been to study the influence of gene mutations on survival, mutation and recombination. A variety of UV-sensitive, recombination-deficient and meiotic mutants have been isolated and analysed in a range of yeasts and filamentous fungi. Some of these data were reviewed in Catcheside (1974) and further reviews will be found in Fincham *et al.* (1979). Work on *Neurospora crassa* and *Aspergillus nidulans* will be used by way of illustration. Table 2 shows data for six loci in *Neurospora crassa*. Biochemical data (Worthy & Epler, 1973) have shown that strains carrying *uvs2* and *upr1* are unable to excise dimers, and a consequence of this is that there is a high level of UV-induced mutation in these strains presumably because repair of dimers occurs by error-prone repair (Schroeder, 1975). A third mutant *uvs3* has a high

Table 2. *Genetic loci in* Neurospora crassa *showing increased sensitivity to ultraviolet radiation (Schroeder, 1975)*

Locus	Spontaneous	UV-induced	Mitotic recombinations or deletions	Dimer excision
*uvs*1	Low	Low	–	Low
*uvs*2	Normal	High	Normal	Absent
*uvs*3				
(*nuh*-4)	×10	None	High	Low
*uvs*5	Normal	Low	Low-normal	Normal
*uvs*6	–	Normal	High	Normal
mei-3	–	–	High	–

Table 3. *Genetic loci in* Aspergillus nidulans *affecting sensitivity to ultraviolet radiation*[a]

Locus	Spontaneous mutation	Mitotic crossing-over	Process assumed to be affected
*uvs*A	–	–	–
*uvs*B	×1.7	High	Excision repair
*uvs*C	×8.7	Low	Recombination repair
*uvs*D	×1.8	High	Excision repair
*uvs*E	×4	Low	Recombination repair
*uvs*F	–	High	Excision repair

[a] See Clutterbuck (1974) for explanation of gene symbols, and see text for further references.

incidence of mitotic crossing-over or deletions and no UV-induced mutation (Schroeder, 1970). Little work has been done on characterizing the enzymes affected in these strains, but Käfer & Fraser (1979) have reported a nuclease halo mutant *nuh*-4 which lacked endonuclease activity (Fraser, 1979). This mutant may be an allele of *uvs*3. A further link between UV sensitivity and recombination has been found by Newmeyer & Galeazzi (1978) who have reported a mutant *mei*-3 in which meiosis is completely blocked and in which there is increased instability of non-tandem duplications.

A range of *uvs* mutants has also been isolated in *Aspergillus nidulans* (Shanfield & Käfer, 1969; Jansen, 1970*a*, *b*; Fortuin, 1971*b*; Jansen, 1972; Table 3). Spontaneous mutation in most of these mutants is increased. Excision repair is assumed to be defective in *uvs*B and *uvs*D as mitotic crossing-over is high, presumably because unexcised dimers stimulate genetic exchange. *uvs*C and *uvs*E strains show low mitotic crossing-over and here recombination-type repair processes are thought to have been lost. *uvs* strains have also been shown to affect the stability of non-tandem duplication strains (Burr, Palmer & Roper, 1971). The influence of *uvs*A on the survival of duplication strains has also been studied (Mayerfeld & Roper, 1978).

There are thus some parallels between fungal and bacterial mutants which suggest that similar processes may be operating. Very few enzymes have been characterized although in *Ustilago maydis* the following are now known.

(1) Single-strand specific nuclease affecting mitotic recombination (Holloman & Holliday, 1973).
(2) DNA polymerase: *pol*1–1 mutants are UV-sensitive with reduced mutation (Jeggo & Banks, 1975).
(3) Single-strand DNA specific protein (Banks, pp. 133–58).

In yeast, a large number of *rad* (radiation sensitive) mutants are known (Game, 1975; Haynes, 1975) but mention will only be made of *rad*-6. This locus appears to control an error-prone repair process as UV-induced mutation is completely blocked although mitotic gene conversion is ten-fold higher (Lawrance & Christensen, 1976; Kern & Zimmerman, 1978). SOS repair in bacteria is inducible by UV and there is some evidence for inducible processes in yeast and fungi. Fabre & Roman (1977) have shown that the stimulus for recombination can be transmitted from irradiated haploid cells to a diploid strain by mating to produce a triploid. Holliday (1975) has also presented evidence for UV-inducible recombination in *Ustilago maydis*. Ferguson & Cox (1980) have found evidence for UV-inducible repair in *Saccharomyces cerevisiae* by the use of split doses. In parallel experiments 200 J m^{-2} of UV were given either as one dose or as two doses of 100 J m^{-2} separated by 8 h under non-growing conditions. Survival in the split dose experiments was 1% and for the single dose survival was 0.06%, even after liquid holding recovery.

Conclusion

The influence of UV radiation on yeast and fungal nuclei is now well characterized in terms of survival, mutation and recombination of the cell. Biochemical analysis of repair and recombination processes has lagged behind physiological and genetic data, but on the whole bacterial systems have provided useful models for the interpretation of fungal data. It appears likely, however, that there will be differences between prokaryotes and eukaryotes particularly as chromatin has been reported to affect repair (Hanawalt *et al.*, 1979). The presence of an intact nuclear membrane throughout mitosis in some fungi is also likely to influence repair. Super coiling of DNA in chromosome organization and a variety of DNA-specific enzymes, such as topoisomerases, may also be expected to affect repair processes. It may prove possible to utilize recombinant DNA techniques to clone genes involved in repair processes so that proteins can be isolated in sufficient quantities for study of repair *in vitro*.

References

Bainbridge, B. W. (1971). Macromolecular composition and nuclear division during spore germination in *Aspergillus nidulans*. *Journal of General Microbiology*, **66**, 319–25.

Bainbridge, B. W. (1976). Estimation of the generation time and peripheral growth zone of *Aspergillus nidulans* and *Alternaria solani* hyphae from radial growth rates and ranges in apical cell length. *Journal of General Microbiology*, **97**, 125–7.

Burr, K. W., Palmer, H. M. & Roper, J. A. (1971). Mitotic non-conformity in *Aspergillus nidulans*: the effects of reduced DNA repair. *Heredity*, **27**, 487.

Catcheside, D. G. (1974). Fungal genetics. *Annual Review of Genetics*, **8**, 279–300.

Chanet, R., Heude, M. & Moustacchi, E. (1974). Variations in UV-induced lethality and 'petite' mutagenesis in synchronous cultures of *Saccharomyces cerevisiae*. II. Responses of radiosensitive mutants to lethal damage. *Molecular and General Genetics*, **132**, 23–30.

Chanet, R., Williamson, D. H. & Moustacchi, E. (1973). Cyclic variations in killing and 'petite' mutagenesis induced by ultraviolet light in synchronized yeast strains. *Biochimica et Biophysica Acta*, **324**, 290–9.

Clutterbuck, A. J. (1970). Synchronous nuclear division and septation in *Aspergillus nidulans*. *Journal of General Microbiology*, **60**, 133–5.

Clutterbuck, A. J. (1974). *Aspergillus nidulans*. In *Handbook of Genetics*, vol. 1: *Bacteria, Bacteriophages and Fungi*, ed. R. C. King, pp. 447–510. New York: Plenum Press.

Davies, P. J., Tippins, R. S. and Parry, J. M. (1978). Cell-cycle variations in the induction of lethality and mitotic recombination after treatment with UV and nitrous acid in yeast, *Saccharomyces cerevisiae*. *Mutation Research*, **51**, 327–46.

Fabre, F. & Moustacchi, E. (1973). Removal of pyrimidine dimers in cells of *Schizosaccharomyces pombe* mutated in different repair pathways. *Biochimica et Biophysica Acta*, **312**, 617–25.

Fabre, F. & Roman, H. (1977), Genetic evidence for inducibility of recombination competence in yeast. *Proceedings of the National Academy of Sciences, USA*, **74**, 1667–71.

Ferguson, L. R. & Cox, B. S. (1980). The role of dimer excision in liquid-holding recovery of UV-irradiated haploid yeast. *Mutation Research*, **69**, 19–41.

Fiddy, C. & Trinci, A. P. J. (1976). Mitosis, septation, branching and the duplication cycle in *Aspergillus nidulans*. *Journal of General Microbiology*, **97**, 169–84.

Fincham, J. R. S., Day, P. R. & Radford, A. (1979). *Fungal Genetics*. Oxford: Blackwell Scientific Publications.

Fortuin, J. J. H. (1971*a*). Another two genes controlling mitoic intragenic recombination and recovery from UV damage in *Aspergillus nidulans*. III. Photoreactivation of UV damage in *uvs*D and *uvs*E mutants. *Mutation Research*, **13**, 131–6.

Fortuin, J. J. H. (1971*b*). Another two genes controlling mitotic intragenic recombination and recovery from UV damage in *Aspergillus nidulans*. I. UV sensitivity, complementation and location of six mutants. *Mutation Research*, **11**, 149–62.

Fraser, M. J. (1979). Alkaline deoxyribonucleases released from *Neurospora crassa* mycelia: two activities not released by mutants with multiple sensitivities to mutagens. *Nucleic Acids Research*, **6**, 231–46.

Game, J. C. (1975). Radiosensitive mutants of yeast. In *Molecular Mechanisms for Repair of DNA*, part B, ed. P. C. Hanawalt & R. B. Setlow, pp. 541–4. New York: Plenum Press.

Ganesan, A. K. (1974). Persistence of pyrimidine dimers during post-replication repair in ultraviolet light-irradiated *Escherichia coli* K12. *Journal of Molecular Biology*, **87**, 103–19.

Gealt, M. A., Sheir-Neiss, G. & Morris, N. R. (1976). The isolation of nuclei from the filamentous fungus *Aspergillus nidulans*. *Journal of General Microbiology*, **94**, 204–10.

Grivell, A. R. & Jackson, J. F. (1968). Thymidine kinase: evidence for its absence from *Neurospora crassa* and some other microorganisms, and the relevance of this to the specific labelling of deoxyribonucleic acid. *Journal of General Microbiology*, **54**, 307–17.

Hanawalt, P. C., Cooper, P. K., Ganesan, A. K. & Smith, C. A. (1979). DNA repair in bacteria and mammalian cells. *Annual Review of Biochemistry*, **48**, 783–836.

Haynes, R. H. (1975). DNA repair and the genetic control of radiosensitivity in yeast. In *Molecular Mechanisms for Repair of DNA*, part B, ed. P. C. Hanawalt & R. B. Setlow, pp. 529–40. New York: Plenum Press.

Holliday, R. (1971). Biochemical measure of the time and frequency of radiation-induced allelic recombination in *Ustilago*. *Nature, London*, **232**, 233–6.

Holliday, R. (1974). *Ustilago maydis*. In *Handbook of Genetics*, vol. 1: *Bacteria, Bacteriophages and Fungi*, ed. R. C. King, pp. 575–95. New York: Plenum Press.

Holliday, R. (1975). Further evidence for an inducible repair system in *Ustilago maydis*. *Mutation Research*, **29**, 149–53.

Holloman, W. K. & Holliday, R. (1973). Studies of a nuclease from *Ustilago maydis*. I. Purification, properties and implications in recombination of the enzyme. *Journal of Biological Chemistry*, **248**, 8107–13.

Hundert, P., Koltin, Y., Stamberg, J. & Wertzberger, R. (1978). Repair of UV-induced damage in wild type and mutant strains of *Schizophyllum commune*. *Mutation Research*, **50**, 157–62.

Jagger, J. (1967). *Introduction to Research in Ultraviolet Photobiology*. Englewood Cliffs: Prentice-Hall.

Jagger, J. (1977). Phototechnology and biological experimentation. In *The Science of Photobiology*, ed. K. C. Smith, pp. 1–26. New York: Plenum Press.

Jansen, G. J. O. (1970*a*). Survival of *uvs*B and *uvs*C mutants of *Aspergillus nidulans* after UV-irradiation. *Mutation Research*, **10**, 21–32.

Jansen, G. J. O. (1970*b*). Abnormal frequencies of spontaneous mitotic recombination in *uvs*B and *uvs*C mutants of *Aspergillus nidulans. Mutation Research*, **10**, 33–41.

Jansen, G. J. O. (1972). Mutator activity in *uvs* mutants of *Aspergillus nidulans. Molecular and General Genetics*, **116**, 47–50.

Jeggo, P. A. & Banks, G. R. (1975). DNA polymerase of *Ustilago maydis*: partial characterization of the enzyme and a *pol*-1 mutation. *Molecular and General Genetics*, **142**, 209–24.

Käfer, E. (1977). Meiotic and mitotic recombination in *Aspergillus* and its chromosomal aberrations. *Advances in Genetics*, **19**, 33–131.

Käfer, E. & Fraser, M. (1979). Isolation and genetic analysis of nuclease halo (*nuh*) mutants of *Neurospora crassa. Molecular and General Genetics*, **169**, 117–27.

Kern, R. & Zimmerman, F. K. (1978). The influence of defects in excision and error-prone repair on spontaneous and induced mitotic recombination and mutation in *Saccharomyces cerevisiae. Molecular and General Genetics*, **161**, 81–8.

King, R. C. (1974) (ed.). *Handbook of Genetics*, vol. 1: *Bacteria, Bacteriophages and Fungi*. New York: Plenum Press.

Lawrence, C. W. & Christensen, R. (1976). UV mutagenesis in radiation-sensitive strains of yeast. *Genetics*, **82**, 207–32.

Marmur, J. J. (1961). A procedure for the isolation of deoxyribonucleic acid from microorganisms. *Journal of Molecular Biology*, **3**, 208–18.

Mayerfeld, J. H. & Roper, J. A. (1978). Dose effects of the *uvs*A$^+$ gene product in duplication strains of *Aspergillus nidulans. Mutation Research*, **49**, 19–26.

Morris, N. R. (1978). Preparation of large molecular weight DNA from the fungus *Aspergillus nidulans. Journal of General Microbiology*, **106**, 387–9.

Newmeyer, D. & Galeazzi, D. R. (1978). A meiotic UV-sensitive mutant that causes deletion of duplications in *Neurospora. Genetics*, **89**, 245–69.

Parry, J. M. & Cox, B. S. (1968). The effects of dark holding and photoreactivation on ultraviolet light-induced mitotic recombination and survival in yeast. *Genetical Research, Cambridge*, **12**, 187–98.

Reynolds, R. J. (1978). Removal of pyrimidine dimers from *Saccharomyces cerevisiae* nuclear DNA under non-growth conditions as detected by a sensitive enzymatic assay. *Mutation Research*, **50**, 43–56.

Schroeder, A. L. (1970). Ultraviolet-sensitive mutants of *Neurospora*. I. Genetic basis and effect on recombination. *Molecular and General Genetics*, **107**, 291–304.

Schroeder, A. L. (1975). Genetic control of radiation sensitivity and DNA repair in *Neurospora*. In *Molecular Mechanisms for Repair of DNA*, Part B, ed. P. C. Hanawalt & R. B. Setlow, pp. 567–79. New York: Plenum Press.

Scott, B. R. & Anderson, T. (1971). The random (non-specific) forward mutational response of gene loci in *Aspergillus conidia* after photosensitisation to near ultraviolet light (365 nm) by 8 methoxypsoralen. *Mutation Research*, **12**, 29–34.

Scott, B. R., Pathak, M. A. & Mohn, G. R. (1976). Molecular and genetic basis of furacoumarin reactions. *Mutation Research*, **39**, 29–74.

Shanfield, B. and Käfer, E. (1969). UV-sensitive mutants increasing mitotic crossing over in *Aspergillus nidulans. Mutation Research*, **7**, 485–7.

Smith, K. C. (1977*a*) (ed.). *The Science of Photobiology*. New York: Plenum Press.

Smith, K. C. (1977*b*). Ultraviolet radiation effects on molecules and cells. In *The Science of Photobiology*, ed. K. C. Smith, pp. 113–42. New York: Plenum Press.

Smith, K. C. (1977*c*). New topics in photobiology. In *The Science of Photobiology*, ed. K. C. Smith, pp. 397–417. New York: Plenum Press.

Unrau, P. (1975). The excision of pyrimidine dimers from the DNA of mutant and wild-type strains of *Ustilago. Mutation Research*, **29**, 53–65.

Unrau, P., Wheatcroft, R. & Cox, B. S. (1972). Methods for the assay of ultraviolet light-induced pyrimidine dimers in *Saccharomyces cerevisiae*. *Biochimica et Biophysica Acta*, **269**, 311–21.

Unrau, P., Wheatcroft, R., Cox, B. & Olive, T. (1973). The formation of pyrimidine dimers in the DNA of fungi and bacteria. *Biochimica et Biophysica Acta*, **312**, 626–32.

Wickner, R. B. (1974). Mutants of *Saccharomyces cerevisiae* that incorporate deoxythymidine-5'-monophosphate into deoxyribonucleic acid *in vivo*. *Journal of Bacteriology*, **117**, 252–60.

Williamson, D. H., Moustacchi, E. & Fennell, D. (1971). A procedure for rapidly extracting and estimating the nuclear and cytoplasmic DNA components of yeast cells. *Biochimica et Biophysica Acta*, **238**, 369–74.

Witkin, E. M. (1976). Ultraviolet mutagenesis and inducible repair in *Escherichia coli*. *Bacteriological Reviews*, **40**, 869–907.

Worthy, T. E. & Epler, J. L. (1973). Biochemical basis of radiation sensitivity in mutants of *Neurospora crassa*. *Mutation Research*, **19**, 167–73.

The synthesis of chromosomal replicons in yeast

D. H. WILLIAMSON AND L. H. JOHNSTON

National Institute for Medical Research, The Ridgeway, Mill Hill, London NW7 1AA, UK

Introduction

The nucleus of the yeast, *Saccharomyces cerevisiae*, has a haploid DNA content of about 9×10^9 daltons (*c.* 0.03 pg) (Hartwell, 1970). This is distributed amongst 17 chromosomes ranging in size from 10^8 to 10^9 daltons, the average being around 6×10^8 daltons (Petes *et al.*, 1973). In terms of DNA therefore, the yeast chromosome is remarkably small (Fig. 1), being only about one-third the size of the *Escherichia coli* chromosome and two orders of magnitude smaller than an average human chromosome (calculated from data in Holliday, 1970). This disparity in size might at first sight suggest that the structural organization of yeast and human chromosomes would have little in common. In fact the converse is true; the more understanding we gain, the more common features are revealed, and it becomes increasingly realistic to suppose that study of the control and execution of chromosomal replication in the yeast cell will provide meaningful pointers to the way these activities are accomplished in higher eukaryotes. In this article we

E. coli

Yeast

Human

Fig. 1. Comparison of the lengths of the DNA molecules of average chromosomes, drawn to scale.

shall attempt briefly to survey current knowledge of DNA synthesis in yeast, paying particular regard to the duplication of the basic replication unit, the replicon.

The replicon
Size, distribution and rate of growth
Replication of chromosomal DNA in all types of cell almost invariably proceeds bi-directionally away from points known as 'origins of replication' or 'replicators'. The complete stretch of DNA that is replicated from a single origin is known as the replicon (Fig. 2), a term first coined by Jacob, Brenner & Cuzin (1963). In prokaryotes there is normally only one origin, the DNA molecule that comprises the prokaryotic genome thus constituting a single replicon. Eukaryotes on the other hand always contain many replicons. There must of course be at least one per chromosome, but in practice there are always many more. Hand (1978) suggests there may be 10^4 replicons in the cells of higher eukaryotes, which could mean around 200 on an average human chromosome. These numbers are necessarily rather vague, since they are based on estimates of average replicon size, and there is considerable variation in size both within and between cells of different species. Globally, the range may reasonably be put at between 10 μm and 500 μm (Hand, 1978; Yurov, 1980). Intra-species variation is seen not only in the spread of sizes within a given population; there is also considerable evidence that the average sizes of replicons in the cells of a given species may change in response to environmental conditions, and may vary in different tissues depending on the state of differentiation of the cells in question (Hand, 1978). There are also indications that in higher eukaryotes at least, changes in the overall rate of synthesis of DNA are achieved more often by a change in replicon size than by a change in the rate of replication fork movement, good examples being

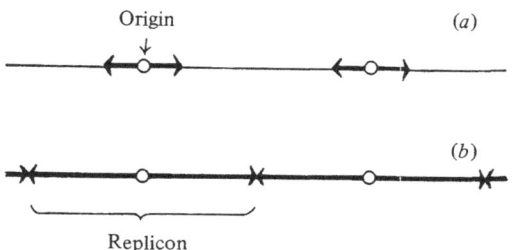

Fig. 2. Diagrammatic illustration of a pair of adjacent eukaryotic replicons, (*a*) early in replication, (*b*) completed.

found in the observation of Callan (1972) and Blumenthal, Kriegstein & Hogness (1973). This variability in replicon size must mean that replication does not necessarily require activation of all potential origins.

The rate of growth of replicons (equal to twice the rate of movement of replication forks) also shows considerable variation within species, even within individual cells (Yurov, 1980), and changes in rate have also been noted at different stages of the S phase (Painter & Schaefer, 1971; Housman & Huberman, 1975). The average rate of fork movement in eukaryotic replicons is usually not more than 2 μm min.$^{-1}$ In contrast, replication forks in *Escherichia coli* move ten times faster (Cairns, 1963).

The size and number of replicons in yeast clearly put this organism in the eukaryotic bracket. Using electron microscopy, Petes & Newlon (1974) and Newlon *et al.* (1974) found multiple replicons between 10 μm and 40 μm in length, with an average size around 30 μm (6 × 10^7 daltons). On this basis, it may be calculated that the genome contains around 150 average replicons and, depending on its size, a chromosome may carry between two and fifteen replicons, the average chromosome having perhaps ten.

Using a 'pulse/chase' labelling protocol, Petes & Williamson (1975) showed that replication was bi-directional, and that the replication forks moved at around 0.7 μm min^{-1} at 24 °C, a rate clearly comparable with that seen in other eukaryotes. A similar low figure was also reported for another diploid strain by Johnston & Williamson (1978), this time on the basis of alkaline sedimentation studies.

More recently, a somewhat different situation has been reported by Rivin & Fangman (1980*a*, *b*) in a haploid strain growing (at 30 °C) in semi-defined media. When ammonium was supplied as sole nitrogen source, a mean generation time of 2 h was achieved, and the S phase of the cell lasted about 1 h. Substitution of ammonia by proline increased the generation time to about 400 min and the S phase to about 200 min. Using DNA fibre autoradiography, it was observed that replicons were about the same size in both sorts of culture, the average being, as in the earlier studies of Petes & Williamson (1975), around 30 μm. Surprisingly, however, the rate of movement of the replication forks was not a constant, but varied inversely with the generation time. Even more surprisingly, in the fastest growing cells (on ammonium), the replication forks moved at about 2 μm min^{-1}, about three times the rate previously observed by Petes & Williamson (1975) in a diploid, and dropped to

0.56 μm min^{-1} in the slowest growing cells (on proline). The significance of these observations in relation to the control of replicon synthesis in the cell cycle is discussed further below.

Assembly of replicons

The molecular events involved in the synthesis of replicons from nucleoside triphosphates are illustrated diagrammatically in Fig. 3. The scheme shown is necessarily hypothetical and is based on generally accepted concepts derived in the main from studies on bacteria (for a review see Johnston, Bonhoeffer & Symmons, 1979). It is centred on two basic principles. The first is that all known DNA polymerases can act only in one direction, from 5′ to 3′; as a consequence one of the daughter strands at the replication fork can in principle be made continuously (the 'leading' strand) but the other ('lagging' strand) has to be made discontinuously. Secondly, DNA polymerases all require a primer, that is a polynucleotide chain with a free 3′-OH end, hydrogen-bonded to the template. It is generally supposed that this takes the form of a short piece of RNA laid down by RNA polymerase(s). Having served its function as primer, this RNA is later removed, the gap thus produced is filled in by DNA polymerase activity, and finally the resulting nick between adjacent single-strand daughter molecules is sealed by the action of DNA ligase. In addition, various less obvious activities are certainly required, for instance, topoisomerases to permit unwinding of the DNA ahead of the replication fork, proteins involved in the separation of single strands, and 'proof-reading' (3′-exonuclease) activity to increase fidelity. Further proteins may be needed to ensure correct termination of the completed replicon, involving at the very least a ligase activity to join adjacent daughter strands at the junction

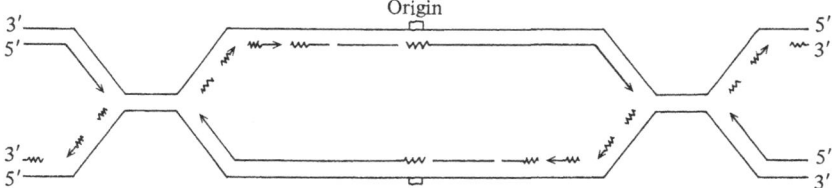

Fig. 3. Schematic replicon, illustrating some basic features of replicon biosynthesis as discussed in the text. Replication starts at the region marked 'origin' and a replication fork proceeds away from this site in each direction. The arrows show the direction of chain growth at each growing point, and wavy lines indicate a requirement for an RNA primer. The diagram illustrates approaching forks of adjacent replicons.

between replicons. Moreover, specific proteins, as yet unidentified in eukaryotes, may be required to recognize the origin and trigger its activation. It should be pointed out that with the exception of certain viruses there is little firm evidence in eukaryotes that either origins or termini are specified solely by particular DNA sequences. A terminus may simply be a region on a DNA molecule where the replication forks of two adjacent replicons meet. Likewise, the sites at which replication is activated might be determined by some structural features of the chromosomes, chromatin or nuclear matrix (Pardoll, Vogelstein & Coffey, 1980). However, it seems difficult to believe that DNA sequence plays no part at all, particularly in determining the sites of origins and, as pointed out below, putative origin sequences are currently being studied in a number of laboratories using recombinant DNA technology.

In the following section we outline the currently scanty knowledge concerning the enzymes and mutants relevant to the process of DNA synthesis in yeast.

Replication enzymes and mutants in yeast

There are three DNA polymerizing activities in yeast. One is apparently mitochondrial (Wintersberger & Blutsch, 1976), but the others are of nuclear origin and are known either as polymerases A and B (Wintersberger, 1974a) or I and II (Chang, 1977). Both of the nuclear polymerases are of high molecular weight, greater than 10^5 daltons, and yeast appears to entirely lack a low molecular weight enzyme comparable with the β-polymerase of mammalian cells (Wintersberger, 1974b). In fact the yeast polymerases do not seem to be closely related to the polymerases of either higher eukaryotes or of prokaryotes. However, polymerase I does bear some resemblance to the polymerase α which in higher cells is believed to be responsible for DNA replication (Wintersberger, 1974a; Chang, 1977; Wintersberger, 1977), while polymerase II, in possessing an associated 3′-exonuclease activity (Helfman, 1973; Chang, 1977; Wintersberger, 1978), resembles polymerases from the smut fungus *Ustilago maydis* (Yarranton & Banks, 1977) and from certain prokaryotes.

The physiological roles of the two nuclear enzymes in yeast are uncertain. For the reason outlined above, it is expected that the enzyme responsible for replicative synthesis would be able to use an RNA primer, and of the two yeast enzymes, only polymerase I has this property (Plevani & Chang, 1977; Wintersberger, 1978). Furthermore, when an extract of growing cells was compared with one from a stationary phase culture, polymerase I had a higher specific activity than polymerase II (Wintersberger, 1974a). On the other hand, only polymerase II has the associated 3'-exonuclease activity required for excising mismatched 3'-nucleotides, the 'proof-reading' property expected of a replicative polymerase. However, this last situation is not without precedent, since no purified polymerases from higher eukaryotes possess any exonuclease activity, and in these cells proof-reading functions presumably reside in different polypeptides that are separated from the polymerase during purification. This same argument may be applicable to yeast polymerase I.

Further information concerning replication enzymes is sketchy. A yeast protein has been partially purified which stimulates DNA synthesis on single-stranded DNA in the absence of added ribonucleoside triphosphates (Chang, Lurie & Plevani, 1978). Although the basis of this stimulation is not understood, it is interesting that yeast polymerase I and mammalian α-polymerases are stimulated much more than yeast polymerase II and mammalian β-polymerases.

An untwisting (nicking-closing) enzyme has also been partly purified from yeast (Durnford & Champoux, 1978). This enzyme is capable of removing both positive and negative superhelices from DNA. In principle it could remove the positive superhelices generated by strand separation at the replication forks, but whether or not it actually does have this function is not known. DNA-binding proteins have recently been isolated from yeast; they are discussed by Banks in an earlier chapter (pp. 133–58).

The removal of the presumed RNA primers used by the replicative polymerase requires a special activity known as RNase H, and a number of such activities has been described in yeast (Wyers *et al.*, 1976). However, the true role of these enzymes *in vivo* is not yet known.

The general uncertainty and lack of information revealed in the preceding paragraphs underscores the need for further study, and it is remarkable that with one exception none of the enzymes referred to above has so far been assigned to a structural gene. The one exception is the DNA ligase. This activity was detected in crude extracts of wild-type strains and was shown by Johnston & Nasmyth (1978) to be coded by the *cdc9* locus, mutations in which confer a cell cycle phenotype (Culotti & Hartwell, 1971). The role of the ligase is discussed further below. Another cell cycle mutant which is apparently defective in DNA chain elongation is *cdc21*, which has been shown to have a lesion in the structural gene for thymidylate synthetase (Game, 1976; Bisson & Thorner, 1977).

A couple of other cell cycle mutants have fairly dramatic effects on DNA replication and may be more or less directly involved with the process. One is *cdc*8, which rapidly ceases synthesis of both nuclear and mitochondria DNA (Hartwell, 1973; Newlon & Fangman, 1975) when cells are switched to the non-permissive temperature; so far the defective gene product has not been identified. The other mutant is *cdc*7 (Hartwell, 1973) which has an execution point just before the initiation of the S phase. It is specific for nuclear (as opposed to mitochondrial) DNA synthesis (Newlon & Fangman, 1975) and has properties suggestive of a role in the initiation of S, but not in chain elongation. Again, the gene product concerned is so far unidentified.

A few other cell cycle mutants, namely *cdc*2, *cdc*6 (Hartwell, 1976) and *cdc*40 (Kassir & Simchen, 1978) have been implicated in DNA synthesis in a more subtle way, in the sense that although they seem to make what approaches a full round of bulk DNA synthesis at the restrictive temperature, 'reciprocal-shift' experiments suggest they make DNA that is in some subtle way defective or incomplete (see also Johnston & Williamson, 1978).

Another group of mutants was isolated as a result of a specific search for DNA synthesis mutants by Johnston & Game (1978). These mutants (referred to as *dds* for 'defective DNA synthesis') were thought to have primary defects in DNA synthesis although they showed a variety of phenotypes. For example, at restrictive temperature one mutant accumulated with the dumb-bell terminal morphology characteristic of *cdc* mutants arrested at mitosis, whilst others were defective in the synthesis of RNA as well as DNA. It is worth noting that this study also revealed that several of the *rna* mutants described by Hartwell, McLaughlin & Warner (1970) were grossly defective in DNA as well as RNA synthesis. This makes the nature of the primary lesion a little less clear, and the molecular defects in both the *dds* mutants and these *rna* mutants are not known.

In a further search in our laboratory, several conditional cell cycle mutants blocked at mitosis have been isolated and screened for macromolecular synthesis. Among these are several mutants which are defective in DNA synthesis but proficient in both RNA and protein synthesis, and which complement the known *cdc* mutants (A. Thomas, unpublished observations).

In none of the above mutants, with the exception of *cdc*9, is the

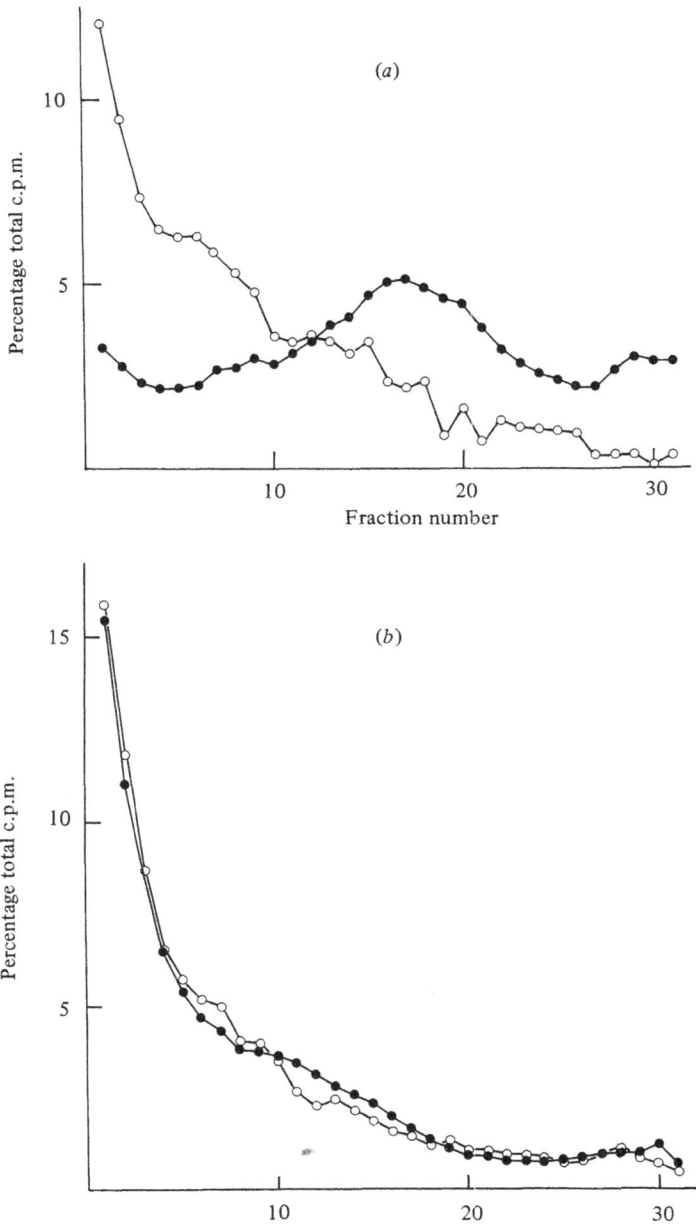

Fig. 4. Alkaline sucrose gradient profiles of single-stranded DNA molecules from log phase cells, pre-labelled by growth for several generations in [¹⁴C]uracil, followed by pulse-labelling in [³H]adenine

molecular nature of the defect known. In part this lack of biochemical information may reflect the comparative inadequacy of existing systems for analysis of DNA replication *in vitro*. These are reviewed briefly in the next section.

Systems for studying DNA synthesis in vitro

A permeabilized cell system devised by Hereford & Hartwell (1971) was later shown by Banks (1973) to synthesize only mitochondrial DNA. Oertel & Goulian (1977) have developed a system based on permeabilized spheroplasts, and Johnston (1979) described a system similar to the cellophane disc procedure devised for bacteria by Schaller *et al.* (1972). Although the *in vivo* defect in *cdc*8 was discernible in the permeabilized spheroplast system, neither this nor the cellophane disc system showed all the properties of *in vivo* replication. For example, they both made only small fragments of DNA and were apparently unable to join them together. They are therefore likely to be of only limited use. In this respect the recent *in vitro* system described by Jazwinski & Edelman (1979) using the 2 μm plasmid as its substrate may be better, since it appears that intact supercoiled molecules are produced.

Replication in vivo

Studies involving the use of alkaline sucrose gradients to follow the synthesis and fate of single-stranded intermediates have been comparatively fruitful. Observations on log phase and synchronized populations (Johnston & Williamson, 1978) and on a temperature-sensitive ligase mutant (Johnston & Nasmyth, 1978) revealed the presence of three broadly defined size classes of single-stranded molecules; small molecules around 10^5 daltons in weight, intermediate molecules in the range 7–60×10^6 daltons, and larger molecules, ranging above 100×10^6 daltons, which began to appear with longer times of labelling, and which probably corresponded to full-size chromosomal daughter strands (Figs. 4 and 5).

The internal precursor pools of yeast make it impossible in this type of

for (*a*), 10 min; (*b*), 40 min. Sedimentation (increasing size) from right to left. Parental ([^{14}C]labelled) molecules are mostly of high molecular weight. In (*a*), the newly made DNA includes replicons and smaller fragments, but when the longer ^{3}H pulse in (*b*) is used, the label appears mostly in high molecular weight form (L. H. Johnston, unpublished). (○) ^{14}C-pre-label; (●) ^{3}H-pulse label.

experiment to 'chase' radioactivity from one class of molecule to another and thus prove a precursor/product relationship unambiguously. However, considered as a whole, these experiments, involving pulse-labelling periods of different lengths, lead to the general idea that as in other organisms (Johnston *et al.*, 1979), replication proceeds via two definable steps. Early in the process, low molecular weight single-stranded fragments appear, as a result either of discontinuous synthesis along at least one daughter strand, or (perhaps additionally) as a secondary consequence of some other phenomenon such as the removal of mis-incorporated uracil by the action of uracil glycosylase (Tye *et al.*, 1977). Whichever mechanism is responsible for generating these frag-

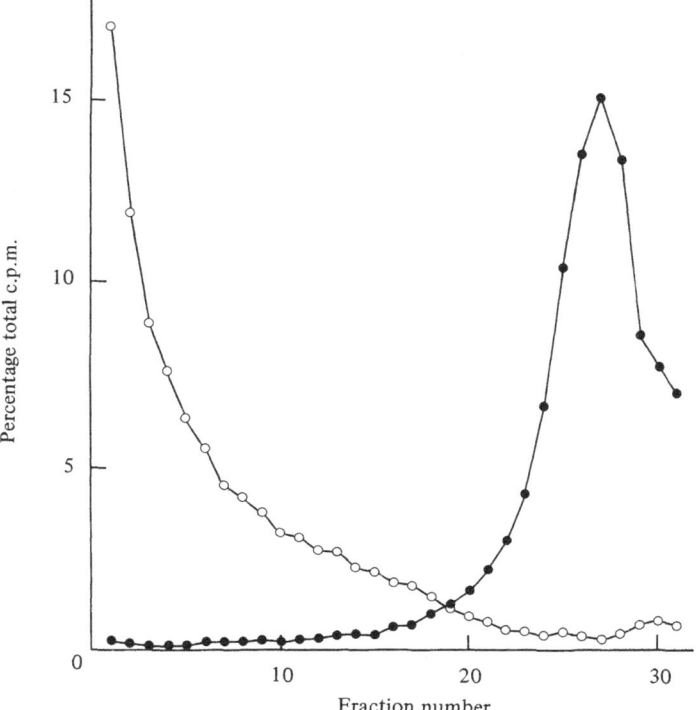

Fig. 5. Alkaline sedimentation analysis of single-stranded DNA molecules from a culture of the ligase mutant *cdc9*. Growing cells pre-labelled with [¹⁴C]uracil at the permissive temperature were incubated for 30 min at the restrictive temperature in the presence of [³H]adenine. Sedimentation from right to left. Parental strands are of normal size, but DNA made at the restrictive temperature in the absence of ligase accumulates as small fragments (L. H. Johnston, unpublished). (○) ¹⁴C-pre-label; (●) ³H-pulse label.

ments in the yeast cells, it is clear that their processing normally involves the *cdc9* ligase (Johnston & Nasmyth, 1978).

The second step in the replicative process involves the growth of the single-stranded molecules, presumably by the joining of the small fragments, and the generation of intermediates in the broad size range of replicons. In the asynchronous cultures, this class of molecule constituted an obvious peak in the alkaline gradients (Fig. 4*a*) which suggests in itself that there may be some delay in the joining of adjacent replicons. However, this delay could not last for very long for it was not measurable in the synchronized cultures, and in these it seemed that most of the replicons must have been joined together at least by the end of the S period.

The experiments with synchronized cultures (Johnston & Williamson, 1978) permitted estimation of the rate of chain growth, and provided ligation of small fragments occurred rapidly, this would essentially reflect the rate of replication fork movement. Discounting the apparently rapid rates observed at the start and end of the S period, which could have been artifactual, the authors concluded that the best estimate they could make was around 0.8 μm min^{-1}. It is not clear if this should be taken as the rate of movement of one or two forks, since it is not known whether the fragments observed in the gradients were growing at both ends or at only one. The latter possibility might occur if a single-stranded daughter strand gap persisted at the origin, a situation which would perhaps arise if, for instance, the origin were protected by some protein to prevent re-initiation. Until this problem is resolved we can only record that the alkaline gradient results showed a rate of chain growth broadly comparable with those determined autoradiographically (Petes & Williamson, 1975; Rivin & Fangman, 1980*b*).

The role of the ligase in replication

It will be apparent from the preceding paragraphs that ligation of single strands plays a key role in the assembly of replicons, and its importance in yeast was clearly demonstrated by the observations of Johnston & Nasmyth (1978) on the conditional mutant *cdc9*. These authors showed that all the DNA made by this mutant at the non-permissive temperature accumulated as small fragments (Fig. 5), and that, as judged by examination of crude extracts, ligase activity was completely absent. From this it may be concluded that the ligase coded by the *cdc9* locus is responsible for joining the single-stranded fragments generated early in the replication process. It now seems likely that the same enzyme has a second important function in replication, namely the joining of adjacent completed replicons.

This conclusion stems from observations (L. H. Johnston, unpublished) on the effect of the *cdc9* mutation on the recovery from treatment with the inhibitor of DNA synthesis, hydroxyurea. Johnston has found that the effects of hydroxyurea are somewhat concentration-dependent, but at a low level the inhibitor reversibly blocks the sealing of adjacent replicons. The mechanism of this phenomenon is not understood, but it has afforded a means of determining if the *cdc9* ligase is required for joining replicons. Cultures of *cdc9* at the permissive temperature were incubated for long enough in the presence of hydroxyurea and [³H]adenine to allow substantial accumulation of labelled replicons, completed but unjoined. The cells were then transferred to fresh medium lacking both hydroxyurea and [³H]adenine, and incubated at either the permissive or non-permissive temperature, before being sampled for analysis on alkaline sucrose gradients. It was found that the joining of the accumulated replicons to form high molecular weight molecules only occurred in the cells incubated (after treatment with hydroxyurea) at the permissive conditions, clearly implicating the *cdc9* ligase in this joining reaction. This conclusion was supported by the observation that a revertant of *cdc9* was able not only to grow at the restrictive temperature but also to join replicons.

Activation of replicons in the cell cycle

Proper duplication of chromosomes in the cell cycle clearly requires co-ordination of the synthesis of replicons, and even in the simplest model there must be some control of the activation of the origins. Assuming the latter occur at fixed sites, it is therefore extremely likely that the genome is replicated in a definite temporal sequence.

Evidence in support of such a defined order of replication in yeast is restricted to a group of experiments based on mutagenesis with nitrosoguanidine, an alkylating agent which, in bacteria, selectively mutagenizes replicating DNA (Guerola, Ingram & Cerda-Olmedo, 1971). This rationale prompted Dawes & Carter (1974), Burke & Fangman (1975) and Kee & Haber (1975) to examine the frequency of mutations at particular loci in cells exposed to nitrosoguanidine at different times in the S period. The overall finding of all three groups was that individual loci were mutated (and by inference therefore were replicating) at specific times in the S period. Using a variant of this technique, Dawes *et al.* (1977) found specific patterns of co-mutation following exposure of growing cells to the same mutagen, and this again was taken as support for a strict control of the timing of replication of different genes in the S phase. There are numerous difficulties in interpreting these experiments, including the possibility that an error-prone repair mechanism might in fact be responsible for fixing the mutations. Nevertheless, the observations argue strongly for some sort of regularity shared by all the cells in the population and, against the background of evidence for ordered replication in other eukaryotes (Mueller & Kajiwara, 1966; Braun & Wili, 1969), it seems reasonable to interpret them in a positive manner.

If we accept, as a first approximation, that there is a defined order of replication, we may then explore the way in which this is achieved. Recent developments in our understanding of the yeast S period have revealed an unexpected degree of complexity, for it is beginning to emerge that different cultures of yeast may structure their S period in fundamentally different ways.

Until recently, the length of the S period in yeast was defined primarily by three studies. Williamson (1965), using whole cell autoradiography, and Slater, Sharrow & Gart (1977), using flow cytofluorimetry agreed in assigning diploid cells an S period occupying around 20–25% of the cell cycle – in absolute terms about 20–30 min. Barford & Hall (1976), also using whole cell autoradiography, arrived at a 40 min S period for cells of a commercial yeast, but since these authors failed to correct for pulse length or for mitochondrial DNA synthesis, their result is likely to be an over-estimate. Accepting therefore the 20–30 min estimate as being broadly correct, we can consider how this relates to the time required for replicon assembly.

From the data of Petes & Williamson (1975) on the growth rate and size of replicons in a diploid (see above), it may be deduced that the average replicon takes about 20 min to be completed, in other words most of the S period. It follows that, as illustrated in Fig. 6a, the majority of replication origins must be activated near the start of the period. Provided origins occur at fixed sites in every cell, such an S period would not be unstructured, but the time of replication of any particular gene would be determined solely by its distance from its origin.

On this basis the yeast S period emerges as a relatively uncomplicated affair, possibly no more than a simple 'start of S' signal being required to activate all the origins with some degree of synchrony. It contrasts strikingly with the situation in higher eukaryotes, in which activation of origins is staggered throughout the S phase (Hand, 1978). However, recent developments are leading us to modify this view, since it now appears that a marked staggering of origin activation can also occur in yeast cells.

This emerges from the observations of Rivin & Fangman (1980*a*, *b*), who examined the S period in a haploid strain growing on semi-defined media at rates governed by different nitrogen sources. They showed that replicon size on all their media was about the same as that previously reported by Petes & Williamson (1975) in a diploid. However the S period was relatively long. At all growth rates it occupied about 50% of the cell cycle, ranging in absolute terms from about 60 to 200 min. Furthermore the rate of movement of replication forks not only varied

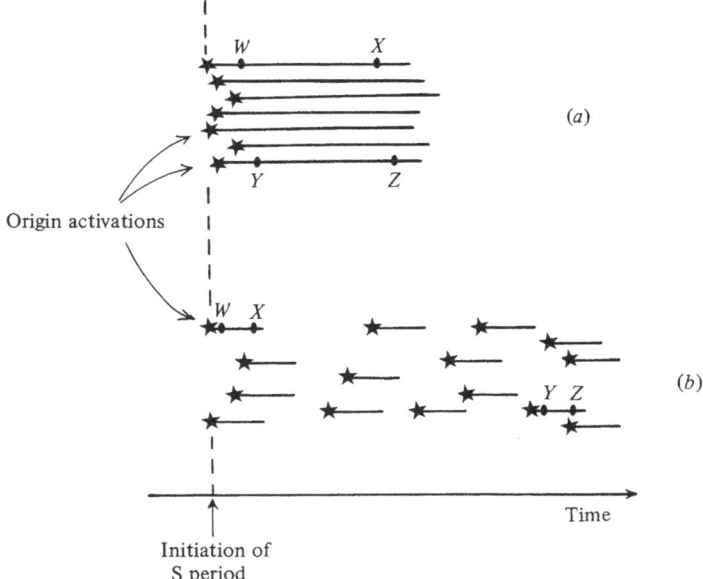

Fig. 6. Two patterns of replicon activation in the yeast cell cycle. (*a*) In some strains (characteristically diploids), most of the origins are activated near the start of S, and completion of the average replicon takes most of the S period. Genes close to origins (*W*, *Y*) replicate early, those further away (*X*, *Z*) replicate later. (*b*) In at least some haploids, the S period has been shown to be relatively prolonged, even though the average replicons, are completed rapidly. In these cases the comparative lengthening of the duration of S is achieved by a staggering of the activation of replication origins. The relative time of replication of particular genes may vary widely in the two different situations.

inversely with the absolute length of S, but in all cases it was sufficiently high that the average replicon would be synthesized in only a fraction of the S period. To take a specific example, cells supplied with ammonium ions as sole nitrogen source had an S period of nearly 60 min, and replication forks moving (bi-directionally) at about 2 μm min^{-1}. The average replicon in these cells would have been completed in 7–8 min, and the only way of stretching out the S period to 60 min would be by staggering the activation of origins, as illustrated in Fig. 6 (*b*).

It is too early yet to pin down the cause of these differences between strains, though it seems likely that a basic effect of mating type or ploidy may be involved. In any event however, these results show that the yeast cell can structure its S period in more than one way, and in this second 'staggered' S period, the order of gene replication is determined as much by the timing of origin activation as by the physical locations of genes on replicons. Of course it must be stressed that there is no direct evidence in these new observations that the origins are in fact activated in a defined sequence, and this must be an early priority for research in this area. Given, however, that such a sequence is demonstrable, we now have in front of us an unrivalled opportunity for detailed dissection of the structure and control of the S period in this eukaryotic organism.

Prospect

From the brief account given in this chapter, it is obvious that we still have much to learn about chromosomal replication in yeast. Our knowledge of the enzymology of the process is fragmentary, and the genetic flexibility of the system has by no means been explored to the full. Likewise, we are obviously only on the threshold of understanding the nature of origins, termini, and the mechanisms that control and co-ordinate replicon duplication.

The answers to some of these outstanding problems may be found relatively soon. Several different laboratories are currently engaged in isolating yeast DNA fragments which enhance the transformation ability of certain plasmids, and may therefore carry replication origins. It is too early yet to be confident about the nature of these fragments, but armed with them, and with our new-found ability, based on recombinant DNA technology, to make probes for specific genes, the time is evidently ripe for rapid progress to be made. The prospect is good.

References

Banks, G. R. (1973). Mitochondrial DNA synthesis in permeable cells. *Nature New Biology*, **245**, 196–9.

Barford, J. P. & Hall, R. J. (1976). Estimation of the length of cell cycle phases from asynchronous cultures of *Saccharomyces cerevisiae*. *Experimental Cell Research*, **102**, 276–84.

Bisson, L. & Thorner, J. (1977). Thymidine-5′-monophosphate-requiring mutants of *Saccharomyces cerevisiae* are deficient in thymidylate synthetase. *Journal of Bacteriology*, **132**, 44–50.

Blumenthal, A. B., Kriegstein, H. J. & Hogness, D. S. (1973) The units of DNA replication in *Drosophila melanogaster* chromosomes. *Cold Spring Harbor Symposia on Quantitative Biology*, **38**, 205–23.

Braun, R. & Wili, H. (1969). Time sequence of DNA replication in *Physarum*. *Biochimica et Biophysica Acta*, **174**, 246–52.

Burke, W. & Fangman, W. L. (1975). Temporal order in yeast chromosome replication. *Cell*, **5**, 263–9.

Cairns, J. (1963). The bacterial chromosome and its manner of replication as seen by autoradiography. *Journal of Molecular Biology*, **6**, 208–13.

Callan, H. G. (1972). Replication of DNA in the chromosomes of eucaryotes. *Philosophical Transactions of the Royal Society, Series B*, **181**, 19–41.

Chang, L. M. S. (1977). DNA polymerases from baker's yeast. *Journal of Biological Chemistry*, **252**, 1873–80.

Chang, L. M. S., Lurie, K. & Plevani, P. (1978). A stimulatory factor for yeast DNA polymerases. *Cold Spring Harbor Symposia on Quantitative Biology*, **43**, 587–95.

Culotti, J. & Hartwell, L. H. (1971). Genetic control of the cell division cycle in yeast. III. Seven genes controlling nuclear division. *Experimental Cell Research*, **67**, 389–401.

Dawes, I. W. & Carter, B. L. A. (1974). Nitrosoguanidine mutagenesis during nuclear and mitochondrial gene replication. *Nature, London*, **250**, 709–12.

Dawes, I. W., MacKinnon, O. A., Ball, D. E., Hardie, I. D., Sweet, O. M., Ross, F. M. & MacDonald, F. (1977). Identifying sites of simultaneous DNA replication in eukaryotes by *N*-methyl-*N*′-Nitro-*N*-nitrosoguanidine multiple mutagenesis. *Molecular and General Genetics*, **152**, 53–7.

Durnford, J. M. & Champoux, J. J. (1978). The DNA untwisting enzyme from *Saccharomyces cerevisiae*. *Journal of Biological Chemistry*, **253**, 1086–9.

Game, J. C. (1976). Yeast cell cycle mutant *cdc21* is a temperature-sensitive thymidylate auxotroph. *Molecular and General Genetics*, **146**, 313–15.

Guerola, N., Ingram, J. L. & Cerda-Olmedo, E. (1971). Induction of closely linked multiple mutations by nitrosoguanidine. *Nature New Biology*, **230**, 122–5.

Hand, R. (1978). Eucaryotic DNA: organisation of the genome for replication. *Cell*, **15**, 317–25.

Hartwell, L. H. (1970). Biochemical genetics of yeast. *Annual Review of Genetics*, **4**, 373–96.

Hartwell, L. H. (1973). Three additional genes required for DNA synthesis in *Saccharomyces cerevisiae*. *Journal of Bacteriology*, **115**, 966–74.

Hartwell, L. H. (1976). Sequential function of gene products relative to DNA synthesis in the yeast cell cycle. *Journal of Molecular Biology*, **104**, 803–17.

Hartwell, L. H., McLaughlin, C. S. & Warner, J. R. (1970). Identification of ten genes that control ribosome formation in yeast. *Molecular and General Genetics*, **109**, 42–56.

Helfman, W. B. (1973). The presence of an exonuclease in highly purified DNA polymerase from baker's yeast. *European Journal of Biochemistry*, **32**, 42–9.

Hereford, L. M. & Hartwell, L. H. (1971). Defective DNA synthesis in permeabilised yeast cells. *Nature New Biology*, **234**, 171–2.

Holliday, R. (1970). The organisation of DNA in eukaryotic chromosomes. In *Organisation and Control in Prokaryotic and Eukaryotic Cells*, ed. H. P. Charles & B. C. Knight, pp. 359–80. Symposium of the Society for General Microbiology 20. Cambridge University Press.

Housman, O. & Huberman, J. A. (1975). Changes in the rate of DNA replication fork movement during S phase in mammalian cells. *Journal of Molecular Biology*, **94**, 173–81.

Jacob, F., Brenner, S. & Cuzin, F. (1963). On the regulation of DNA replication in bacteria. *Cold Spring Harbor Symposia on Quantitative Biology*, **28**, 329–36.

Jazwinski, S. M. & Edelman, G. M. (1979). Replication *in vitro* of the 2 μm DNA plasmid of yeast. *Proceedings of the National Academy of Sciences, USA*, **76**, 1223–7.

Johnston, L. H. (1979). In-vitro DNA synthesis in a concentrated yeast lysate. *Molecular and General Genetics*, **175**, 217–21.

Johnston, L. H., Bonhoeffer, F. & Symmons, P. (1979). The molecular principles and the enzymatic machinery of DNA replication. In *Cell Biology, a Comprehensive Treatise*, vol. 2, ed. D. M. Prescott & L. Goldstein, pp. 59–130. London & New York: Academic Press.

Johnston, L. H. & Game, J. C. (1978). Mutants of yeast with depressed DNA synthesis. *Molecular and General Genetics*, **161**, 205–14.

Johnston, L. H. & Nasmyth, K. A. (1978). *Saccharomyces cerevisiae* cell cycle mutant *cdc9* is defective in DNA ligase. *Nature, London*, **274**, 891–3.

Johnston, L. H. & Williamson, D. H. (1978). An alkaline sucrose gradient analysis of the mechanism of nuclear DNA synthesis in the yeast *Saccharomyces cerevisiae*. *Molecular and General Genetics*, **164**, 217–25.

Kassir, Y. & Simchen, G. (1978). Meiotic recombination and DNA synthesis in a new cell cycle mutant of *Saccharomyces cerevisiae*. *Genetics*, **90**, 49–68.

Kee, S. G. & Haber, J. E. (1975). Cell cycle-dependent induction of mutations along a yeast chromosome. *Proceedings of the National Academy of Sciences, USA*, **72**, 1179–83.

Mueller, G. C. & Kajiwara, K. (1966). Early and late-replicating deoxyribonucleic acid complexes in HeLa nuclei. *Biochimica et Biophysica Acta*, **114**, 108–15.

Newlon, C. S. & Fangman, W. L. (1975). Mitochondrial DNA synthesis in cell cycle mutants of *S. cerevisiae*. *Cell*, **5**, 423–8.

Newlon, C. S., Petes, T. D., Hereford, L. M. & Fangman, W. L. (1974). Replication of yeast chromosomal DNA. *Nature, London*, **247**, 32–5.

Oertel, W. & Goulian, M. (1977). DNA synthesis in permeabilised spheroplasts of *Saccharomyces cerevisiae*. *Journal of Bacteriology*, **132**, 233–46.

Painter, R. B. & Schaefer, A. W. (1971). Variation in the rate of DNA chain growth throughout the S phase in HeLa cells. *Journal of Molecular Biology*, **58**, 289–95.

Pardoll, D. M., Vogelstein, B. & Coffey, D. S. (1980). A fixed site of DNA replication in eucaryotic cells. *Cell*, **19**, 527–36.

Petes, T. D. & Newlon, C. S. (1974). Structure of DNA in DNA replication mutants of yeast. *Nature, London*, **251**, 637–9.

Petes, T. D., Newlon, C. S., Byers, B. & Fangman, W. (1973). Yeast chromosomal DNA: size, structure and replication. *Cold Spring Harbor Symposia on Quantitative Biology*, **38**, 9–16.

Petes, T. D. & Williamson, D. H. (1975). Fiber autoradiography of replicating yeast DNA. *Experimental Cell Research*, **95**, 103–10.

Plevani, P. & Chang, L. M. S. (1977). Enzymatic initiation of DNA synthesis by yeast DNA polymerases. *Proceedings of the National Academy of Sciences, USA*, **74**, 1937–41.

Rivin, C. J. & Fangman, W. L. (1980*a*). Cell cycle phase expansion in nitrogen-limited cultures of *Saccharomyces cerevisiae*. *Journal of Cell Biology*, **85**, 96–107.

Rivin, C. J. & Fangman, W. L. (1980*b*). Replication fork rate and origin activation during the S phase of *Saccharomyces cerevisiae*. *Journal of Cell Biology*, **85**, 108–15.

Schaller, H., Otto, B., Nusslein, V., Huf, J., Herrmann, R. & Bonhoeffer, F. (1972). DNA replication *in vitro*. *Journal of Molecular Biology*, **63**, 183–200.

Slater, M. L., Sharrow, S. O. & Gart, J. J. (1977). Cell cycle of *Saccharomyces cerevisiae* in populations growing at different rates. *Proceedings of the National Academy of Sciences, USA*, **74**, 3850–4.

Tye, B., Nyman, P., Lehman, I. R., Hochhauser, S. & Weiss, B. (1977). Transient accumulation of Okazaki fragments as a result of uracil incorporation into nascent DNA. *Proceedings of the National Academy of Sciences, USA*, **74**, 154–7.

Williamson, D. H. (1965). The timing of deoxyribonucleic acid synthesis in the cell cycle of *Saccharomyces cerevisiae*. *Journal of Cell Biology*, **25**, 517–28.

Wintersberger, E. (1974*a*). DNA polymerases from yeast. Further purification and characterisation of DNA-dependent DNA polymerases A and B. *European Journal of Biochemistry*, **50**, 41–7.

Wintersberger, U. (1974*b*). Absence of a low molecular weight DNA polymerase from nuclei of the yeast *S. cerevisiae*. *European Journal of Biochemistry*, **50**, 197–202.

Wintersberger, E. (1977). DNA-dependent DNA polymerases from eukaryotes. *Trends in Biochemical Sciences*, **2**, 58–61.

Wintersberger, E. (1978). Yeast DNA polymerases: antigenic relationship, use of RNA primer and associated exonuclease activity. *European Journal of Biochemistry*, **84**, 167–72.

Wintersberger, U. & Blutsch, H. (1976). DNA polymerase from yeast mitochondria. *European Journal of Biochemistry*, **68**, 199–206.

Wyers, F., Huet, J., Sentenac, A. & Fromageot, P. (1976). Role of DNA·RNA hybrids in eukaryotes. Characterisation of yeast ribonucleases H1 and H2. *European Journal of Biochemistry*, **69**, 385–95.

Yarranton, G. T. & Banks, G. R. (1977). A DNA polymerase from *Ustilago maydis*. Evidence for proof-reading by the associated $3' \rightarrow 5'$ deoxyribonuclease activity. *European Journal of Biochemistry*, **77**, 521–7.

Yurov, Y. B. (1980). Rate of DNA replication fork movement within a single mammalian cell. *Journal of Molecular Biology*, **136**, 339–42.

Genetic regulation of RNA and protein patterns in the monokaryon–dikaryon transition

J. G. H. WESSELS, J. J. M. DONS, J. H. C. HOGE,
J. SPRINGER, O. M. H. DE VRIES and
A. ZANTINGE

Department of Developmental Plant Biology, Biological Centre, University of Groningen, Haren, The Netherlands

Introduction

This review is mainly concerned with work done in our laboratory during the past three years with the aim of studying the levels of regulation of genome expression in a simple eukaryotic organism. The choice of the monokaryon–dikaryon transition in *Schizophyllum commune* Fr. as a model system was made because (i) the genetic elements controlling cell morphogenesis (the incompatibility factors) were well known and (ii) the genomic complexity of this fungus was expected to be low compared with higher eukaryotes, which would facilitate analysis of the system.

Genetic elements and cell differentiation

The genetic elements involved in the monokaryon–dikaryon transition in *Schizophyllum commune* have been studied extensively by Raper and his students. The results of these studies have been excellently reviewed (Raper, 1966; Raper, J. R. & Raper, C. A., 1973; Raper, 1978) and just a few remarks will suffice here to provide a background for the present purpose (Fig. 1).

The presence in a homokaryon of one wild-type allele for each of the two loci, α and β, for both the A factor and the B factor, results in a mycelium with monokaryotic morphology, i.e. a mycelium that exclusively consists of uninucleate cells separated by simple septa. For convenience this will be called the A-off B-off state of morphology. Mating of such monokaryons may result in the introduction in a common cytoplasm of different alleles of the loci of the incompatibility factors (heterokaryon), and it is found that a difference in only one of the two loci of an incompatibility factor suffices to elicit a morphogenetic response. Allelic differences in both the *A* factor and the *B* factor result in the formation of a mycelium with dikaryotic morphology, distinguished by binucleate cells and clamp connections at each septum separating the cells (*A*-on *B*-on state of morphology). If an allelic difference occurs only in the *B* factor a heterokaryon is formed in which the processes of septal dissolution and nuclear migration (normally transient in the formation of the dikaryon) are constitutive (*A*-off *B*-on). If an allelic difference occurs only in the *A* factor, other

Heterokaryon	Homokaryon	Sequence	Morphology
	Ax Bx	*A*-off *B*-off	
Ax Bx/Ax By	*Ax Bmut*	*A*-off *B*-on	
Ax Bx/ Ay Bx	*Amut Bx*	*A*-on *B*-off	
Ax Bx/Ay By	*Amut Bmut*	*A*-on *B*-on	

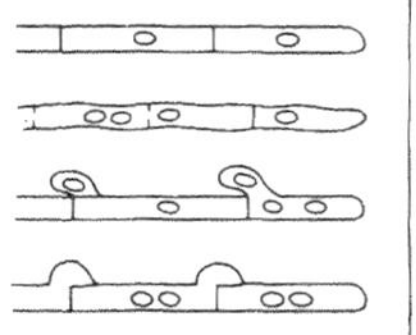

Fig. 1. Mycelial morphologies as determined by different alleles of the incompatibility factors in heterokaryons or by primary mutations in these alleles in homokaryons.

processes related to dikaryon formation, e.g. initiation of clamp connections, become operative. Septal dissolution and nuclear migration do not occur, however, and the initiated clamps fail to fuse, resulting in a mycelium bearing pseudoclamps (*A*-on *B*-off).

As indicated in Fig. 1 the different morphogenetic sequences can also be switched on in homokaryons by a primary mutation in one of the loci of the incompatibility factors. Such rare mutations co-ordinately express a number of morphogenetic functions, mutations for which can be found outside the incompatibility loci. With the incompatibility loci as the controlling elements, these mutations apparently affect the controlled part of the system. However, also within the incompatibility loci, mutations are known that seem to affect only certain parts of the regulated sequences (secondary mutations). This led to the conclusion that each incompatibility locus is composed of a recognition subunit (determining allelic specificity) and a regulatory subunit containing information pertaining to the controlled sequences (Raper, J. R. & Raper, C. A., 1973). If the recognition subunits are thought to be analogous to genetic elements that can be affected by external stimuli (sensors), then the incompatibility loci very much resemble the hypothetical integrator genes envisaged by Georgiev (1969) and Britten

& Davidson (1969) for the control of cytodifferentiation in metazoa. Thus, a study of the incompatibility system at the molecular level would offer unique opportunities to unravel details of such regulatory circuits because in this fungal system (i) the different cell types can be grown in isolation and mutations in regulatory genes do not upset development as would be the case in multicellular organisms (Davidson & Britten, 1973) and (ii) with a low genetic complexity and only four different cell types involved, the regulatory circuits would be expected to be relatively simple.

If the incompatibility loci are considered to be regulatory genes that are not essentially different from the integrator genes envisaged for the regulation of development in higher eukaryotes, their unique character obviously lies in the recognition subunits that determine the interaction of different alleles. Several models have been proposed to explain how recognition of the products of different alleles might occur (Kuhn & Parag, 1972; Ullrich, 1978; Raper, 1978) but none of these models has been put through biochemical tests. In fact it is still difficult to reconcile all properties of the system with known properties of either nucleic acids or proteins. However, the genetic evidence (Raper, J. R. & Raper, C. A., 1973) at least strongly suggests that the product of an incompatibility allele is inactive by itself but is converted into an active regulatory substance by interaction with the product of another allele, or by a rare mutation (positive regulation).

Biochemical indices of cell differentiation

Ideally a study on the regulation of morphological differentiation between cell types should start with identifying the molecular basis of morphogenesis. In this regard particular attention has been paid to the *B* sequence of morphogenesis because the *B*-on phenotype has a number of peculiarities which immediately suggest the type of biochemical changes involved. Among these are the constitutive degradation of septa, which permit nuclei to migrate, and the irregularly bulged hyphae, which suggest weakened walls. This has initiated a long-term study on the chemical and morphological structure of hyphal walls and septa of *Schizophyllum commune*, including the degradative processes involved. This work was recently reviewed in detail (Wessels, 1978; Wessels & Sietsma, 1979). Briefly, the results indicate that in the *B*-on phenotype a number of hydrolases are very active among which a crucial role is played by an enzyme (R-glucanase) that hydrolytically solubilizes the $\beta(1-3)/\beta(1-6)$-glucan component of wall and septum. Differences in the construction of lateral wall and septum have been shown to be responsible for the preferential dissolution of septa and a model has been proposed suggesting that both the activity of R-glucanase and resistance of septa against lysis are regulated by the incompatibility factors. Very little is known about biochemical processes related to other morphogenetic sequences, e.g. formation of clamp connections. In the absence of more information on proteins that can be specifically assigned to morphological changes, it would be useful to know how many proteins are regulated by the incompatibility factors to obtain an impression of the extensiveness of the regulated system.

Regulation of protein patterns

In an early study Raper & Esser (1961) demonstrated that extracts of two compatible coisogenic monokaryons were apparently identical in antigenic properties, whereas those of the dikaryon constituted by mating these monokaryons showed antigenic differences. Extending these studies, Wang & Raper (1969, 1970) compared protein and isozyme patterns by one-dimensional electrophoresis. The patterns from monokaryons were identical but markedly different from those of other cell types; for example, isozyme patterns of 14 out of 15 common enzymes. However, Ullrich (1977) has not been able to reproduce this result. In fact he did not see any difference in isozyme patterns of four enzymes previously shown to be very different.

Therefore we undertook to compare the protein patterns of two compatible coisogenic monokaryons (*A41 B41* and *A43 B43*), the dikaryon constituted by mating these monokaryons, and a coisogenic series of homokaryotic developmental mutants carrying a primary mutation in either the *A* factor or the *B* factor (*A41 Bmut* and *Amut B41*) or in both factors (*Amut Bmut*) (de Vries, Hoge & Wessels, 1980). Young growing mycelia with fully expressed phenotypes were briefly labelled with [^{35}S]methionine and the radioactive proteins were extracted under conditions avoiding proteolysis and other artifacts and analysed on two-dimensional gels, using isoelectrofocusing in the first direction and sodium dodecyl sulphate (SDS) electrophoresis in the second direction. This enabled comparison of about 700 polypeptides.

If the presence or absence of a specific spot is taken as a criterion, then the differences in protein patterns of the various mycelial types can be summarized as in Fig. 2. Analysis of identical extracts revealed differences in patterns of less than 0.5%. Extracts from coisogenic monokaryons differed by 2.3%, which is much less than the difference

(11%) between these monokaryons and a non-isogenic monokaryon (*A51 B51*) which is morphologically identical. The differences between the coisogenic monokaryons and the dikaryon amounted to 7%. The *Amut Bmut* homokaryon had a protein pattern that was intermediate between those of the coisogenic monokaryons and the dikaryon. This may be related to the fact that it did not express the dikaryotic morphology in the present experiment, but rather made monokaryotic and pseudoclamped cells.

When the total number of mutual differences between the coisogenic monokaryons and the derived dikaryon (i.e. 59 out of 710 spots

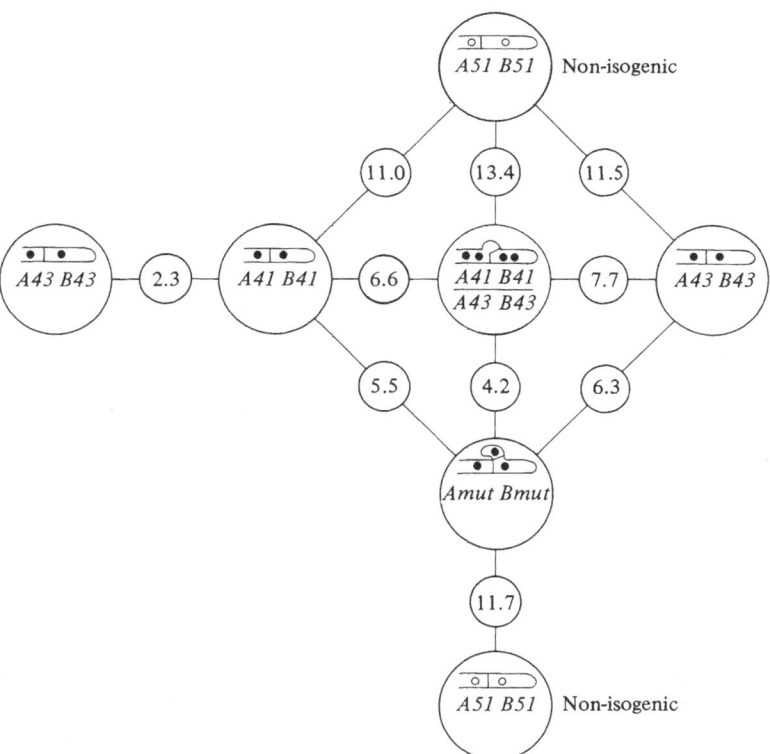

Fig. 2. Diagram of qualitative differences (expressed as percentages of the total number (710) of spots examined) between the spectra of protein synthesis of various strains analysed by two-dimensional electrophoresis of urea-Nonidet protein extracts. The various strains were: two coisogenic monokaryons 4–39 (*A41 B41*) and 4–40 (*A43 B43*), the derived dikaryon (*A41 B41/A43 B43*), the double-factor mutant (*Amut Bmut*), and the non-isogenic monokaryon 4–50 (*A51 B51*). (From de Vries, Hoge & Wessels, 1980.)

analysed) was examined it was found that 23 spots specifically appeared in the pattern of the dikaryon whereas 20 spots specifically disappeared from the pattern. Of the proteins switched off in the dikaryon, the majority could still be detected in the developmental mutants. This is to be expected since all developmental mutants formed monokaryotic cells in addition to the cell types specified by the mutations. As shown in Fig. 3, of the 23 proteins specifically detected in the dikaryon, 12

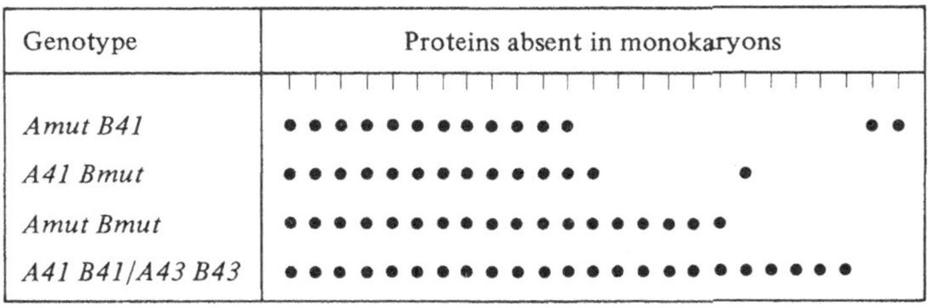

Genotype	Proteins absent in monokaryons
Amut B41	• • • • • • • • • • • • • •
A41 Bmut	• • • • • • • • • • • • • •
Amut Bmut	• • • • • • • • • • • • • • • • • •
A41 B41/A43 B43	• •

Fig. 3. The presence in developmental mutants of proteins specifically synthesized in the dikaryon.

proteins were also found in the *Amut B41* and *A41 Bmut* mycelia which differed by only four proteins. This was unexpected because of the vastly different morphogenetic sequences switched on in these mutants. The concerted action of the mutations in both the *A* factor and the *B* factor (*Amut Bmut*) appeared to switch on five more dikaryotic proteins but four were still missing, probably due to incomplete expression of the dikaryotic phenotype in this double mutant. Apart from two proteins specifically present in the *Amut B41* mycelium, none of the developmental mutants showed proteins not present in the dikaryon.

In this analysis we compared only about 700 sulphur-containing proteins with isoelectric points between 5 and 7 and apparent molecular weights between 17 000 and 130 000 which were synthesized at relatively rapid rates. It is thus impossible to extrapolate the percentage of polypeptides (6%, cf. Table 1), apparently under control of the *A* factor and the *B* factor, to the whole population of polypeptides which may comprise 10 000 different species (see below). Therefore it was considered necessary to look for differences in the complex class of RNA encompassing both the abundant RNA species – probably directing synthesis of the proteins observed *in vivo* – as well as rare RNA species, to see whether control by the incompatibility factors occurred by regulation of the amount of specific RNAs or otherwise. For this analysis a study of the organization of the genome was also necessary and a summary of the findings is presented below.

Organization of the genome of *Schizophyllum commune*

In *Schizophyllum commune* nuclear DNA (nDNA) and mitochondrial DNA (mtDNA) can easily be distinguished on CsCl gradients because of their widely different G + C contents (Dons, de Vries & Wessels, 1979). The G + C content of nDNA was 57%, calculated from its buoyant density and melting curve. For mtDNA a G + C content of 22% was calculated from the derivative melting curve.

Mitochondrial DNA

About half of the mtDNA melted sharply at low temperature, indicating the occurrence of very rich A + T stretches (A + T content 84%) in the mtDNA. However, because the distribution of mtDNA in isopycnic CsCl gradients was unimodal, these A + T stretches are probably interspersed. This suggests that the mtDNA of *Schizophyllum commune* has a similar organization to that found in the mtDNAs of *Saccharomyces cerevisiae* and *Euglena gracilis* (Bernardi, 1976).

We also determined the molecular weight of the mtDNA of *Schizophyllum commune* by digesting the DNA with the restriction endonucleases Eco R1 (11 fragments), Hind III (10 fragments) and Xba I (10 fragments), and adding up the fragment lengths after electrophoresis of the digest (J. Groffen, unpublished). A molecular weight of 31×10^6 was found which is intermediate between values found for *Neurospora crassa* (40×10^6; Terpstra, Holtrop & Kroon, 1977) and *Aspergillus nidulans* (22×10^6; Lopez-Perez & Turner, 1975).

Complexity of nuclear DNA

Nuclear DNA, labelled with ^{32}P, was sheared to an average fragment length of 400–600 base pairs and melted. Reassociation of the single-stranded fragments was followed at 25 °C below the melting temperature by hydroxyapatite chromatography (Dons *et al.*, 1979). An analysis of the reassociation curves (Fig. 4) revealed a genome size of 22.8×10^9 daltons which is within the range of 6–30×10^9 found for other fungi (Ohja & Dutta, 1977). In *Schizophyllum commune* nDNA, three kinetic components could be distinguished: zero-time double-stranded DNA (2%), repetitive DNA (7%), and single-copy DNA (91%) with a kinetic complexity of 20.7×10^9 daltons. This means that the nDNA of *Schizophyllum commune* is 7.4 times more complex than the DNA of *Escherichia coli.*

Absence of interspersed repetitive DNA sequences

Several lines of evidence indicated that the repetitive DNA sequences were clustered and not interspersed with single copy sequences (Dons & Wessels, 1980). For instance, large fragments of nDNA (7700 base pairs) were dissociated and reassociated at low C_0t. With such DNA fragments one would expect interspersed repetitive sequences to reassociate, leaving the single copy DNA as single-stranded tails attached to the double-stranded repetitive DNA. Digestion of these single strands with S1 nuclease would then leave the double-stranded regions of repetitive DNA, the size of which could be determined by gel filtration. As shown in Fig. 5, such a procedure detects the presence of short (300 base pairs), interspersed, repetitive sequences in calf DNA but fails to detect such sequences in *Schizophyllum commune* DNA.

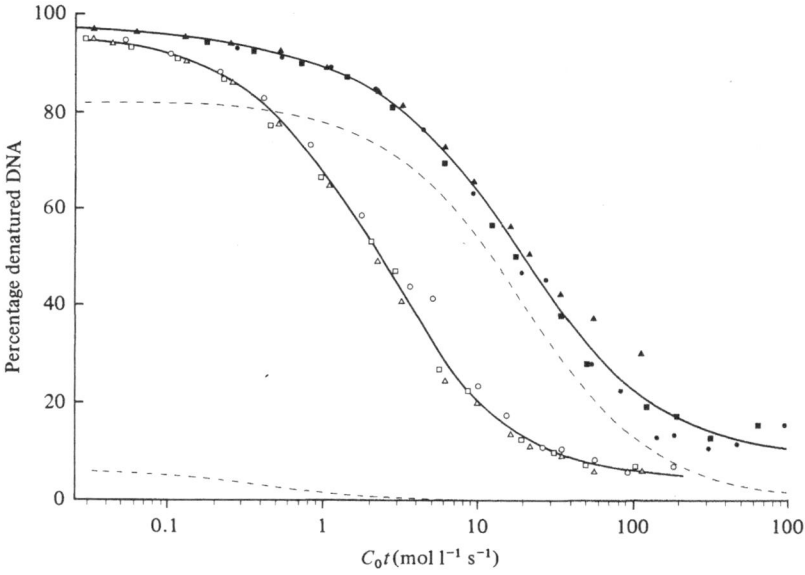

Fig. 4. Reassociation kinetics of *S. commune* nuclear DNA. Nuclear *S. commune* [32P]DNA (closed symbols) was reassociated in the presence of *E. coli* [3H]DNA (open symbols); the best least square solution to the data yielding two second-order components for *S. commune* DNA (– – –) and one second-order component for *E. coli* DNA. C_0t is expressed as the product of DNA concentration (mol nucleotides l^{-1}) and time (s). (From Dons *et al.*, 1979.)

Since it is known that ribosomal RNA (rRNA) cistrons generally occur as clustered repetitive sequences, the size and multiplicity of these cistrons was determined.

Ribosomal RNA cistrons

Denatured labelled nDNA was hybridized with a saturating amount of rRNA and remaining single strands of DNA were digested with S1 nuclease. Assuming asymmetric transcription, 2.2% (0.5×10^9

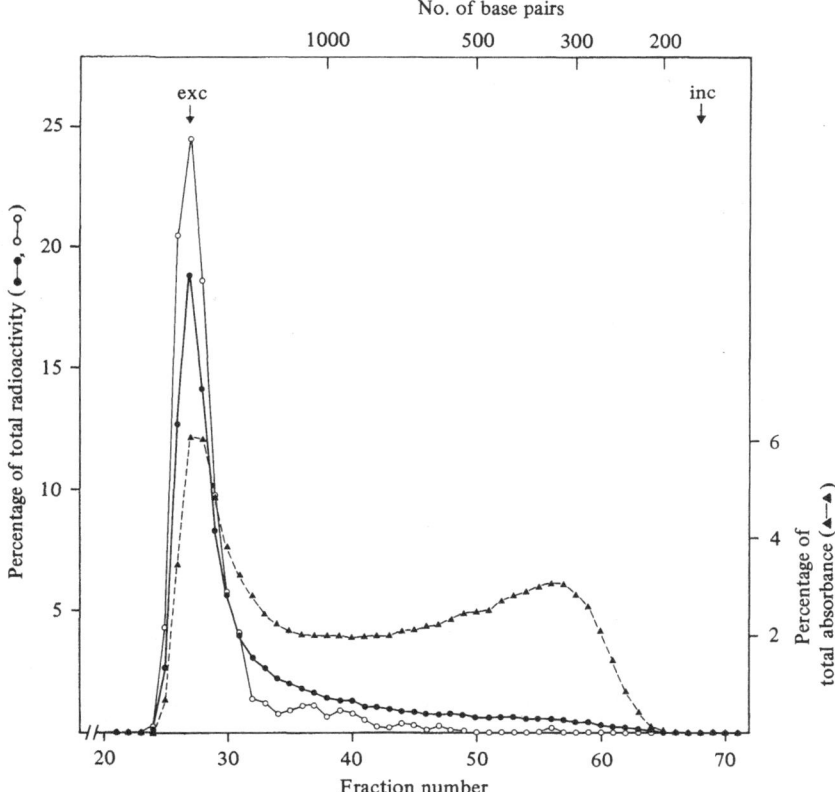

Fig. 5. Size distribution of *S. commune* repetitive duplexes. 7.7 kilobase-pair fragments of nuclear [^{32}P]DNA were reassociated to C_0t 0.145 and treated with S1 nuclease. S1-resistant duplexes were collected by hydroxyapatite binding and chromatographed on Agarose A50 (●——●). ○——○, elution pattern of duplexes bound to hydroxyapatite without prior S1 nuclease treatment; ▲– – –▲, elution pattern of S1-resistant repetitive DNA from calf thymus, reassociation to C_0t 2. exc., position of the excluded volume of the column; inc., position of the inclusion marker [^{3}H]leucine. (From Dons & Wessels, 1980.)

daltons) of the DNA was found to account for the sequences present in mature rRNAs. The total molecular weight of these rRNAs (25 S, 18 S, 5.8 S, and 5 S rRNA) was estimated as 2.14×10^6, indicating that the rRNA cistrons are repeated in the genome approximately 120 times (Dons & Wessels, 1980).

To estimate the size of the rRNA cistron, high molecular weight nDNA (25×10^6) was cleaved with the restriction enzymes Eco R1 and Hind III. Three prominent bands, which hybridized with rRNA, were seen on the gels. Their total molecular weight was 11.9×10^6, indicating that the rRNA cistrons are about 2.8 times the size of mature rRNA, and thus occupy about 6.2% of the nuclear genome. From hybridizations between nDNA and total RNA it was concluded that another 0.6% probably codes for repetitive sequences such as transfer RNA (Dons & Wessels, 1980).

Taken together these results indicate that in *Schizophyllum commune*, short interspersed sequences of repetitive DNA do not exist or are extremely rare. All repetitive DNA is clustered and consists of rRNA genes, tRNA genes, and possibly a few structural genes. Similar conclusions have now been reached for the nuclear genomes of *Saccharomyces cerevisiae* (Lauer, Roberts & Klotz, 1977), *Aspergillus nidulans* (Timberlake, 1978), and *Neurospora crassa* (Krumlauf & Marzluf, 1979). All these studies lead to the conclusion that the organization of these fungal genomes is fundamentally different from those of higher eukaryotes. In *Schizophyllum commune* the absence of interspersed repetitive sequences makes it less likely that the incompatibility genes operate according to the Britten–Davidson model of regulation.

Complex class of RNA in *Schizophyllum commune*

Poly(A)-containing RNA was isolated from total RNA by binding to oligo(dT)-cellulose and its complexity was determined and compared with that of total RNA (Zantinge, Dons & Wessels, 1979). The size of the poly(A) tails was found to be small (number-average size: 33 nucleotides) and 2.5% of the total RNA contains such poly(A) tracts. These poly(A)-containing RNAs have a number-average length of 1100 nucleotides, which is in good agreement with the number-average molecular weight (34 000) of proteins synthesized *in vitro* in a cell-free system encoded by total RNA of *Schizophyllum commune* (de Vries, Hoge & Wessels, 1980).

To obtain a value for the complexity of poly(A)-containing RNA, it was hybridized to single-copy DNA. For this purpose single-copy DNA was isolated from total DNA by first removing the rapidly reassociating component (repetitive DNA) and then labelled *in vitro* to high specific activity (gap translation). Denatured single-copy [3H]DNA was then incubated with an excess of poly(A)-containing RNA at various RNA : DNA ratios and at various times the amount of hybrids was determined by removing remaining single-stranded DNA with S1 nuclease (Fig. 6). On the average 13% of the single-copy DNA could hybridize to poly(A)-containing RNA. Taking into account the degree of hybridization (80%) obtainable with nick-translated DNA, and assuming asymmetric transcription, about 32% of single-copy DNA sequences are transcribed in *Schizophyllum commune*. With a single-copy DNA complexity of 20.7×10^9 (3.4×10^7 base pairs) this would

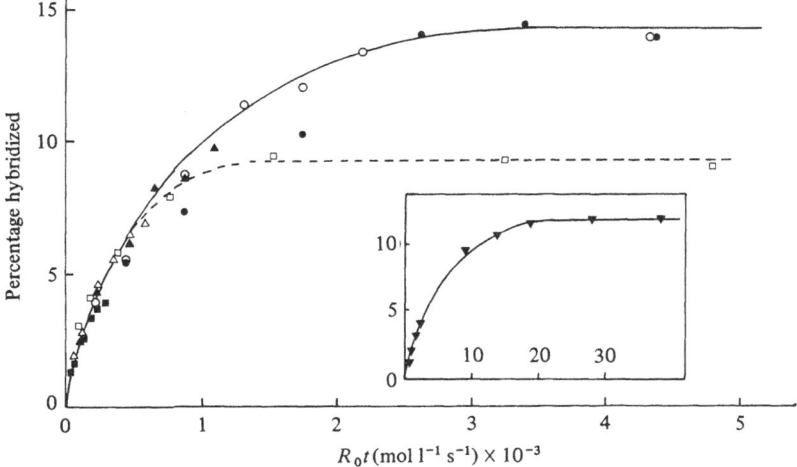

Fig. 6. Hybridization of poly(A)-containing RNA to single-copy [3H]DNA at different RNA : DNA ratios. Increasing amounts of poly(A)-containing RNA were hybridized to single-copy [3H]DNA with the following RNA : DNA ratios: 6600 (■), 13 200 (△), 25 000 (▲), 50 000 (○) and 100 000 (●). To show saturation at a low RNA : DNA ratio and high R_0t values one series of hybridizations was performed with an RNA : DNA ratio of 6300 (□). For each incubation time the amount of DNA/DNA hybrid was subtracted from the total amount of duplexes. R_0t values are related to the concentration of poly(A)-containing RNA and equivalent R_0t values. The inset shows hybridization of total RNA to single-copy [3H]DNA at an RNA : DNA ratio of 1.1×10^6. DNA/DNA renaturation was negligible; 1.4% S1 nuclease resistance of single-copy [3H]DNA at $t = 0$ was subtracted at each R_0t value. (From Zantinge, Dons & Wessels, 1979.)

indicate the presence of about 10 000 different, average-sized poly(A)-containing RNAs.

The inset in Fig. 6 shows that with hybridization of total RNA to single-copy DNA a similar hybridization level was reached as with poly(A)-containing RNA, but hybridization was ten times slower. This suggests that total RNA contains about 10% RNA sequences hybridizable to single-copy DNA with a complexity essentially the same as poly(A)-containing RNA. Since 2.5% of the RNA is polyadenylated, this means that one out of four individual RNA sequences contains a poly(A)-tract and that there does not exist a separate class of non-polyadenylated RNAs among the complex RNAs. In addition, polysomal mRNA attained the same hybridization level as total RNA and poly(A)-containing RNA (results not shown). Apparently all complex RNA is also involved in translation. As a consequence there appears to exist no class of large heterogeneous nuclear RNA (hnRNA) as in higher eukaryotes, a conclusion also reached for *Achlya ambisexualis* (Timberlake, Shumark & Goldberg, 1977; Rozek, Orr & Timberlake, 1978) and *Saccharomyces cerevisiae* (Hereford & Rosbash, 1977).

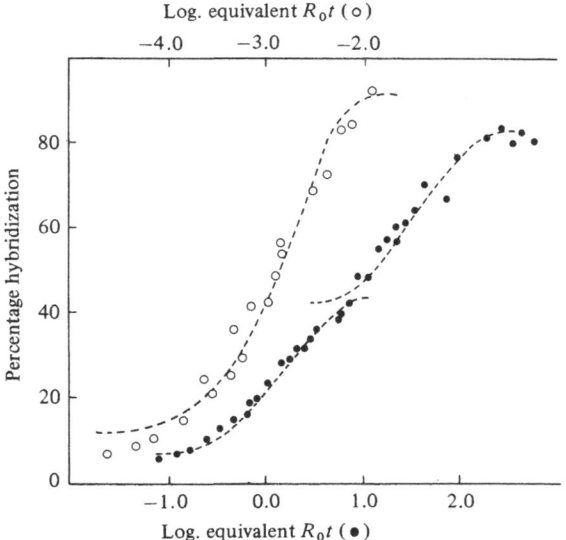

Fig. 7. Complexity of poly(A)-containing RNA from *S. commune* as measured by RNA/cDNA hybridization (●), with globin mRNA as a standard (○). The dashed curves are based on computer analysis assuming two kinetic components in the RNA from *S. commune*. R_0t is expressed as the product of the concentration of poly(A)-containing RNA (mol nucleotides l^{-1}) and time (s).

The complexity of poly(A)-containing RNA of *Schizophyllum commune* was also determined by hybridizing this RNA to its copy DNA (cDNA) prepared with reverse transcriptase. Fig. 7 shows the kinetics of this hybridization compared to that of pure globin mRNA to its cDNA. In contrast to results obtained with poly(A)-containing RNA from higher eukaryotes there were no clear-cut abundancy classes, but because the hybridization took place over about four R_0t orders of magnitude a two-component computer analysis was applied to resolve the data (Fig. 7). Based on this kinetic analysis there is a class of about 600 different abundant poly(A)-containing RNAs accounting for 50% of the mass whereas the other 50% of the mass comprises 13 000 different rare poly(A)-containing RNAs. This total complexity approximates the value of 10 000 as revealed by excess hybridization to gap-translated single-copy DNA (Fig. 6). If it is considered that in a monokaryotic cell the RNA : DNA ratio is about 75 and that 10% of the RNA consists of complex RNA sequences, each sequence being polyadenylated in 25% of the cases, it can be calculated that on average the 600 abundant RNA sequences and the 13 000 rare RNA sequences are represented in the cell by 400 and 20 copies, respectively. However, as indicated above, there is a continuum between the two classes of abundance thus defined.

Comparison of complex class RNA in monokaryon and dikaryon
Comparison of poly(A)-containing RNA by hybridization to single-copy DNA

Poly(A)-containing RNA was isolated from actively growing mycelia of two coisogenic monokaryons of *Schizophyllum commune* and from the derived dikaryon. The RNAs were then hybridized in excess with nick-translated single-copy DNA (cf. Fig. 6) either separately or in combinations. After digesting the remaining single-stranded material with S1 nuclease, and correcting for S1 nuclease resistance at zero time and for renaturation of single-copy [³H]DNA, about 13% of the DNA was found in RNA/DNA hybrids in all cases with no significant differences in hybridization levels (Zantinge *et al.*, 1979). This method which could indicate qualitative differences in rare poly(A)-containing RNAs thus fails to demonstrate any difference between monokaryon and dikaryon. However, the inevitable large errors in determining hybridization values precluded demonstration of qualitative differences in RNA sequences within about 1% of the hybridization values, involving about 1000 RNA sequences. Nevertheless this result contrasts with the remarkably large differences in RNA sequences found using a similar method in various developmental stages of *Dictyostelium discoideum* (Firtel, 1972) and *Neurospora crassa* (Dutta & Chaudhury, 1975) involving 38% and 30% of the genome, respectively.

Comparison of poly(A)-containing RNA and total RNA by RNA/cDNA hybridization

Homologous RNA/cDNA hybridizations were performed with cDNAs prepared on poly(A)-containing RNA from coisogenic monokaryons and the derived dikaryon. The hybridization curves were all identical to that shown for one of the monokaryons in Fig. 7. To reveal differences in RNA sequences, heterologous hybridizations were carried out, i.e. using cDNA synthesized on poly(A)-containing RNA from a monokaryon and hybridizing with poly(A)-containing RNA from the dikaryon and vice versa. The final levels of hybridization (80%) were indistinguishable from those in the homologous hybridizations, but the curves shifted to higher R_0t values, indicating quantitative differences in poly(A)-containing RNAs in monokaryon and dikaryon.

To detect any qualitative differences in poly(A)-containing RNAs, non-hybridized cDNA remaining after a heterologous hybridization was re-isolated. Such cDNA would be enriched with any cDNA specific for the cell type from which it originated. Subsequent hybridization of this cDNA with poly(A)-containing RNA from the two cell types should then show higher hybridization levels with the homologous RNA than with the heterologous RNA, if differences exist. This was indeed observed; both the monokaryon and the dikaryon contained cDNA specific to cell type amounting to 1–1.5% of the cDNA mass. If these differences are assigned to the rare RNA sequences, this means that each cell type would contain 250–400 specific poly(A)-containing RNAs.

Although previously no indications were found for the existence of a separate class of non-polyadenylated RNAs, this did not rule out the possibility that the observed difference in cDNAs derived from the two cell types was due to the presence or absence of poly(A) tails on 250–400 specific RNA sequences. Therefore, hybridizations on enriched cDNA were carried out as above but now using total RNA (including non-polyadenylated RNA), instead of poly(A)-containing RNA, to drive the reaction. Surprisingly, no differences could now be detected between the complex RNAs from monokaryon and dikaryon. Taking into account the errors involved and assigning any undetectable differences to the class of rare sequences it could be calculated that the RNA sequences in monokaryon and dikaryon are qualitatively the same for at least 99%; no conclusions are possible for about 100 rare RNA sequences.

Although polysomal mRNA (derived from polysomes containing 8–12 ribosomes per mRNA) was not compared in heterologous hybridizations with cDNA, it was found in homologous hybridizations with cDNA that this polysomal mRNA encompasses virtually the entire complexity of poly(A)-containing RNA. This agrees with the earlier mentioned results obtained by hybridization to single-copy DNA.

Comparison of complex RNA by cell-free translation

Although the methods of RNA/single-copy DNA and RNA/ cDNA hybridizations failed to demonstrate differences in RNAs from monokaryon and dikaryon, these methods are less suitable for comparing abundant RNA sequences. Yet, a comparison of abundant RNA sequences is important in view of the likelihood that proteins visualized on gels after labelling *in vivo* are encoded by these abundant RNA sequences. For the same reason it was expected that the spectrum of abundant RNA sequences could be analysed by examining the products obtained by translation of RNA *in vitro*.

Total RNA from *Schizophyllum commune* was translated in a mRNA-dependent cell-free system derived from wheat germ or rabbit reticulocytes (de Vries, Hoge & Wessels, 1980). The products, labelled with [^{35}S]methionine, were fractionated two-dimensionally by iso-electrofocusing/SDS electrophoresis in the same way as proteins labelled *in vivo*. Both cell-free systems synthesized about 600 distinct polypeptides, a number that agrees with the number of abundant average-sized (1100 nucleotides) RNA sequences as determined by RNA/cDNA hybridization. The protein patterns obtained with the two different cell-free systems were very similar but vastly different from those obtained after labelling *in vivo*; only 40% of the spots coincided on the gels. This does not negate the value of *in vitro* translation in determining the occurrence of individual abundant RNA sequences. It rather indicates the importance of other factors, in addition to the mere presence of mRNAs, that determine the appearance of specific proteins *in vivo*, e.g. post-translational modifications, including cleavage and turnover of proteins.

Table 1. *Differences in protein species between the parental monokaryons and the dikaryon*

Classification of differences[a]			Number of differences		
			Total RNA		Polysomal RNA
Monokaryon	Monokaryon	Dikaryon	*In vivo*	*In vitro*	*In vitro*
(strain 4–39)	(strain 4–40)	(4–39/4–40)	(710)[b]	(550)[b]	(455)[b]
+	+	−	20	1	3
−	−	+	23	3	2
+	−	+	11	1	0
−	+	+	3	4	0
+	−	−	1	4	0
−	+	−	1	1	0

[a] Presence (+) or absence (−) of a particular protein species.
[b] Total number of proteins analysed.

When total RNA preparations from two unrelated monokaryons, showing 11% difference in their protein patterns, were compared by translation *in vitro*, a 24% difference was found in the translatable RNAs. When, however, total RNA preparations from the coisogenic monokaryons and the derived dikaryon were compared in this way, they were found to be near identical. As shown in Table 1, the number of proteins specifically absent (20) or specifically present (23) in the dikaryon cannot be accounted for by the differences in the RNAs coding for proteins *in vitro*. In fact, translation of the RNAs *in vitro* did not reveal any significant differences between the RNAs from monokaryons and the dikaryon because the absolute error in comparing protein patterns for the same protein extract is about 0.5%.

In the absence of differences in abundant RNA sequences in total RNA, the differences between proteins synthesized *in vivo* by monokaryons and the dikaryon might conceivably be explained by differences in the selection or exclusion of certain RNA sequences for translation. To test this possibility, RNA was isolated from polysomes carrying 8–12 ribosomes per mRNA and translated *in vitro*. As shown in Table 1 no differences could be detected between the polysomal mRNAs of monokaryons and the dikaryon. Significantly, the protein patterns obtained by translation of polysomal mRNA and total RNA were also nearly identical, indicating that all mRNAs present are actually being translated in the cells.

Conclusion

The monokaryon and the dikaryon of *Schizophyllum commune* are very similar in the proteins they synthesize; of the 700 proteins analysed only about 20 could be assigned specifically to the dikaryon whereas about 20 other proteins were present in the monokaryon but absent in the dikaryon. The regulation of this pattern by the *A* factor and the *B* factor is apparent because a homokaryotic mutant carrying mutations in both the *A* factor and the *B* factor not only exhibited a similar morphology but also a similar protein pattern as the normal heterokaryotic dikaryon. Although the *A* factor and the *B* factor each control quite different morphogenetic sequences it was surprising to find that the majority of the regulated proteins is similarly controlled by both factors.

Although the incompatibility factors control morphogenetic sequences, and thus may be compared with regulatory genes that control cytodifferentiation in multicellular organisms, the present analysis indicates that regulation of protein patterns in fungi may be achieved in different ways. Among the striking differences in fungi is the apparent absence of interspersed repetitive DNA and heterogeneous nuclear RNA. This would immediately exclude the existence of regulatory mechanisms for regulating concentrations of specific mRNAs, as proposed for higher eukaryotes (Britten & Davidson, 1969; Davidson & Britten, 1979). In relation to the two morphologically different cells examined in the present study, we were unable to detect any variations in RNA sequences that could explain the differences in protein patterns. Even at the level of polysomal mRNA no differences were found, and it appears that all the 10 000 or so different RNA sequences transcribed in these cells are actually used in translation.

The apparent absence in the monokaryon and dikaryon of a difference in the transcription of structural genes coding for abundant and rare RNA sequences agrees with recent findings in metazoa, where such genes were found to be transcribed constitutively (cf. Davidson & Britten, 1979). To explain differences in polysomal mRNAs in different cell types of these organisms, regulation of processing of the primary transcripts of structural genes is inferred. In the present fungal system, however, polysomal mRNAs of the two cell types appeared identical and no indications for RNA processing were found. It is thus concluded that the observed differences in protein patterns of the two cell types are generated at the post-translational level. The nature of such post-translational modifications and the way in which their occurrence is regulated by the incompatibility factors remain obscure. A lead may be that the proteins specifically present in the dikaryon are, on average, larger and more acid than those specifically present in the monokaryon (de Vries *et al.*, 1980).

Finally it should be stressed that the observed absence of differences in RNA only refers to the differentiation in monokaryotic and dikaryotic cell type. Preliminary experiments have revealed that differences in RNA between monokaryon and dikaryon occur when the latter is allowed to form fruiting bodies in surface cultures.

Acknowledgements. The present investigation has been carried out under the auspices of the Netherlands Foundation for Fundamental Biological Research (BION) and with financial aid from the Netherlands Organization for the Advancement of Pure Research (ZWO).

References

Bernardi, G. (1976). The mitochondrial genome of yeast: organization and recombination. *Genetics and Biogenesis of Chloroplasts and Mitochondria*, ed. Th. Bücher, pp. 503–10. Amsterdam: Elsevier.

Britten, R. J. & Davidson, E. H. (1969). Gene regulation for higher cells: a theory. *Science*, **165**, 349–57.

Davidson, E. H. & Britten, R. J. (1973). Organization, transcription, and regulation in the animal genome. *Quarterly Review of Biology*, **48**, 565–613.

Davidson, E. H. & Britten, R. J. (1979). Regulation of gene expression: possible role of repetitive sequences. *Science*, **204**, 1052–9.

de Vries, O. M. H., Hoge, J. H. C. & Wessels, J. G. H. (1980). Translation of RNA from *Schizophyllum commune* in a wheat germ and rabbit reticulocyte cell-free system: comparison of in-vitro and in-vivo products after two-dimensional gel electrophoresis. *Biochimica et Biophysica Acta*, **607**, 373–8.

de Vries, O. M. H., Hoge, J. H. C. & Wessels, J. G. H. (1980). Regulation of the pattern of protein synthesis in *Schizophyllum commune* by the incompatibility genes. *Developmental Biology*, **74**, 22–36.

Dons, J. J. M. & Wessels, J. G. H. (1980). Sequence organization of the nuclear DNA of *Schizophyllum commune*. *Biochimica et Biophysica Acta*, **607**, 385–96.

Dons, J. J. M., de Vries, O. M. H. & Wessels, J. G. H. (1979). Characterization of the genome of the basidiomycete *Schizophyllum commune*. *Biochimica et Biophysica Acta*, **563**, 100–12.

Dutta, S. K. & Chaudhury, R. K. (1975). Differential transcription of nonrepeated DNA during development of *Neurospora crassa*. *Developmental Biology*, **43**, 35–41.

Firtel, R. A. (1972). Changes in the expression of single-copy DNA during development of the cellular slime mold *Dictyostelium discoideum*. *Journal of Molecular Biology*, **66**, 363–77.

Georgiev, G. P. (1969). On the structural organization of operon and the regulation of RNA synthesis in animal cells. *Journal of Theoretical Biology*, **25**, 473–90.

Hereford, L. M. & Rosbash, M. (1977). Number and distribution of polyadenylated RNA sequences in yeast. *Cell*, **10**, 453–62.

Krumlauf, R. & Marzluf, G. A. (1979). Characterization of the sequence complexity and organization of the *Neurospora crassa* genome. *Biochemistry*, **18**, 3705–13.

Kühn, J. & Parag, Y. (1972). Protein-subunit aggregation model for self-incompatibility in higher fungi. *Journal of Theoretical Biology*, **35**, 77–91.

Lauer, G. D., Roberts, T. M. & Klotz, L. C. (1977). Determination of the nuclear DNA content of *Saccharomyces cerevisiae* and implications for the organization of DNA in yeast chromosomes. *Journal of Molecular Biology*, **114**, 507–26.

Lopez-Perez, M. J. & Turner, G. (1975). Mitochondrial DNA from *Aspergillus nidulans*. *FEBS Letters*, **58**, 159–63.

Ohja, M. & Dutta, S. K. (1977). Nuclear control of differentiation. In *The Filamentous Fungi*, ed. J. E. Smith & D. R. Berry, vol. 3, pp. 8–27. London: Edward Arnold.

Raper, C. A. (1978). Control of development by the incompatibility system in basidiomycetes. In *Genetics and Morphogenesis in the Basidiomycetes*, ed. M. N. Schwalb & P. G. Miles, pp. 3–29. London & New York: Academic Press.

Raper, J. R. (1966). *Genetics of Sexuality in Higher Fungi*. New York: Ronald Press.

Raper, J. R. & Esser, K. (1961). Antigenic differences due to the incompatibility factors in *Schizophyllum commune*. *Zeitschrift Vererbungslehre*, **92**, 439–44.

Raper, J. R. & Raper, C. A. (1973). Incompatibility factors: regulatory genes for sexual morphogenesis in higher fungi. *Brookhaven Symposia for Biology*, **25**, 19–38.

Rozek, C. E., Orr, W. C. & Timberlake, W. E. (1978). Diversity and abundance of polyadenylated RNA from *Achlya ambisexualis*. *Biochemistry*, **17**, 716–22.

Terpstra, P., Holtrop, M. & Kroon, A. M. (1977). A complete cleavage map of *Neurospora crassa* mtDNA obtained with endonucleases Eco R1 and Bam HI. *Biochimica et Biophysica Acta*, **475**, 571–88.

Timberlake, W. E. (1978). Low repetitive DNA content in *Aspergillus nidulans*. *Science*, **202**, 973–75.

Timberlake, W. E., Shumard, D. S. & Goldberg, R. B. (1977). Relationship between nuclear and polysomal RNA populations of *Achlya*. *Cell*, **10**, 623–32.

Ullrich, R. C. (1977). Isozyme patterns and cellular differentiation in *Schizophyllum*. *Molecular and General Genetics*, **156**, 157–61.

Ullrich, R. C. (1978). On the regulation of gene expression: incompatibility in *Schizophyllum*. *Genetics*, **88**, 709–22.

Wang, C. S. & Raper, J. R. (1969). Protein specificity and sexual morphogenesis in *Schizophyllum commune*. *Journal of Bacteriology*, **99**, 291–97.

Wang, C. S. & Raper, J. R. (1970). Isozyme patterns and sexual morphogenesis in *Schizophyllum commune*. *Proceedings of the National Academy of Sciences, USA*, **66**, 882–9.

Wessels, J. G. H. (1978). Incompatibility factors and the control of biochemical processes. In *Genetics and Morphogenesis in the Basidiomycetes*, ed. M. N. Schwalb & P. G. Miles, pp. 81–104. London & New York: Academic Press.

Wessels, J. G. H. & Sietsma, J. H. (1979). Wall structure and growth in *Schizophyllum commune*. In *Fungal Walls and Hyphal Growth*, ed. J. H. Burnett & A. P. J. Trinci, pp. 27–48. British Mycological Symposium 2. Cambridge University Press.

Zantinge, B., Dons, J. J. M. & Wessels, J. G. H. (1979). Comparison of poly(A)-containing RNAs in different cell types of the lower eukaryote *Schizophyllum commune*. *European Journal of Biochemistry*, **101**, 251–60.

Co-ordination of transcription with translation in yeast

S. G. OLIVER

Applied Molecular Biology Group, Department of Biochemistry, University of Manchester Institute of Science and Technology, Manchester M6O 1QD, UK

Introduction

The nucleus contains the bulk of the genetic information of the fungal cell and therefore plays the dominant role in controlling growth and development. However, it cannot direct these activities without receiving information from the cytoplasm as to the current metabolic state of the cell. Such nucleo-cytoplasmic interactions are critical in determining gene activity and among the most important of them is the regulation of the rate of RNA synthesis in the nucleus to match the demands of protein synthesis in the cytoplasm. The cost, in both energetic and material terms, of synthesizing RNA is so high that the co-ordination of transcription with translation is one of the central homeostatic controls of the cell.

The usual strategy employed to study these general controls on RNA synthesis is to change the organism's nutrient environment, such that an increase or decrease in protein synthetic rate is brought about. The manner in which the rate and pattern of RNA synthesis is changed in response to this is then analysed. This account will therefore be divided into two parts corresponding to the two types of nutritional transition studied: starvation and shift-up.

Starvation experiments

Studies with *Escherichia coli* have led to the definition of a classic response, the stringent response, which occurs in auxotrophic bacteria when they are deprived of required amino acids. The response of the cell to its inability to carry out net protein synthesis any longer is to reduce drastically its rate of RNA synthesis. The minimal requirements for a stringent response mechanism can be outlined as follows:

SIGNAL ⟶ TARGET ⟶ EFFECT

There has to be some *signal* which the cell monitors to determine it is being starved. In the case of amino acid starvation this signal could be the absence of the required amino acid in the cytoplasmic pool, the decrease in charged tRNA, the increase in uncharged tRNA or the reduction in the rate of protein synthesis itself. This signal must have its impact on some *target*, the RNA polymerase complex or the promoter sites within the DNA, in order to elicit the *effect* of reducing the rate of RNA synthesis. The simplest mechanism of stringent control that could be postulated would therefore be one in which the uncharged tRNA that accumulates during amino acid starvation binds to the RNA polymerase complex and prevents it from synthesizing RNA. Studies *in vitro* (Bremer, Yegian & Konrad, 1965; Spassky *et al.*, 1979) have demonstrated that tRNA binds to and inhibits *Escherichia coli* RNA polymerase. However there is little or no difference between the binding coefficients of charged and uncharged tRNA, and therefore this process cannot mediate the stringent response. Experiments *in vivo* with cells having elevated amounts of tRNA (Ezekiel & Valulis, 1966) have confirmed that direct interaction between tRNA and RNA polymerase cannot be the mechanism of stringent control.

It has emerged (for reviews see Edlin & Broda, 1968 and Nomura, Tissières & Lengyel, 1974) that the mechanism of stringent control in *Escherichia coli* is a complex two-stage system:

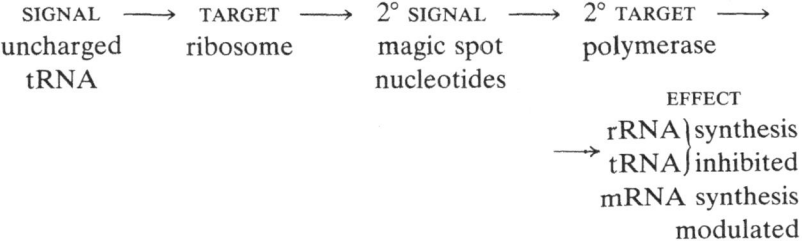

The signal of starvation, uncharged tRNA, binds to the acceptor site of the ribosome and stimulates a ribosomal protein (the stringent factor) to synthesize a series of unusual guanosine nucleotides – the 'magic spots' ppGpp, pppGpp and pGpp (Cashel & Gallant, 1974; Pao & Gallant, 1979). It is these spots, the secondary signals, which apparently direct the *E. coli* RNA polymerase complex to reduce the rate of synthesis of rRNA and tRNA and to change its pattern of mRNA synthesis. The synthesis of some messages, such as those for ribosomal proteins (Dennis & Fiil, 1979) is repressed, whereas that of others such as those encoding amino acid biosynthetic enzymes is stimulated (Stephens, Artz & Ames, 1975).

The situation in *Escherichia coli* has been discussed in some detail, not because it provides a good model for the control of RNA synthesis in the fungi (current data suggest that it does not), but because an appreciation of the reasons for the use of this two-level control system by *E. coli* may give us some clue as to the alternative mechanisms open to eukaryotic cells such as the fungi. The use of secondary signals or messengers is one which is familiar from the hormone control systems of higher organisms. There, such systems are used to amplify a weak signal, to transmit a signal to a remote target site or to enable a single signal to elicit a pleiotropic response from the body. It seems likely that *E. coli* also uses a system of secondary signals (the magic spot nucleotides) to elicit a pleiotropic response. A more direct form of control is denied it due to the bacterium's structural and functional simplicity. In *E. coli* there is no separation of the site of transcription from the site of translation, and there is only a single RNA polymerase complex which must synthesize all three classes of RNA. The experimental organism used in our studies on the stringent response in the fungi, *Saccharomyces cerevisiae,* will now be discussed in this context.

Yeast – structural and functional organization

The yeast cell contains a number of quantitatively and qualitatively distinct genomes. The nuclear genome, 2 μ-circular DNA and the double-stranded RNA (dsRNA) genomes associated with the yeast virus-like particle (VLP) all use the cytoplasmic ribosomes to express their genes. The mitochondrial genome uses its own specific translation system for gene expression.

The nuclear genome of yeast has a total weight of *c.* 10^{10} daltons of DNA. This genome is divided into 17 chromosomes which have been defined by linkage studies. In contrast to the genome of *Escherichia coli*, the genes of the yeast chromosomes are not clustered according to their functions. There are no operons in yeast, and complex loci, such as *his* 4 on chromosome III, have been shown to encode single, multifunctional polypeptides (Keesey, Bigelis & Fink, 1979). The genes that encode tRNA molecules, which have been defined by suppressor mutations, are also scattered throughout the yeast genetic map. The only set of clustered genes is that encoding rRNA. There are *c.* 140 rRNA genes (Schweizer, MacKechnie & Halvorson, 1969) and they represent the only highly reiterated gene set in the yeast chromosomes.

The 140 genes are clustered together in an extensive tandem array (Cramer *et al.*, 1977; Petes, Hereford & Skryabin, 1978; Kaback & Davidson, 1980) and this array appears to be located on a single chromosome, number XII (Petes, 1979*a, b*). These rRNA genes are also physically separated from the rest of the nucleus in a nucleolus-like structure (Gordon, 1977). The yeast nucleolus is a rather ill-defined organelle, however there is some preliminary evidence for the membrane association of the rRNA genes (Walmsley & Oliver, 1980).

In addition to this structural compartment of the yeast genome there is also a functional compartmentation of transcriptional activity. The yeast VLP and the mitochondrion both have their own specific, and rather simple, RNA polymerase enzymes (Scragg, 1976; Herring & Bevan, 1977). The yeast nucleus contains three RNA polymerase complexes which each have different transcriptional specificities. RNA polymerase I is responsible for the synthesis of the 35 S precursor ribosomal RNA from which the 25 S, 18 S and 5.8 S mature rRNA species are derived by endonucleolytic cleavage (Udem & Warner, 1972). RNA polymerase II is responsible for the transcription of the structural genes into mRNA. There are at least two forms of this enzyme (IIA and IIB; Buhler *et al.*, 1976) but the functional significance of this is not understood. Finally, RNA polymerase III is responsible for the synthesis of the tRNAs and of the 5 S rRNA which is transcribed from the opposite strand of the DNA duplex to the 35 S pre-rRNA (Maxam *et al.*, 1977).

The greater structural and functional complexity of the yeast genomes and their transcriptional apparatuses may mean that more flexible and direct means of controlling RNA synthesis are available to yeast cells than can be achieved by simpler, prokaryotic organisms. The effect of starvation on the synthesis of RNA from the different genomes of yeast will now be considered.

Starvation studies with yeast

The existence of a stringent response in the yeast, *Saccharomyces cerevisiae* has been demonstrated by studying the level of total RNA synthesis in auxotrophic strains starved for required amino acids (Schmidt *et al.*, 1956; Ycas & Brawerman, 1957; McLaughlin, Magee & Hartwell, 1969; Wehr & Parks, 1969; Roth & Dampier, 1972). The overall rate of RNA synthesis in amino acid-starved yeast falls to less than one-third of that in control cells (Oliver & McLaughlin, 1977). However, the severity of the restriction of RNA synthesis and the time lag before it is exerted depends on the pool size of the amino acid starved for. A rapid shut-down in RNA synthesis can be obtained by starving for tyrosine (Shulman, Sripati & Warner, 1977) or leucine (Clare & Oliver, 1979, and unpublished data) or by starving for more than one amino acid at once (Roth & Dampier, 1972; Oliver & McLaughlin, 1977). A stringent response may also be exerted in yeast by switching a mutant strain which has a temperature-sensitive aminoacyl-tRNA synthetase to its restrictive temperature (McLaughlin *et al.*, 1969; Oliver & McLaughlin, 1977). In this case the stringent response may be exerted in a rich medium which contains all the amino acids required by the cells. This result leads to the important conclusion that it cannot be the absence of amino acids *per se* that signals the stringent response in yeast. The organism must monitor starvation by measuring tRNA charging or the rate of protein synthesis itself.

An early result which implied that tRNA charging was the important step was the observation that the protein synthesis inhibitor cyclohexi-mide stimulated RNA synthesis in starved yeast cells (de Kloet, 1966; Foury & Goffeau, 1973; Gros & Pogo, 1974). However, this finding seemed to contradict the observation that the same inhibitor reduced the rate of RNA synthesis in control (unstarved) cells (Roth & Dampier, 1972). This problem was not resolved until the effect of the stringent response on the synthesis of different classes of RNA mole-cules was studied.

The regulation of nuclear RNA synthesis

The detailed study of how the synthesis of RNAs encoded in the yeast nucleus responds to amino acid starvation has revealed two important differences between yeast and *Escherichia coli*. The first is that the direct inhibition of protein synthesis by the use of drugs such as cycloheximide results in a reduction in the rate of RNA synthesis (Roth & Dampier, 1972; Shulman *et al.*, 1977; Kelker & Pogo, 1980). This result implies that it is the reduction in the rate of protein synthesis itself, and not the level of uncharged tRNA, which signals the starved condition in yeast cells.

The paradoxical effect of cycloheximide in apparently causing the relaxation of RNA synthesis in starved cells is explained by the second important finding: that the stringent control of rRNA and tRNA synthesis in yeast is markedly non-co-ordinate. The initial response of the yeast cell to amino acid starvation is to repress the synthesis of the ribosomal RNAs whilst allowing tRNA synthesis to continue (Shulman *et al.*, 1977; Oliver & McLaughlin, 1977; Kelker & Pogo, 1980). A similar response is elicited by raising a temperature-sensitive aminoacyl-tRNA synthetase mutant to its restrictive temperature (Oliver & McLaughlin, 1977). Protein synthesis inhibitors act at the ribosome in starved cells to stimulate the synthesis of tRNA alone (Oliver & McLaughlin, 1977); there is little or no effect on rRNA synthesis (Oliver & McLaughlin, 1977; Kelker & Pogo, 1980). This effect of translation inhibitors on tRNA synthesis is not seen in the aminoacyl-tRNA synthetase mutant at the restrictive temperature (Oliver & McLaughlin, 1977) and this implies that these drugs act by promoting the trickle-charging of tRNA. Direct experiments (Clare & Oliver, 1980, and unpublished) have confirmed that translation inhibitors do promote trickle-charging of tRNA in starved yeast cells. Edeine and 2-(4-methyl-2,6-dinitroanilino)-*N*-methyl propionamide, two inhibitors of an early step in initiation, which did not stimulate RNA synthesis in starved cells (Oliver & McLaughlin, 1977), were nevertheless found to promote trickle-charging (Clare & Oliver, 1980, and unpublished). It may be that these two inhibitors have some non-specific effect on RNA synthesis or the result might suggest that the 80 S ribosome–mRNA complex has some essential role to play in the regulation of RNA synthesis.

The distinct behaviour of tRNA synthesis in starved cells leads us to the conclusion that it is the subject of distinct and separate controls. Its synthesis continues in starved cells and is stimulated by the promotion of trickle-charging due to the action of protein synthesis inhibitors at the ribosome. Our working model of the control of tRNA synthesis is therefore one of autoregulation by charged tRNA through a positive feed-back loop. The effect of this regulatory scheme is promoter-specific rather than polymerase-specific since 5 S rRNA and tRNA are synthesized non-co-ordinately (Oliver & McLaughlin, 1977), although they are both made by RNA polymerase III. This last point has been contested by Kelker & Pogo (1980) who find that the synthesis of 5 S rRNA and tRNA respond co-ordinately during starvation.

The search for a mechanism of rRNA control
The discovery that the guanosine polyphosphates acted as pleiotropic effectors in the control of RNA synthesis in *Escherichia coli* led to a search for similar molecules in yeast and other fungi. Many of the early experiments in this area involved determining whether ribosomes from various organisms could produce ppGpp either on their own or when complemented with *E. coli* stringent factor or ribosomal

subunits. Ribosomes from yeast, *Chlamydomonas reinhardtii* and mammalian cells were used in the *in vitro* system. The general conclusion was that ribosomes from the eukaryotic cytoplasm were unable to produce ppGpp but that organelle ribosomes (from chloroplasts and mitochondria) could synthesize the 'magic spot' (Richter, 1973; Sy *et al.*, 1974).

Experiments *in vivo* have yielded similar results. Neither *Saccharomyces cerevisiae* (Kudrna & Edlin, 1975, 1980; Lusby & McLaughlin, unpublished) or *Neurospora crassa* (Alberghina *et al.*, 1973; Buckel & Bock, 1973) produce ppGpp. A temperature-sensitive aminoacyl-tRNA synthetase mutant of yeast also failed to produce the 'magic spot' compounds at the restrictive temperature (E. W. Lusby & S. G. Oliver, unpublished). Pao *et al.* (1977) demonstrated the synthesis of ppGpp in heat-shocked yeast. The production of the compound was inhibited by oxytetracycline and it was therefore concluded that ppGpp was produced by mitochondrial ribosomes. This is supported by the failure to demonstrate ppGpp synthesis in petite yeast (J. R. Warner, personal communication).

There is some evidence that the production of highly phosphorylated nucleotides may be associated with differentiation in the lower eukaryotes. Klein (1974) demonstrated that the cellular slime mould, *Dictyostelium discoideum* produced ppGpp during the starvation-induced transition from the amoeboid to the grex form. Work from LéJohn's laboratory (e.g. LéJohn *et al.*, 1975) has suggested that the production of diguanosine nucleotides is associated with sporulation in *Achlya*, although this has been disputed (Warner *et al.*, 1977). Finally Rhaese's group have shown that the production of adenosine tetra-, penta- and hexaphosphates accompanies ascospore formation in *Saccharomyces cerevisiae* (Rhaese *et al.*, 1979).

If there are no low molecular weight pleiotropic effectors involved in the stringent control of transcription in yeast and other fungi, we must search for some other mechanism. Any hypothesis should preferably account for the non-coordinate nature of the control and also for the fact that the direct inhibition of protein synthesis, as well as amino acid starvation, provokes the stringent response. An attractive idea (Shulman & Warner, 1978) was that the reduction in the rate of rRNA synthesis was caused by the deficiency in new ribosomal proteins which would result from the absence of protein synthesis. It was possible to test this idea directly by using a yeast mutant, ts-368, which is temperature-sensitive for the synthesis of ribosomal proteins (Hartwell,

McLaughlin & Warner, 1970). However, it was found that whilst the 35 S pre-rRNA was not processed to the mature species at the restrictive temperature in this mutant, its rate of synthesis was undiminished (Shulman & Warner, 1978). Thus the synthesis of the ribosomal precursor RNA is not dependent on the synthesis of ribosomal proteins nor on its own further processing.

I would suggest that the idea that tRNA is directly involved in the control of rRNA synthesis is worthy of examination. The direct interaction of tRNA with the RNA polymerase complex has been demonstrated in *Escherichia coli* systems but has been shown to be of no physiological significance (see above). However, a sufficiently discriminatory form of control is possible in the eukaryotes due to the functional compartmentation of RNA synthetic capactiy. Thus tRNA could inhibit the synthesis of the 35 S pre-rRNA by RNA polymerase I whilst having no effect on (or even stimulating) the activity of RNA polymerases II and III. Separation of the protein synthetic and regulatory functions of tRNA might be achieved by having a precursor molecule or a processing fragment responsible for regulation. The direct involvement of tRNA in the repression of rRNA synthesis would also provide an explanation for the continued synthesis of tRNA during starvation.

The regulation of double-stranded RNA synthesis

The double-stranded (ds) RNA molecules associated with the yeast VLP are interesting in the context of regulation since they represent independent genomes outside the nucleus and use the same translation system as the nuclear genome (see review by Wickner, 1979). We (Clare & Oliver, 1979) have studied the synthesis of P1 or L dsRNA, the 3.4×10^6 mol wt (Holm *et al.*, 1978) genome of the VLP (Hopper *et al.*, 1977). During amino acid starvation the pattern of P1 dsRNA synthesis is similar to that of rRNA. The rate of synthesis is markedly reduced and is not stimulated by the addition of the translation inhibitor cycloheximide. Moreover, the cycloheximide-induced inhibition of protein synthesis in control cells also causes a reduction in the rate of P1 dsRNA synthesis.

Although the response of P1 dsRNA synthesis and rRNA synthesis to the reduction in protein synthetic rate, brought about by either amino acid starvation or inhibitor treatment, are very similar it should not be assumed that the control mechanisms are the same. It is possible that the continual synthesis of dsRNA replication enzymes is required as pre-existing molecules are removed into completed virions. Alternatively, if the dsRNA is itself replicated within an intact VLP (see Buck & Ratti, 1975) then the continual synthesis of host cell proteases may be required to release the daughter genomes from the virus-like particle.

The regulation of mitochondrial RNA synthesis

The transcription and translation of mitochondrial DNA is carried out by specific systems located within the mitochondrion and separated from the rest of the cell by the inner and outer mitochondrial membranes. However, the mitochondrial RNA polymerase and all but one of the mitochondrial ribosomal proteins (see review by Borst & Grivell, 1978) are synthesized, not in the mitochondrion, but in the cytoplasm. This, together with the fact that mitochondrial biogenesis is a co-operative activity of both the nuclear and mitochondrial genomes, suggests that nuclear and mitochondrial transcription must be integrated in some way.

Ray & Butow (1979a, b) have made a very careful study of the regulation of mitochondrial rRNA synthesis in yeast. They have found that it is very tightly co-ordinated with nuclear rRNA synthesis. It is under stringent control from amino acid starvation and also when cytoplasmic, but not mitochondrial, protein synthesis is inhibited by cycloheximide. Chloramphenicol, a specific inhibitor of mitochondrial protein synthesis, did not significantly reduce the rate of mitochondrial RNA synthesis. Whilst this is a similar response to that of *Escherichia coli*, it was found that chloramphenicol did not stimulate the phenotypic relaxation of mitochondrial rRNA synthesis in starved cells. This result probably indicates that the mitochondrial and cytoplasmic amino acid pools are in rapid equilibrium such that it is impossible to build up a pool of charged tRNA exclusively within the mitochondrion. The effects of tetracycline (which prevents the binding of tRNA to the ribosome) on mitochondrial rRNA synthesis in starved cells might prove informative, especially in view of the results of Pao *et al.* (1977) on ppGpp synthesis (see above).

Ray & Butow (1979b) did, however, manage to discover conditions under which the synthesis of mitochondrial and cytoplasmic rRNAs was uncoupled. If the initiation of cytoplasmic protein synthesis is inhibited using the temperature-sensitive mutant ts-187 (Hartwell & McLaughlin, 1969) then mitochondrial rRNA synthesis continued at *c.* 75% of the level found in control cells. Indeed, using previously published data (Hartwell & McLaughlin, 1968; McLaughlin *et al.*, 1969), these workers were able to point to an inverse correlation between the proportion of cytoplasmic ribosomes found in polysomes and the rate of mitochondrial rRNA synthesis. There is some evidence (Oliver & McLaughlin, 1977) that the rRNA–ribosome complex is involved in the regulation of tRNA synthesis and the two phenomena may be interrelated, representing different facets of an overall control system.

Shift-up studies with yeast

A complementary way of using nutritional transitions to examine the co-ordination of transcription with translation is to follow how the cell adjusts to conditions that induce an increase in its rate of growth and division. Early experiments of this sort with yeast (Wehr & Parks, 1969) employed ill-defined transitions from poor to rich media. More recently Waldron (1977) has studied the shift-up produced by adding a mixture of L-amino acids to a culture which contained proline as nitrogen source. The amino acid catabolic enzymes are not repressed in such a medium and others (Ludwig, Oliver & McLaughlin, 1977) have preferred to use a regime in which amino acids are added to a minimal medium containing a repressing concentration of ammonium so that the added amino acids do not act as an extra source of nitrogen. This type of shift produced a 19% increase in growth rate, from a doubling-time $(T_d) = 3$ h $(\mu = 0.23$ h$^{-1})$ to a $T_d = 2.4$ h $(\mu = 0.29$ h$^{-1})$. This increase was not observed until some 120 min after amino acids were added to the minimal medium.

The accumulation of the different classes of macromolecules was followed during this transitional period between the addition of amino acids and an observable increase in growth rate. The rate of protein accumulation was found to increase by 19% immediately after the addition of amino acids while there was a lag of 15–20 min before any increase in RNA accumulation was observed. This result indicates that there is spare translational capacity available at the lower growth rate and that an increase in ribosome number is not required to increase the rate of protein synthesis. This agrees with the finding that ribosome efficiency in yeast is lower at slower growth rates (Boehlke & Friesen, 1975; Waldron & Lacroute, 1975). The rate of DNA accumulation also increased by c. 20% but this occurred much later, at 100–120 min after the addition of amino acids. It may be that the net rate of DNA synthesis by a yeast cell can be adjusted only at a particular point in the cell cycle (see Hartwell *et al.*, 1974; Nurse, next chapter) and that this determines the lag before an increased rate of cell division is recorded.

The pattern of RNA synthesis during a shift-up in yeast is more complex than that for DNA or protein synthesis. Similar data for RNA accumulation have been obtained with both the proline → amino acids (Waldron, 1977) and the ammonium → amino acids (Ludwig *et al.*, 1977) shifts. In the latter experiments there was a 15–20 min lag before any increase in the rate of RNA accumulation was seen. The rate then increased by 74% over that in the control culture. This very high rate of RNA accumulation was maintained until 120 min after the addition of amino acids. The rate was then reduced to only 19% above that found in the control culture. Thus after 120 min the rates of culture growth, DNA, RNA and protein accumulation had all increased by the same proportion and the culture can be considered to have achieved a new state of balanced growth.

The perplexing feature of the pattern of macromolecular synthesis seen during the shift-up transition from a lower to a higher rate of balanced growth is that there is an overproduction of RNA during the transition, even though the cell has an adequate supply of ribosomes to increase immediately its rate of protein synthesis. In our studies (Ludwig *et al.*, 1977) we examined the effect of the shift-up on the rate of synthesis of the different classes of stable RNA molecules. The dramatic increase in RNA accumulation during the transition was found to be solely due to the synthesis of the rRNA species derived from the 35 S precursor molecule (25 S, 18 S and 5.8 S rRNAs). The rate of synthesis of 5 S rRNA and of tRNA increased much later and to a lesser extent. The differential control of the synthesis of these two classes of molecule may readily be achieved in yeast since they are synthesized by different RNA polymerase complexes; RNA polymerase I produces the 35 S pre-rRNA, and RNA polymerase III is responsible for the synthesis of both tRNA and 5 S rRNA. Thus the functional compartmentation of RNA synthetic capacity again permits greater flexibility of control.

The question remains as to why there is this great increase in the rate of 35 S pre-rRNA synthesis during the shift-up transition. Ribosome supply itself is not limiting the cell's ability to shift to the higher growth rate. Indeed, the overproduction of rRNA is confined to the species derived from the 35 S precursor and does not include 5 S rRNA, which is also essential to the construction of a functional ribosome. This suggests that the overproduction of 35 S pre-rRNA during the shift-up transition has some regulatory function and that the synthesis or processing of the precursor is an important pre-requisite for progress through the cell cycle. In this way the rate of synthesis of the 35 S pre-rRNA or the cellular concentration of the species derived from it might control the rate of cell growth and division.

The idea that pre-rRNA synthesis has a regulatory function in yeast cell growth and division has been pioneered and promoted by the work of Johnston and his collaborators (see Nurse, pp. 331–45 for an alternate view of their work). They have found that a number of inhibitors, such as the zinc chelators *o*-phenanthroline and 8-hydroxyquinoline (Johnston & Singer, 1978), the methionine analogue ethionine (Singer, Johnston & Bedard, 1978) and the DNA gyrase inhibitor (Sugino *et al.*, 1977) nalidixic acid (Singer and Johnston, 1979), have their immediate effect on the synthesis of rRNA and also cause the transient arrest of cells in the G1 phase of the cell cycle. They have also found that the temporary inhibition of rRNA synthesis induced by temperature shock in yeast (Warner & Udem, 1972) is also accompanied by G1 arrest. This temperature shock effect is a complication since it means that the further investigation of the role of pre-rRNA in cell-cycle control will have to involve the isolation of conditional lethal mutants, dependent on some condition other than high temperature for the expression of the mutant phenotype.

Conclusions

The co-ordination of transcription with translation in eukaryotes such as the fungi is dominated by the compartmentation of RNA synthetic capacity within the cell. There are two forms of compartmentation. (1) Structural compartmentation comprises the separation of the site of transcription from the site of translation by the nuclear membrane and also the subdivision of the nucleus itself into nucleolus and nucleoplasm. (2) Functional compartmentation is brought about by having three RNA polymerase complexes within the nucleus which each have different transcriptional specificities.

Studies of the effect of nutritional transitions have shown that these divisions permit fungi to have a very flexible means of regulating RNA synthesis. The greater structural and functional complexity of the fungi may mean that they are able to use much more direct systems of control than are the bacteria. They may allow the RNA molecules themselves to be important regulatory components, determining not only their own rate of synthesis but also that of the growth and division of the entire cell.

Acknowledgement. Work in my own laboratory has been supported by SRC grants GR/A/39752 and GR/B/26114 and by a Scientific Investigations grant from the Royal Society.

References

Alberghina, F. A. M., Schiaffonati, L., Zardi, L. & Sturani, E. (1973). Lack of guanosine tetraphosphate accumulation during inhibition of RNA synthesis in *Neurospora crassa*. *Biochimica et Biophysica Acta*, **312**, 435–9.

Boehlke, K. W. & Friesen, J. D. (1975). Cellular content of ribonucleic acid and protein in *Saccharomyces cerevisiae* as a function of growth rate. *Journal of Bacteriology*, **121**, 429–39.

Borst, P. & Grivell, L. A. (1978). The mitochondrial genome of yeast. *Cell*, **15**, 705–23.

Bremer, H., Yegian, C. & Konrad, M. (1965). Inactivation of purified *Escherichia coli* RNA polymerase by transfer RNA. *Journal of Molecular Biology*, **16**, 94–103.

Buck, K. W. & Ratti, G. (1975). A model for the replication of double-stranded ribonucleic acid mycoviruses. *Biochemical Society Transactions*, **3**, 542–4.

Buckel, P. & Bock, A. (1973). Lack of accumulation of unusual guanosine nucleotides upon amino acid starvation in two eukaryotic organisms. *Biochimica et Biophysica Acta*, **324**, 184–7.

Buhler, J. M., Iborra, F., Sentenac, A. & Fromageot, P. (1976). The presence of phosphorylated subunits in yeast RNA polymerases A and B. *FEBS Letters*, **71**, 37–41.

Cashel, M. & Gallant, J. (1974). Cellular regulation of guanosine tetraphosphate and pentaphosphate. In *Ribosomes*, ed. M. Nomura, A. Tissières & P. Lengyel, pp. 735–45. New York: Cold Spring Harbor Laboratory.

Clare, J. J. & Oliver, S. G. (1979). The regulation of RNA synthesis in yeast. IV. Synthesis of double-stranded RNA. *Molecular and General Genetics*, **171**, 161–6.

Clare, J. J. & Oliver, S. G. (1980). Protein synthesis inhibitors promote the trickle-charging of tRNA in starved yeast cells. (Abstract.) *Transactions of the British Mycological Society*, in press.

Cramer, J. H., Farrelly, F. W., Barnitz, J. & Rownd, R. H. (1977). Construction and restriction endonuclease mapping of hybrid plasmids containing *Saccharomyces cerevisiae* ribosomal DNA genes. *Molecular and General Genetics*, **151**, 229–44.

DeKloet, S. R. (1966). Ribonucleic acid synthesis in yeast. The effect of cycloheximide on the synthesis of ribonucleic acid in *Saccharomyces carlsbergensis*. *Biochemical Journal*, **99**, 566–81.

Dennis, P. P. & Fiil, N. P. (1979). Transcriptional and post-transcriptional control of RNA polymerase and ribosomal protein genes cloned on composite ColEl plasmids in the bacterium *Escherichia coli*. *Journal of Biological Chemistry*, **254**, 7540–47.

Edlin, G. & Broda, P. (1968). Physiology and genetics of the ribonucleic acid control locus of *Escherichia coli*. *Bacteriological Reviews*, **32**, 206–26.

Ezekiel, D. H. & Valulis, B. (1966). Control of ribonucleic acid synthesis in *Escherichia coli* cells with altered transfer RNA concentration. *Biochimica et Biophysica Acta*, **129**, 123–39.

Foury, F. & Goffeau, A. (1973). Stimulation of RNA synthesis in yeast by cycloheximide and 3′,5′-cyclic AMP. *Nature New Biology*, **245**, 44–7.

Gordon, C. N. (1977). Chromatin behaviour during the mitotic cell cycle of *Saccharomyces cerevisiae. Journal of Cell Science,* **84**, 81–93.

Gros, K-J. & Pogo, A. O. (1974). Control mechanisms of ribonucleic acid synthesis in eukaryotes. *Journal of Biological Chemistry,* **249**, 568–76.

Hartwell, L. H., Culotti, J., Pringle, J. R. & Reid, B. J. (1974) Genetic control of the yeast cell cycle. *Science,* **183**, 46–51.

Hartwell, L. H. & McLaughlin, C. S. (1968). Temperature-sensitive mutants of yeast exhibiting a rapid inhibition of protein synthesis. *Journal of Bacteriology,* **96**, 1664–71.

Hartwell, L. H. & McLaughlin, C. S. (1969). A mutant of yeast apparently defective in the initiation of protein synthesis. *Proceedings of the National Academy of Sciences, USA,* **62**, 468–74.

Hartwell, L. H., McLaughlin, C. S. & Warner, J. R. (1970). Identification of ten genes that control ribosome formation in yeast. *Molecular and General Genetics,* **109**, 42–56.

Herring, A. J. & Bevan, E. A. (1977). Yeast virus-like particles possess a capsid-associated single-stranded RNA polymerase. *Nature, London,* **268**, 464–6.

Holm, C. A., Oliver, S. G., Newman, A. M., Holland, L. E., McLaughlin, C. S., Wagner, E. K. & Warner, R. C. (1978). The molecular weight of yeast P1 double-stranded RNA. *Journal of Biological Chemistry,* **253**, 8332–6.

Hopper, J. E., Bostian, K. A., Rowe, L. B. & Tipper, D. J. (1977). Translation of the L-species dsRNA genome of the killer-associated virus-like particles of *Saccharomyces cerevisiae. Journal of Biological Chemistry,* **252**, 9010–17.

Johnston, G. C. & Singer, R. A. (1978). RNA synthesis and the control of cell division in the yeast, *S. cerevisiae. Cell,* **14**, 951–8.

Kaback, D. & Davidson, N. (1980). Organization of the ribosomal RNA gene cluster in the yeast *Saccharomyces cerevisiae. Journal of Molecular Biology,* **138**, 745–54.

Keesey, J. R., Jr., Bigelis, R. & Fink, G. R. (1979). The product of the *his 4* gene cluster in *Saccharomyces cerevisiae. Journal of Biological Chemistry,* **254**, 7727–37.

Kelker, H. C. & Pogo, A. O. (1980). The stringent and relaxed phenomena in *Saccharomyces cerevisiae. Journal of Biological Chemistry,* **255**, 1526–35.

Klein, C. (1974). Presence of magic spots in *Dictyostelium discoideum. FEBS Letters,* **38**, 149–52.

Kudrna, R. & Edlin, G. (1975). Nucleotide pools and regulation of RNA synthesis in yeast. *Journal of Bacteriology,* **121**, 740–2.

LéJohn, H. B., Cameron, L. E., McNaughton, D. R. & Klassen, G. R. (1975). Diguanosine nucleotides in fungi that regulate RNA polymerases isolated and partially characterised. *Biochemical and Biophysical Research Communications,* **66**, 460–7.

Ludwig, J. R., II, Oliver, S. G. & McLaughlin, C. S. (1977). The regulation of RNA synthesis in yeast. II. Amino acids shift-up experiments. *Molecular and General Genetics,* **158**, 117–22.

Lusby, E. W., Jr. & McLaughlin, C. S. (1980). The effect of amino acid starvation on a major, acid-soluble compound in *Saccharomyces cerevisiae. Molecular and General Genetics,* **179**, 699–701.

Maxam, A. M., Tizard, R., Skryabin, K. G. & Gilbert, W. (1977). Promoter region for yeast 5 S rRNA, *Nature, London,* **267**, 643–45.

McLaughlin, C. S., Magee, P. T. & Hartwell, L. H. (1969). Role of isoleucyl transfer ribonucleic acid synthesis and enzyme repression in yeast. *Journal of Bacteriology,* **109**, 773–79.

Nomura, M., Tissières, A. & Lengyel, P. (eds.) *Ribosomes.* New York: Cold Spring Harbor Laboratory.

Oliver, S. G. & McLaughlin, C. S. (1977). The regulation of RNA synthesis in yeast. I. Starvation experiments. *Molecular and General Genetics,* **154,** 145–53.

Pao, C. C., Paietta, J. & Gallant, J. (1977). Synthesis of guanosine tetraphosphate (magic spot I) in *Saccharomyces cerevisiae. Biochemical and Biophysical Research Communications,* **74,** 314–22.

Pao, C. C. & Gallant, J. (1979). A new nucleotide involved in the stringent response in *Escherichia coli.* Guanosine 3′-diphosphate -2′- monophosphate. *Journal of Biological Chemistry,* **254,** 688–92.

Petes, T. D. (1979a). Yeast ribosomal genes are located on chromosome XII. *Proceedings of the National Academy of Sciences, USA,* **76,** 410–14.

Petes, T. D. (1979b). Meiotic mapping of yeast ribosomal deoxyribonucleic acid on chromosome XII. *Journal of Bacteriology,* **138,** 185–92.

Petes, T. D., Hereford, L. M. & Skryabin, K. G. (1978). Characterization of two types of yeast ribosomal DNA genes. *Journal of Bacteriology,* **134,** 295–305.

Ray, D. B. & Butow, R. A. (1979a). Regulation of mitochondrial ribosomal RNA synthesis in yeast. I. In search of a relaxation of stringency. *Molecular and General Genetics,* **173,** 227–38.

Ray, D. B. & Butow, R. A. (1979b). Regulation of mitochondrial ribosomal RNA synthesis in yeast. II. Effects of temperature-sensitive mutants defective in cytoplasmic protein synthesis. *Molecular and General Genetics,* **173,** 239–47.

Rhaese, H. J., Schekel, R., Groscurth, R. & Stomminger, G. (1979). Studies on the control of development. Highly phosphorylated nucleotides (HPN) are correlated with ascospore formation in *Saccharomyces cerevisiae. Molecular and General Genetics,* **170,** 57–65.

Richter, D. (1973). Formation of guanosine tetraphosphate (magic spot I) in homologous and heterologous systems. *FEBS Letters,* **34,** 291–4.

Roth, R. M. & Dampier, C. (1972). Dependence of ribonucleic acid synthesis on continuous protein synthesis in yeast. *Journal of Bacteriology,* **109,** 773–9.

Schmidt, G., Seraidarian, K., Greenbaum, L. M., Hickey, M. & Thanhauser, S. J. (1956). The effect of certain conditions on the formation of purines and ribonucleic acid in baker's yeast. *Biochimica et Biophysica Acta,* **20,** 135–49.

Schweizer, E., MacKechnie, C. & Halvorson, H. O. (1969). The redundancy of ribosomal and transfer RNA genes in *Saccharomyces cerevisiae. Journal of Molecular Biology,* **40,** 261–77.

Scragg, A. H. (1976). The isolation and properties of a DNA-directed RNA polymerase from yeast mitochondria. *Biochimica et Biophysica Acta,* **442,** 331–42.

Shulman, R. W., Sripati, C. E. & Warner, J. R. (1977). Non-coordinated transcription in the absence of protein synthesis in yeast. *Journal of Biological Chemistry,* **252,** 1344–9.

Shulman, R. W. & Warner, J. R. (1978). Ribosomal RNA transcription in a mutant of *Saccharomyces cerevisiae* defective in ribosomal protein synthesis. *Molecular and General Genetics,* **161,** 221–3.

Singer, R. A. & Johnston, G. C. (1979). Nalidixic acid causes a transient G1 arrest in the yeast *Saccharomyces cerevisiae. Molecular and General Genetics,* **176,** 37–39.

Singer, R. A., Johnston, G. C. & Bedard, D. (1978). Methionine analogues and cell division regulation in the yeast *Saccharomyces cerevisiae. Proceedings of the National Academy of Sciences, USA,* **73,** 6083–7.

Spassky, A., Busby, S. J. W., Danchin, A. & Bac, H. (1979). On the binding of tRNA to *Escherichia coli* RNA polymerase. *European Journal of Biochemistry,* **99,** 187–201.

Stephens, J. C., Artz, S. W. & Ames, B. N. (1975). Guanosine
5'-diphosphate-3'-diphosphate (ppGpp): positive effector for the histidine
operon transcription and general signal for amino acid control. *Proceedings of
the National Academy of Sciences, USA*, **72**, 4389–93.

Sugino, A., Peebles, C. L., Kreuzer, K. N. & Cozzarelli, N. R. (1977). Mechanism of
action of nalidixic acid: purification of *Escherichia coli nal A* gene product and
its relationship to DNA gyrase a novel nicking-closing enzyme. *Proceedings of
the National Academy of Sciences, USA*, **74**, 4767–71.

Sy, J., Nam-Hai, C., Ogawa, Y. & Lipmann, F. (1974). Ribosome-specificity for the
formation of guanosine polyphosphates. *Biochemical and Biophysical Research
Communications*, **56**, 611–16.

Udem, S. A. & Warner, J. R. (1972). Ribosomal RNA synthesis in *Saccharomyces
cerevisiae. Journal of Molecular Biology*, **65**, 227–42.

Waldron, C. (1977). Synthesis of ribosomal and transfer ribonucleic acids in yeast during a
nutritional shift-up. *Journal of General Microbiology*, **96**, 215–21.

Waldron, C. & Lacroute, F. (1975). Effect of growth rate on the amounts of ribosomal
and transfer ribonucleic acids in yeast. *Journal of Bacteriology*, **122**, 855–65.

Walmsley, R. M. & Oliver, S. G. (1980). Some evidence for the association of yeast
ribosomal RNA genes with membranous material. *Society for General
Microbiology Quarterly*, **7**, 96.

Warner, A. H., Thomas, D. ds., Shridhar, V. & McCurdy, H. D. (1977). The
diaguanosine nucleotides: do they exist in aquatic fungi? *Canadian Journal of
Biochemistry*, **55**, 841–6.

Warner, J. R. & Udem, S. A. (1972). Temperature-sensitive mutations affecting ribosome
synthesis in *Saccharomyces cerevisiae. Journal of Molecular Biology*, **65**,
243–57.

Wehr, C. T. & Parks, L. W. (1969). Macromolecular synthesis in *Saccharomyces
cerevisiae* in different growth media. *Journal of Bacteriology*, **98**, 458–66.

Wickner, R. B. (1979). The killer double-stranded RNA plasmids of yeast. *Plasmid*, **2**,
303–22.

Ycas, M. & Brawerman, G. (1957). Interrelationships between nucleic acid and protein
synthesis in microorganisms. *Archives of Biochemistry and Biophysics*, **68**,
118–29.

Genetic control of the yeast cell cycle: a reappraisal of 'start'

PAUL NURSE

School of Biological Sciences, University of Sussex, Brighton BN1 9QG, UK

Introduction

The nucleus undergoes two dramatic events during the cell cycle. The first is S phase when the DNA undergoes replication, and the second is nuclear division when the replicated chromosomes are segregated between the two daughter nuclei. The control of these nuclear activities has been subjected to extensive genetic analysis in the budding yeast *Saccharomyces cerevisiae* (Hartwell, 1974; Hartwell, 1978; Simchen, 1978). These studies have led to the concept of 'start' as the point in the cell cycle at which cells become committed to S phase and nuclear division, and where the overall control of the nuclear and cell division cycles takes place. Because of the importance of this concept, in this chapter I want to reappraise 'start' in the light of more recent work, dealing mainly with *Saccharomyces cerevisiae* but also briefly with the fission yeast *Schizosaccharomyces pombe*.

Outline of the 'start' concept

'Start' is the earliest gene-controlled event of the cell cycle, and is the point at which the cell becomes committed to the mitotic cell cycle (Hartwell, 1974). Before a cell traverses 'start' it monitors various conditions such as the presence of nutrients, the attainment of a critical cell size, the absence of conditions promoting conjugation or sporulation, and the completion of mitosis from the previous cycle. If these conditions are satisfactory the cell will begin a new mitotic cycle, but if they are not the cell will undertake an alternative developmental pathway such as conjugation, sporulation or entry into stationary phase.

Several lines of evidence have contributed to this formulation of the 'start' concept. The gene *cdc* 28 has been shown to be involved in the traverse of 'start'. Its function is required during G1 before the initiation of DNA replication can take place, and it sequences before the action of any other *cdc* genes that function during G1 (Hartwell, 1973; Hereford & Hartwell, 1974). When a temperature-sensitive mutant of *cdc* 28 is incubated at the restrictive temperature, the cells arrest unbudded, and with an unduplicated spindle plaque (Byers & Goetsch, 1975). Temperature-sensitive mutants of other *cdc* genes which act during G1 arrest beyond this stage, having undergone bud emergence and spindle plaque duplication. These observations indicate that *cdc* 28 functions in the earliest known gene-controlled event of the cell cycle. As a consequence, this event has been called 'start'.

Reid & Hartwell (1977) have shown that cells arrested at the *cdc* 28 step can still undergo conjugation whereas cells arrested at later *cdc* gene steps are unable to do so. This result is consistent with the fact that the mating hormone α-factor arrests cells at the same stage of the cell cycle as does the mutant of *cdc* 28 (Hereford & Hartwell, 1974). α-factor is involved in blocking conjugating cells early in G1 so that the cells synchronize their cycles before conjugation (Bücking-Throm *et al.*, 1973). Therefore cells leave the mitotic cycle to undergo conjugation prior to traverse of the *cdc* 28 step. In addition, cells leave the mitotic cycle to enter stationary phase prior to traverse of 'start' (Johnston, Pringle & Hartwell, 1977). Cells deprived of nutrients become blocked in cell cycle progress before bud emergence and before completion of the *cdc* 28 function. These experiments suggest that the *cdc* 28 function acts at a point of commitment in the cell cycle. Once beyond this point cells must complete a further mitotic cycle before undergoing conjugation or entering stationary phase.

The role of cell size in the traverse of 'start' has been shown using a population of nitrogen-starved cells of various sizes which had been reinoculated into fresh medium (Johnston, Pringle & Hartwell, 1977). Complete of the *cdc* 28 function and bud-emergence were both used as markers of 'start', and only took place when the cell had attained a critical size. Smaller cells took longer before they produced a bud because they had to grow for a longer time before they reached this critical size. As well as attaining a critical size, cells must have completed the mitosis of the previous cycle before traversing 'start', since cells blocked in mitosis are unable to undergo any further bud emergence or DNA replication (Culotti & Hartwell, 1971; Hartwell *et al.*, 1974).

Conditions influencing 'start'
Cell size

The influence of cell size on 'start' has been further investigated by reducing the growth rate of cells, which results in a fall of cell size at division (Hartwell & Unger, 1977; Jagadish, Lorincz & Carter, 1977; Carter & Jagadish, 1978). Despite the reduced cell size at division, cell size at bud emergence is almost unchanged, suggesting that a critical cell size must be attained before 'start' can take place. In the slow-growing cells the phase of the cell cycle before 'start' becomes extended. This was shown using the timing of S phase (Carter & Jagadish, 1978; Jagadish & Carter, 1978) and the completion of the *cdc* 28 and α-factor functions (Jagadish & Carter, 1977), as markers for traverse of 'start'. The extension of the cell cycle before 'start' at slow growth rates can be neatly explained if it is assumed that a critical size is required for traverse of 'start' and that a constant time has to elapse between 'start' and cell division (Hartwell & Unger, 1977; Jagadish & Carter, 1977). At slow growth rates the cell divides at a smaller size because it grows less during the constant time between 'start' and cell division. As a consequence it has to grow more in the next cell cycle to attain the critical size required for 'start', and therefore spends a larger proportion of the cell cycle in the period before 'start'.

The explanation just given for the extension of the cell cycle before 'start' at slow growth rates can also account for the accumulation of cells before 'start' when they are deprived of nutrients. Under these conditions the growth rate will fall, cell size at division will be reduced, and cells will accumulate before 'start', since they are too small to traverse this event. Therefore, there is no need to propose a signal at 'start' for monitoring nutrients. However, some more recent results indicate that nutritional level may directly interact with the size control at 'start'. Careful measurements at bud emergence have shown that at fast growth rates cell volume is about 40% larger than at slow growth rates (Lorincz & Carter, 1979; Johnston *et al.*, 1979). In addition, cells shifted from rich to poor medium are advanced into budding, as would be predicted if cell size at bud emergence were to be reduced at slower growth rates. A cell-cycle model generated from a detailed study of mean cell size, of length of budded phase and of generation time, also predicts an increase in cell size at bud-emergence at fast growth rates (Tyson, Lord & Wheals, 1979). All these experiments can be understood in terms of a nutritional modulation of the cell size control acting at 'start' (Fantes &

Nurse, 1977). At fast growth rates the size is modulated upwards and at slow ones it is modulated downwards.

An interesting experiment performed by Shilo, Simchen & Pardee (1978) may also be illuminating about nutritional modulation of the cell size control. Cells of a histidine auxotroph were accumulated before 'start' by histidine starvation. Limiting concentrations of histidine were then added back to the cells and the subsequent rates of bud emergence and protein synthesis were followed. Significant levels of protein synthesis took place in the absence of bud emergence, suggesting that attainment of a critical cell size was insufficient for bud emergence. It is possible that the unusual nutritional conditions present in this experiment may have disturbed the usual nutritional modulation system, so that the cell size required for 'start' was increased, and as a consequence bud emergence could not take place.

Nutrients

As explained in the previous section, the monitoring of nutrients at 'start' is not necessary to explain the accumulation of cells before 'start' in starvation conditions. However, there must be some signal responsible for the nutritional modulation of the cell size control. Therefore a number of experiments designed to investigate the signal monitoring nutrients may still be useful for analysing the nutritional modulation signal which interacts with the cell size control at 'start'. These experiments are considered here.

Cells starved of sulphate arrest at 'start' (Pringle & Maddox, cited in Hartwell, 1974). Unger & Hartwell (1976) have studied the sulphate monitoring signal by blocking cells at different steps in the sulphate assimilation pathway, between sulphate and methionyl transfer RNA. These authors hoped to find a block in the pathway where arrest at 'start' did not take place even though cells were failing to assimilate sulphate, since they could have then concluded that the signal monitoring sulphate for 'start' was generated before that block. However, wherever the block was imposed in the pathway, cells arrested at 'start'. As a consequence, the signal for 'start' must have been generated at or beyond the formation of methionyl transfer RNA. These results led Unger & Hartwell (1976) to suggest that the signal for monitoring all nutrients was generated at the level of protein synthesis.

In another study Wolfner *et al.* (1975) have proposed that a gene *tra* 3 functions as a 'sensor' of amino acid metabolism and acts as a component of the signal monitoring nutrients at 'start'. Mutants in *tra* 3 are derepressed for various enzymes of amino acid biosynthesis and are also temperature sensitive for growth, accumulating before 'start' at the restrictive temperature. This apparently dual role led the authors to suggest that the *tra* 3 gene product relays information on metabolic activity to 'start'. However, an alternative interpretation is that the arrest at 'start' is very much a secondary consequence of the *tra* 3 mutation. In this mutant the rates of RNA and protein synthesis are very much reduced at the restrictive temperature (Wolfner *et al.*, 1975), suggesting that some nutritional component is in short supply. If growth is limited by growing cells on poor medium then cells accumulate before 'start' (Jagadish & Carter, 1977; Hartwell & Unger, 1977). Therefore, the most likely explanation of the behaviour of *tra* 3 mutants is that the rates of growth and macromolecular synthesis are reduced because some nutritional component is limiting, and as a secondary effect cells accumulate before 'start'. For this reason the arguments in favour of *tra* 3 being specifically involved in monitoring nutrients at 'start' are not strong.

A third group has attempted to implicate the rate of production of ribosomal precursor RNA in the nutritional signal at 'start' (Johnston & Singer, 1978; Singer, Johnston & Bedard, 1978). RNA synthesis inhibitors, *o*-phenanthroline and ethionine, caused cells to accumulate before 'start', and the authors claim that this is a direct consequence of a reduction in the rate of synthesis of ribosomal precursor RNA. Unfortunately these experiments are difficult to interpret because the inhibitors may be having a number of effects on the cell. For example, the synthesis of other RNA species may also have been inhibited, but information on this point is lacking. Also, there are some reductions in the rate of protein synthesis, albeit less than that seen for RNA, which could be affecting cell cycle progress (see also Oliver, previous chapter).

A rather different approach to the problem of monitoring nutrients has been taken by Shilo, Simchen & Pardee (1978). Cells of various auxotrophs were accumulated before 'start' by starving them of their nutritional requirements. The nutritional requirements were then added back at limiting concentrations, and the relationship between the rate of bud-emergence as a monitor of traverse of 'start', and the concentration of the requirement, was analysed on a Hill plot. The Hill coefficients for adenine, methionine and histidine were all similar at 2.4, which led the authors to postulate that there is a common intermediate, made up of the three requirements, that regulates the traverse of 'start'. I do not find this argument very convincing since it assumes that the reaction limiting the rate of bud emergence is the same for the three nutritional requirements, even though this may not be the case. The fact that the maximum rates of budding obtainable for the three auxotrophs are different (between 0.37 and 0.87 h^{-1}), suggests that the limiting reactions are different. In addition, simply showing that the three requirements have similar Hill coefficients does not establish that the requirements all assemble to make a common intermediate.

None of these experiments is particularly informative about the signal for the nutritional modulation of the size control at 'start'. The major problem is that treatments which slow down growth or reduce nutrient availability are likely to have many effects on the cell and it is difficult to establish which one is important for the signal.

α-factor

α-factor is a short polypeptide (Duntze *et al.*, 1973; Stötzler, Kiltz & Duntze, 1976) that arrests mating-type cells in G1 (Bücking-Throm *et al.*, 1973). The major question of interest is how this hormone interacts with the cell to cause it to arrest at 'start'. The number of cellular sites to which α-factor binds has been investigated by Uden & Finkelstein (1978). They found that the order of the reaction between α-factor and cycle arrest was one, indicative of a single saturatable cellular site. It has been suggested that α-factor interaction with the cell involves breakdown of the α-factor by a protease located in the cell membrane (Maness & Edelman, 1978). The breakdown of the α-factor either results in hormone fragments entering the cell to act as triggers, or in allosteric changes of the protease which directly influence the cell. This hypothesis was based on the observations that α-factor is broken down by *a* cells, and that preincubation of *a* cells with the protease inhibitor trasyol blocks the action of α-factor. However, Finkelstein & Strausberg (1979) have disagreed with the hypothesis. They found that another protease inhibitor chloroquine, which inhibits the breakdown of α-factor, does not block the action of α-factor, but enhances it. Therefore, it is unlikely that α-factor breakdown is required for the α-factor arrest of cells at 'start'. If this is the case, it is not clear what the breakdown of α-factor is for. One reason might be to reduce the concentration of α-factor (Finkelstein & Strausberg, 1979), since conjugating cells in normal physiological conditions may require only an elongation of the phase before 'start' prior to conjugation. If conjugation does not take place relatively rapidly in these conditions, the

α-factor is broken down and the cells are released from their arrest at 'start'.

The relation between α-factor and the release of storage compounds from the vacuole has been investigated by Sumrada & Cooper (1978). Addition of α-factor to *a* cells results in the induction of allophanate hydrolase. A similar induction is seen when cells are starved of ammonia, which the authors postulate is the result of storage allantoin and arginine being released from the vacuole into the cytoplasm. They further postulate that the α-factor effect is the result of the hormone also causing the release of allantoin and arginine from the vacuole. The relevance of these observations to normal progress through the cell cycle is not clear. It would seem more reasonable that a cell would mobilize its storage compounds *after* it had traversed 'start' rather than *before* since they could then contribute to the completion of the cycle already in progress. Perhaps this mobilization of storage compounds before 'start' is related more to the onset of conjugation than to progress through the mitotic cycle.

Structure of 'start'

Folded chromosomes can be isolated from budding yeast by gentle lysis of cells (Pinon & Salts, 1977). The changes observed in these folded chromosomes in various situations related to the cell cycle have been very informative about the detailed structure of 'start'. The folded chromosomes change their sedimentation profile as cells progress through the cycle, enabling characteristic G1 and G2 forms to be identified. Stationary phase cells starved of nitrogen have a different G_0 form which is not seen in the normal cell cycle (Pinon, 1978). This suggests that stationary phase cells are off the normal cell cycle. α-factor-arrested cells have yet another form of folded chromosome which shows a heterogeneous sedimentation pattern (Pinon & Pratt, 1979). We can conclude that the terminal phenotypes of α-factor-arrested cells and nitrogen-starved cells are different even though both treatments should result in arrest at 'start'. Pinon & Pratt (1979) also showed that α-factor-arrested cells can enter stationary phase and vice versa, without going through a further mitotic cycle. The easiest explanation of these data is that cells leave the cycle at 'start' but then progress along different pathways according to the condition used to arrest the cells. These pathways are reversible and so if the conditions are changed the cell can move about between various states within a 'start' area (Pinon & Pratt, 1979, and Fig. 1). If the cell is an *a*/α diploid then the stationary phase pathway can be extended into sporulation, since nitrogen-free medium induces these cells to sporulate and their folded chromosomes take on a G0 form before progressing to meiosis (Pinon, 1979).

These experiments suggest that 'start' has a complex structure and should be considered more of a 'start' area in which cells can shift between different developmental states. If this is the case then 'start' mutants may not arrest cells at a single 'start' position located directly in the mitotic cycle, but could arrest cells on one of the reversible pathways leading to conjugation sporulation or stationary phase. The folded chromosomes of mutants of two genes *cdc* 25 and *cdc* 28 which arrest at 'start' are not very informative on this point since they are unstable (Pinon, 1979). This is unlike the forms seen in stationary phase, α-factor-arrested, G1-arrested or sporulating cells, and may be due to some abnormal development of the folded chromosomes seen only in the 'start' mutants. However, some experiments concerned with the initiation of sporulation may be relevant in this context (Shilo, Simchen & Shilo, 1978). *cdc* 25 mutants were shown to initiate sporulation in rich medium which normally prevents it from taking place. One explanation for this is that *cdc* 25 mutants may be derepressed for sporulation at the restrictive temperature, so that this takes place when the cells should

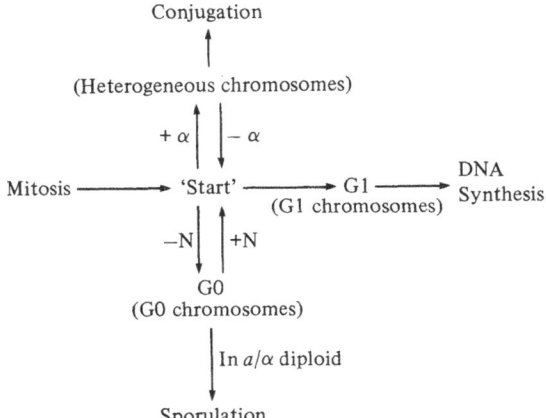

Fig. 1. Various states within the 'start' area. Addition of α-factor causes cells to leave the mitotic cycle and to progress towards conjugation. The folded chromosomes become heterogeneous. Deprivation of nitrogen causes cells to enter G0 and if they are a/α diploids progress towards sporulation. The folded chromosomes take on a G0 form. (Based on Figure given in Pinon & Pratt, 1979.)

normally be undergoing vegetative growth. The mutation may mimic poor nutritional conditions, a possibility supported by its poor growth at the restrictive temperature (Johnston, Pringle & Hartwell, 1977). If the mutant cell is a haploid it will leave the mitotic cycle at 'start' and progress along the pathway leading to stationary phase. If the mutant cell is an a/α diploid it can continue on into sporulation even if it is growing in rich medium.

Commitment

It has been shown that cells arrested after 'start' in the cell cycle are unable to conjugate (Reid & Hartwell, 1977), and it has usually been assumed that such cells will also be unable to sporulate (Hartwell, 1974). This assumption has been questioned by Hirschenberg & Simchen (1977) who tested the ability of cells arrested at various *cdc* steps to undergo sporulation. They found that arrested *cdc* 4 cells could sporulate without proceeding through an intervening mitotic cell cycle first. Since the *cdc* 4 function sequences after that of *cdc* 28, this means that cells can traverse 'start' without being committed to the mitotic cell cycle. It is interesting that the behaviour of arrested *cdc* 4 cells with respect to conjugation is not known since the *cdc* 4 function is required for conjugation to take place (Reid & Hartwell, 1977). It is possible that the *cdc* 4 function should also be included in the 'start' area if cells are not yet fully committed at this stage of the cell cycle.

'Start' as a rate-limiting step

All steps influence the rate at which cell division takes place, but some steps may have more effect than others. These will act as major rate-controlling steps for progress towards cell division. Johnston *et al.* (1977) have proposed that traverse of 'start' is the major rate-limiting step of the cell cycle. This was based on the observations that a critical cell size is necessary before 'start' can take place, and that the time required for a doubling in mass is longer than that required for completing the events of the mitotic cell cycle. As a consequence, small daughter cells take longer to complete their cell cycle than do large ones, since the small cells have to grow much longer to attain the critical size required for 'start', and it is the need for this growth that is the major rate-controlling step of the cell cycle. Therefore traverse of start can be considered rate-controlling or rate-limiting. In slow-growing chemostat cultures, traverse of 'start' is even more rate-controlling, since newly born cells are smaller and the cell cycle becomes much extended before 'start'. On the other hand, in the larger rapidly growing mother cells, the need to attain a critical size before traverse of 'start' may be less important than the rate-controlling step, since the cells begin the cell cycle almost big enough to undergo 'start'.

Shilo, Shilo & Simchen (1976) have also come to the conclusion that traverse of 'start' is rate-limiting. They followed the time course of bud emergence in cells which had been released from blocks at 'start' imposed by *cdc* mutants or α-factor. They found that the proportion of unbudded cells fell exponentially, indicating that recovery from the block at 'start' required traverse of some event which was probabilistic in nature. That is, the recovery cells passed through some state which they left in a probabilistic fashion. Once cells were in this state they all had an equal probability of leaving it and progressing on to bud-emergence. A similar state is an important element of the transition probability model for cell-cycle control (Smith & Martin, 1973). In this model the transition from an A-state to a B-phase was proposed to be probabilistic, and to act as the major rate-limiting step of the cell cycle. On the basis of their results Shilo *et al.* (1976) suggested that traverse of 'start' was probabilistic and was equivalent to the A-state/B-phase transition. Some criticisms have been made of this suggestion, because of certain internal inconsistencies with the data and because of difficulties of being sure that the fall in unbudded cells really is exponential (Nurse & Fantes, 1977; Wheals, 1977). However, if 'traverse' of start is equivalent to the A-state/B-phase transition, then this is clearly very important for an understanding of the 'start' control. It may suggest that 'start' involves components or structures present in only low amounts within the cell.

Mechanism of 'start'

Little is known about the mechanism of 'start'. Since the duplication of the spindle plaque is blocked in the *cdc* 28 mutant, it is possible that the 'start' function may involve the spindle plaque structure. Indirect support for this possibility comes from Dutcher & Hartwell (1978) who have shown that *cdc* 28 mutants have a *kar* phenotype. *kar* mutants are defective in nuclear fusion (Conde & Fink, 1976), and this fusion process requires participation of the spindle plaques (Byers & Goetsch, 1975). Also the dominance properties of the *kar* phenotype can be interpreted in terms of the *cdc* 28 gene product participating in a nucleus-limited structure, an obvious candidate being the spindle plaque. In this context it is of considerable interest that centrioles, which can be considered as analogues of spindle plaques, have been implicated in cell-cycle control in mammalian cells (Tucker, Pardee & Fujiwara, 1979; Brooks, Bennett & Smith, 1980).

One possible approach to investigating the molecular mechanism of 'start' is to use an *in vitro* assay for the initiation of DNA synthesis (Jazwinski & Edelman, 1976). Such studies have shown that cell extracts of *cdc* 28 cannot support initiation but that this deficiency can be made up by wild-type components with a molecular weight of over 50 000. In principle, this system could be used to assay and purify the deficiency of the *cdc* 28 mutant.

'Start' in fission yeast

Some preliminary experiments are suggestive that the 'start' control may also be relevant to the fission yeast *Schizosaccharomyces pombe* (P. Nurse & Y. O. Bissett, unpublished). Cells arrested at two *cdc* gene steps, *cdc* 10 and *cdc* 2, can undergo conjugation, whereas cells arrested at other *cdc* gene steps are unable to do so. *cdc* 10 is required for DNA replication (Nurse, Thuriaux & Nasmyth, 1976); *cdc* 2 was originally described as being required only for mitosis, but more recent results indicate that it is required for DNA replication as well (P. Nurse & Y. O. Bissett, unpublished). Only those *cdc* 2 cells arrested at the DNA replication block point can undergo conjugation. Therefore there appears to be a stage in G1 of *Schizosaccharomyces pombe* defined by *cdc* 10 and *cdc* 2 which is analogous to 'start' in *Saccharomyces cerevisiae*. At this point cells are not yet committed to the mitotic cycle but can still undergo the alternative developmental pathway of conjugation. Cell size may also play a role in the *Schizosaccharomyces pombe* 'start' since the cell must attain a critical size before DNA replication can take place (Nurse, 1975; Nurse & Thuriaux, 1977; Nasmyth, Nurse & Fraser, 1979).

The *cdc* 2 gene function has also been implicated in the control initiating mitosis (Thuriaux, Nurse & Carter, 1978; P. Nurse & P. Thuriaux, unpublished). Certain *cdc* 2 alleles result in cells being advanced into mitosis at a small size, indicating that the *cdc* 2 function acts in a major rate-controlling step for mitosis. If *cdc* 2 also controls commitment of the cell to DNA replication, then it has regulatory functions at the two major events of the mitotic cycle: S phase and mitosis. Another common feature of the two controls involved in these two events is the role of cell size. As mentioned above, cell size may play a role in 'start' and in addition it is important for the timing of mitosis (Nurse, 1975; Fantes, 1977; Fantes & Nurse, 1978). In the latter control, the size is also nutritionally modulated, being higher in faster growing cells and lower in slower growing cells (Fantes & Nurse, 1977).

Given these results it is interesting to compare the controls in *Schizosaccharomyces pombe* and *Saccharomyces cerevisiae*. In *Schizosaccharomyces pombe* there are two controls, one analogous to 'start' and one acting over mitosis. Both share at least one common gene, *cdc* 2, and both involve cell mass, shown in one case to be nutritionally modulated. In *Saccharomyces cerevisiae* there is a single control 'start' which also involves cell mass and nutritional modulation. This single control indirectly determines the timing of mitosis since this usually occurs at a fixed time after 'start'. What is intriguing about *Saccharomyces cerevisiae* is that in G1 the cell produces a short mitotic spindle (Byers & Goetsch, 1975). One interpretation of this observation is that mitosis is initiated in G1 after 'start'. If this is correct, then in *Saccharomyces cerevisiae* the 'start' control and mitotic control could have been combined together, whilst in *Schizosaccharomyces pombe* the controls have been separated but share common features and components.

Conclusions

The major conclusions of this chapter can be summarised as follows:

(1) The main features of 'start' remain intact, and it has proved to be a useful conceptual framework for investigating control of the nuclear and cell division cycles in *Saccharomyces cerevisiae*.

(2) Because of the critical cell size required for traverse of 'start' and the constant time period between 'start' and cell division, cells deprived of nutrients will accumulate before 'start'. Therefore a signal monitoring nutrients at 'start' is not necessary to explain the accumulation before 'start' after nutrient deprivation.

(3) There is evidence for nutritional modulation of the cell size control involved at 'start'.

(4) Traverse of 'start' acts as a major rate-limiting step controlling the rate at which the cell progresses through the cell cycle.

(5) 'Start' should be considered more of an area rather than a single point at the beginning of the cell cycle. Different methods of blocking at 'start' arrest cells at different places within this 'start' area.

(6) There is preliminary evidence that the 'start' concept may also

apply to *Schizosaccharomyces pombe*. In this yeast some components of the 'start' control may also play a role in the control initiating mitosis.

Acknowledgements. I should like to thank Bruce Carter, Peter Fantes and Murdoch Mitchison for many stimulating discussions concerning 'start' and the control of the cell cycle.

References

Brooks, R., Bennett, D. & Smith, J. (1980). Mammalian cell cycles need two random transitions. *Cell*, **19**, 493–504.

Bücking-Throm, E., Duntze, W., Hartwell, L. H. & Manney, T. R. (1973). Reversible arrest of haploid yeast cells at the initiation of DNA synthesis by a diffusible sex factor. *Experimental Cell Research*, **76**, 99–110.

Byers, B. & Goetsch, L. (1975). Behaviour of spindles and spindle plaques in the cell cycle and conjugation of *Saccharomyces cerevisiae*. *Journal of Bacteriology*, **124**, 511–23.

Carter, B. L. A. & Jagadish, M. N. (1978). Control of cell division in the yeast *Saccharomyces cerevisiae* cultured at different growth rates. *Experimental Cell Research*, **112**, 373–83.

Conde, J. & Fink, G. (1976). A mutant of *Saccharomyces cerevisiae* defective for nuclear fusion. *Proceedings of the National Academy of Sciences USA*, **73**, 3651–5.

Culotti, J. & Hartwell, L. H. (1971). Genetic control of the cell division cycle in yeast. III. Seven genes controlling nuclear division. *Experimental Cell Research*, **67**, 389–401.

Duntze, W., Stötzler, D., Bücking-Throm, E. & Kalbitzer, S. (1973). Purification and partial characterisation of α-factor, a mating-type specific inhibitor of cell reproduction from *Saccharomyces cerevisiae*. *European Journal of Biochemistry*, **35**, 357–65.

Dutcher, S. & Hartwell, L. H. (1978). The involvement of *cdc* gene products in conjugation. Abstracts of 9th International Conference on Yeast Genetics and Molecular Biology, p. 70. University of Rochester, New York.

Fantes, P. (1977). Control of cell size and cycle time in *Schizosaccharomyces pombe*. *Journal of Cell Science*, **24**, 51–67.

Fantes, P. & Nurse, P. (1977). Control of cell size at division in fission yeast by a growth modulated size control over nuclear division. *Experimental Cell Research*, **107**, 377–86.

Fantes, P. & Nurse, P. (1978). Control of the timing of cell division in fission yeast. *Experimental Cell Research*, **155**, 317–29.

Finkelstein, D. B. & Strausberg, S. (1979). Metabolism of α-factor by *a*-mating type cells of *Saccharomyces cerevisiae*. *Journal of Biological Chemistry*, **254**, 796–803.

Hartwell, L. H. (1973) Three additional genes required for deoxyribonucleic acid synthesis in *Saccharomyces cerevisiae*. *Journal of Bacteriology*, **115**, 966–74.

Hartwell, L. H. (1974). *Saccharomyces cerevisiae* cell cycle. *Bacteriological Reviews*, **38**, 164–98.

Hartwell, L. H. (1978). Cell division from a genetic perspective. *Journal of Cell Biology*, **77**, 627–37.

Hartwell, L. H., Culotti, J., Pringle, J. R. & Reid, B. J. (1974). Genetic control of the cell division cycle in yeast: a model. *Science*, **183**, 46–51.

Hartwell, L. H. & Unger, M. W. (1977). Unequal division in *S. cerevisiae* and its implications for the control of division. *Journal of Cell Biology*, **75**, 422–435.

Hereford, L. M. & Hartwell, L. H. (1974). Sequential gene function in the initiation of *Saccharomyces cerevisiae* DNA synthesis. *Journal of Molecular Biology*, **84**, 445–61.

Hirschenberg, J. & Simchen, G. (1977). Commitment to the mitotic cell cycle in yeast in relation to meiosis. *Experimental Cell Research*, **105**, 245–52.

Jagadish, M. N. & Carter, B. L. A. (1977). Genetic control of cell division in yeast cultured at different growth rates. *Nature, London*, **269**, 145–7.

Jagadish, M. N. & Carter, B. L. A. (1978). Effects of temperature and nutritional conditions on the mitotic cell cycle of *Saccharomyces cerevisiae*. *Journal of Cell Science*, **31**, 71–8.

Jagadish, M. N., Lorincz, A. & Carter, B. L. A. (1977). Cell size and cell division in yeast cultured at different growth rates. *FEMS Microbiology Letters*, **2**, 235–7.

Jazwinski, S. & Edelman, G. (1976). Activity of yeast extracts in cell-free stimulation of DNA replication. *Proceedings of the National Academy of Sciences, USA*, **73**, 3933–6.

Johnston, G. C., Ehrhardt, C. W., Lorincz, A. & Carter, B. L. A. (1979). Regulation of cell size in the yeast *Saccharomyces cerevisiae*. *Journal of Bacteriology*, **137**, 1–5.

Johnston, G. C., Pringle, J. R. & Hartwell, L. H. (1977). Co-ordination of growth with cell division in the yeast *Saccharomyces cerevisiae*. *Experimental Cell Research*, **105**, 79–98.

Johnston, G. C. & Singer, R. A. (1978). RNA synthesis and control of cell division in the yeast *S. cerevisiae*. *Cell*, **14**, 951–8.

Lorincz, A. & Carter, B. L. A. (1979). Control of cell size at bud initiation in *Saccharomyces cerevisiae*. *Journal of General Microbiology*, **113**, 287–95.

Maness, P. F. & Edelman, G. M. (1978). Inactivation and chemical alteration of mating factor alpha by cells and sphaeroplasts of yeast. *Proceedings of the National Academy of Sciences, USA*, **75**, 1304–08.

Nasmyth, K., Nurse, P. & Fraser, R. (1979). The effect of cell mass on the cell cycle timing and duration of S-phase in fission yeast. *Journal of Cell Science*, **39**, 215–33.

Nurse, P. (1975). Genetic control of cell size at cell division in yeast. *Nature, London*, **256**, 547–55.

Nurse, P. & Fantes, P. (1977). Transition probability and cell-cycle initiation in yeast. *Nature, London*, **267**, 647.

Nurse, P. & Thuriaux, P. (1977). Controls over the timing of DNA replication during the cell cycle of fission yeast. *Experimental Cell Research*, **107**, 365–75.

Nurse, P., Thuriaux, P. & Nasmyth, K. (1976). Genetic control of the cell division cycle in the fission yeast *Schizosaccharomyces pombe*. *Molecular and General Genetics*, **146**, 167–78.

Pinon, R. (1978). Folded chromosomes in non-cycling yeast cells: evidence for a characteristic G_0 form. *Chromosoma, Berlin*, **67**, 263–74.

Pinon, R. (1979). A probe into nuclear events during the cell cycle of *Saccharomyces cerevisiae*: studies of folded chromosomes in *cdc* mutants which arrest in G1. *Chromosoma, Berlin*, **70**, 337–52.

Pinon, R. & Pratt, D. (1979). Folded chromosomes of mating-factor arrested yeast cells: comparison with G_0 arrest. *Chromosoma, Berlin*, **73**, 117–29.

Pinon, R. & Salts, Y. (1977). Isolation of folded chromosomes from the yeast *Saccharomyces cerevisiae. Proceedings of the National Academy of Sciences, USA*, **74**, 2850–54.

Reid, B. J. & Hartwell, L. H. (1977). Regulation of mating in the cell cycle of *S. cerevisiae. Journal of Cell Biology*, **75**, 355–65.

Shilo, B., Shilo, V. & Simchen, G. (1976). Cell-cycle initiation in yeast follows first-order kinetics. *Nature, London*, **264**, 767–70.

Shilo, B., Simchen, G. & Pardee, A. B. (1978). Regulation of cell-cycle initiation in yeast by nutrients and protein synthesis. *Journal of Cell Physiology*, **97**, 177–88.

Shilo, V., Simchen, G. & Shilo, B. (1978). Initiation of meiosis in cell cycle initiation mutants of *Saccharomyces cerevisiae. Experimental Cell Research*, **112**, 241–8.

Simchen, G. (1978). Cell cycle mutants. *Annual Review of Genetics*, **12**, 161–91.

Singer, R. A., Johnston, G. C. & Bedard, D. (1978). Methionine analogs and cell division regulation in the yeast *Saccharomyces cerevisiae. Proceedings of the National Academy of Sciences, USA*, **75**, 6083–7.

Smith, J. & Martin, L. (1973). Do cells cycle? *Proceedings of the National Academy of Sciences, USA*, **70**, 1263–7.

Stötzler, D., Kiltz, H. & Duntze, W. (1976). Primary structure of α-factor peptides from *Saccharomyces cerevisiae. European Journal of Biochemistry*, **69**, 397–400.

Sumrada, R. & Cooper, T. G. (1978). Control of vacuole permeability and protein degradation by the cell cycle arrest signal in *Saccharomyces cerevisiae. Journal of Bacteriology*, **136**, 234–46.

Thuriaux, P., Nurse, P. & Carter, B. (1978). Mutants altered in the control coordinating cell division with cell growth in the fission yeast *Schizosaccharomyces pombe. Molecular and General Genetics* **161**, 215–20.

Tucker, R., Pardee, A. & Fujiwara, K. (1979). Centriole ciliation is related to quiescence and DNA synthesis in 3T3 cells. *Cell*, **17**, 527–35.

Tyson, C. B., Lord, P. G. & Wheals, A. E. (1979). Dependency of size of *Saccharomyces cerevisiae* cells on growth rate. *Journal of Bacteriology*, **138**, 92–8.

Uden, M. M. & Finkelstein, D. B. (1978). Reaction order of *Saccharomyces cerevisiae* alpha-factor-mediated cell cycle arrest and mating inhibition. *Journal of Bacteriology*, **133**, 1501–7.

Unger, M. W. & Hartwell, L. H. (1976). Control of cell division in *Saccharomyces cerevisiae* by methionyl-tRNA. *Proceedings of the National Academy of Sciences, USA*, **73**, 1664–8.

Wheals, A. (1977). Transition probability and cell-cycle initiation in yeast. *Nature, London*, **267**, 647.

Wolfner, M., Yep, D., Messenguy, F. & Fink, G. R. (1975). Integration of amino acid biosynthesis into the cell cycle of *Saccharomyces cerevisiae. Journal of Molecular Biology*, **96**, 273–90.

Index of specific names

Subject index

p denotes *passim* (scattered references)